Comparative Legal Frameworks for Pre-Implantation Embryonic Genetic Interventions

Pin Lean Lau

Comparative Legal Frameworks for Pre-Implantation Embryonic Genetic Interventions

 Springer

Pin Lean Lau
Central European University
Budapest, Hungary

ISBN 978-3-030-22307-6   ISBN 978-3-030-22308-3   (eBook)
https://doi.org/10.1007/978-3-030-22308-3

This Springer imprint is published by the registered company Springer Nature Switzerland AG.
The registered company address is: Gewerbestrasse 11, 6330 Cham, Switzerland

Foreword

When the human genome program was completed and we started to understand the genetic makeup of human beings, the theoretical possibility of manipulating the genetic structure also emerged. In the debates on how much and what kind of interventions are ethically acceptable, only a few resulted in widespread agreement, and one of them was on the prohibition of modifying the genome of future generations. In Europe, under Article 13 of the Oviedo Convention, an intervention seeking to alter the human genome may only be undertaken for preventive, diagnostic, or therapeutic purposes and only if its aim is not to change the genome of descendants. The Oviedo Convention is more than 20 years old, but its ethical principles are considered the standard setting even beyond the European continent. With the introduction of preimplantation gene editing, it seems that the distinction between germ line and somatic line is getting blurred. Therefore, it is crucial to explore and take stock of the manifold legal challenges of preimplantation genetic interventions on a global scale.

Pin Lean Lau's book provides an excellent review of this field, and it is unique in that it discusses preimplantation genetic interventions in a multidisciplinary and comparative context. Many layers of analysis complement each other: besides discussing the philosophical understanding of reproduction and enhancement, the book also explores the ethical principles formed in the debates on the status of the human embryo, on abortion, on early prenatal testing technologies, and on genetic interventions, and it engages with various legal theories as well on the fundamental rights and constitutional rights and on the role of regulation. Pin Lean Lau's work also challenges the notion that European or Western ethical principles and moral values are considered universally valid. The traditional focus on European bioethical discourses is reframed within a rich comparative ethical and legal context: the book analyzes a wide landscape of jurisdictions, including Malaysia, Thailand, and Australia, besides the United States and the United Kingdom. As the field of biomedical research itself becomes more and more global and new technological advances are reported increasingly from Asian countries, it is essential that the philosophical, ethical, and legal analysis of genetic research and interventions also develops a comparative on a global focus.

As Sheila Jasanoff stated in her work *Reframing Rights: Bioconstitutionalism in the Genetic Age*, "Two encyclopedic bodies of writing—one social, the other scientific—define the meaning of life in our era. Encompassing, respectively, law and biology, these intertwined, mutually supporting, indeed coproducing textual projects frame the possibilities, limits, rights, and responsibilities of being alive—most especially for the species we call human." In the above quote, Sheila Jasanoff distinguishes between the two bodies of writing, the social and the biological, and claims that the two are now linked more than ever before. Although the two domains, law and science, have developed separately, the new textuality of genetics brings much closer the two disciplines than ever before. Now, we simply cannot escape the multidisciplinary studies in this field.

This multidisciplinary and comparative perspective is clearly shown in the structure of the book: after a rich introduction, it discusses the legacy of eugenics in contemporary law, the legal and ethical debates in embryo selection, the regulatory framework in biomedical technologies, the international biomedical laws in the field of genetic, and also the dynamics of basic constitutional rights in different jurisdictions. The bibliography also reflects this multidisciplinary approach, and the references to the American, British, Australian, Malaysian, and Thai legal doctrines promise to be a rich source of material for further research.

Pin Lean Lau eloquently demonstrates that working in the field of contemporary biomedical law and bioethics requires passion and patience: passion to understand the multiplicity of philosophical, ethical, and legal issues related to genetic research and the technologies of genetic interventions and patience to develop a consistent comparative analysis of a wide variety of ethical values and legal jurisdictions. Without such passion and patience, it would not be possible to work in this dynamic field, where new scientific challenges can shake the existing normative framework every day. The book shows that science, law, and ethics consist of not just dry texts that need to be collected, assessed, and compared, but it could also be formulated eloquently. I believe that it will be an important source book for lawyers, scientists, professors, students, and different kinds of stakeholders within the biomedical industry.

Central European University   Judit Sándor
Budapest, Hungary

Center for Ethics and Law in Biomedicine
Budapest, Hungary

Acknowledgments

My warmest gratitude:

To my mentor, Professor Judit Sandor, without whom this book would not be possible. No words are enough to express my gratitude for your support, kindness, and friendship.

To my mother, Champi, who encouraged me to follow my dreams, spread my wings, and seek my passion beyond my comfort zone and who taught me how to be strong, resilient, and free-spirited and to embrace the world and my own abilities. To my late father, KB, whom I will always love until the end of my days. To my daughter, Emilie, the center of my universe, the only human being who will ever know what my heart sounds like from the inside. You have taught me so much more than I expected to learn, for the love you bestow upon me and the privilege I've been given to be your mother. I love you beyond the infinite universe. To my brother, Chin Wei, whose strong and silent presence throughout reminds me why I am grateful to you.

To my wonderful friends from home and those in my adopted home who have become my family. All of you, whose names are forever etched in my heart, I love that the breadth and depth of oceans and skies between us have never changed what we mean to each other; every one of you beyond the confines of our geographical borders, I am thankful to call you my dearest friends.

To my colleagues in the Central European University Legal Studies Department, my brothers and sisters now, thank you for the love, the camaraderie, the endless flow of coffee and chocolates in our lab, and the fellowship; the privilege has been mine to know each and every one of you. For my doctoral program directors, Professor Csilla Kollonay-Lehoczky and Professor Mathias Möschel, for your words of wisdom, kindness, continuous support and understanding. To Professors Károly Bárd, Renáta Uitz, Tibor Tajti, Eszter Polgári, and Petra Bárd, all of whom have extended advice and support in one way or other. To the effervescent Ms. Lea Tilless, for all your support, advice, easy friendship, and encouragement throughout these years. To Ms. Natalia Nyikes, who always has time to give me hugs and spur me on with words of wisdom and encouragement. To all the other wonderful people at the Central European University, for being part of my daily life and giving me the privilege of knowing you.

To Professor Jane Kaye, Dr. Michael Morrison, Dr. Harriet Teare, Miranda Mourby, Imogen Holbrook, and Fiona Coldwell at the Center for Health, Law and Emerging Technologies (HeLEX), University of Oxford, my sojourn at HeLEX and all your encouragement and scholarly feedback have impacted my research studies in ways that allowed me to explore beyond what I knew, and I am grateful for that.

To Dr. Aurelie Mahalitchmy and Dr. Michael Morrison, whose invaluable advice and feedback have enabled me to undertake a richer and more comprehensive reflection of my research, for your encouragement, support, and continued future collaborations.

To my publisher and editor, Anke Seyfried, who took a chance on my work and has supported and encouraged me throughout, and to Springer International, for making this a reality.

Thank you.

Budapest, Hungary   Pin Lean Lau
May 2019

Contents

Abbreviations

AHEC	Australian Health Ethics Committee
AHRC	Australian Human Rights Commission
AHRD	ASEAN Human Rights Declaration
ARTs	Assisted Reproductive Technologies
ASEAN	Association of Southeast Asian Nations
CEDAW	Convention on the Elimination of All Forms of Discrimination Against Women
CRISPR	Clustered Regularly Interspaced Short Palindromic Repeats
DNA	Deoxyribonucleic Acid
ECHR	European Convention on Human Rights
ECtHR	European Court of Human Rights
EMA	European Medicines Agency
EU	European Union
FSA	Fertility Society of Australia
GDP	Gross Domestic Product
HFEA Act	Human Fertilization and Embryology Authority Act
HFEA	Human Fertilization and Embryology Authority
HLA	Human Leukocyte Antigen
HRA	Human Rights Act 1998
HRBA	Human Rights-Based Approach
HRCM Act	Human Rights Commission of Malaysia Act 1999
IBC	International Bioethics Committee
ICCPR	International Covenant on Civil and Political Rights
ICESCR	International Covenant on Economic, Social and Cultural Rights
ICPD	International Conference on Population and Development
IDHGD	International Declaration on Human Genetic Data 2003
IVF	In Vitro Fertilization
JRC	Joint Research Centre for Science and Policy
LF	Legal Foresighting
NHMRC	National Health and Medical Research Council
NHRC	National Human Rights Committee

NHSCB	National Health Service Commissioning Board
NIH	National Institutes of Health
OECD	Organization for Economic Co-operation and Development
OHCHR	Office of the High Commissioner for Human Rights
PGD	Preimplantation Genetic Diagnosis
PGS	Preimplantation Genetic Screening
RRI	Responsible Research Innovation
RTAC	Reproductive Technology Accreditation Committee
SUHAKAM	Human Rights Commission of Malaysia
UDBHR	Universal Declaration on Bioethics and Human Rights 2005
UDHG	Universal Declaration on the Human Genome and Human Rights 1997
UDHR	Universal Declaration on Human Rights 1948
UN	United Nations
UNESCO	United Nations Educational, Scientific and Cultural Organization
WHO	World Health Organization
WMA	World Medical Association

Chapter 1
Introduction

1.1 Theoretical Background

The last quarter of the twentieth century was a period of remarkable scientific advances. Information technology, nanotechnology, and biotechnology were three particular areas of scientific endeavors that generated intense public interest in the transformative power of science on the course of future human existence.[1] Advances in biotechnology were led by the drive to enhance human life, a concept with immense implications that demonstrated a simultaneous capacity to thrill and to frighten. The moral implications alone are staggering. However, the application of diverse channels for replicating biological substances has sparked global interest based on its systematic and moral inferences.

Advancements in science and technology across the world greatly affect people's lives, in addition to providing alternatives essential to human survival. One of the major developments in science and technology in the last decade is in the area of human genetic engineering. Human genetic engineering is defined as the process of changing the genotype[2] of an individual with an intention to alter the existing phenotype[3] of an adult or child, or choosing a newborn's phenotype.[4] The proponents of human genetic engineering view it as a solution to genetic diseases and an enhancement mechanism of human immunity to viruses or diseases. The propo-

---

[1] Heinemann and Honnefelder (2003), p. 530.

[2] A genotype is *"the genetic contribution to the phenotype"* according to the National Human Genome Research Institute. See also (n 3).

[3] 'National Human Genome Research Institute (NHGRI)' (*National Human Genome Research Institute (NHGRI)*) https://www.genome.gov/. The National Human Genome Research Institute defines a phenotype as "an individual's observable traits, such as height, eye color, and blood type…..Some traits are largely determined by the genotype, while other traits are largely determined by environmental factors."

[4] Lewis (2017).

© Springer Nature Switzerland AG 2019
P. L. Lau, *Comparative Legal Frameworks for Pre-Implantation Embryonic Genetic Interventions*, https://doi.org/10.1007/978-3-030-22308-3_1

nents also believe that further, in-depth research into the area can be essential in not only improving human mental faculties, such as intelligence and memory, but also in changing the metabolism and physical appearance of human beings.[5] It is indisputable scientific knowledge that genes influence a person's general health condition; some proponents further advocate that genes also significantly impact a person's behavior and personality traits.[6] This fact has prompted some researchers to make use of genetic technology to unravel genomes contributions to the aforementioned phenotypes.[7] In the process of seeking to achieve this goal, what is being discovered is a Pandora's Box of other possible applications of genetic engineering technology, particularly for humans, and how these applications may solve some of the genetic related problems that afflict human beings.[8]

One of the applications of human engineering technologies is in the field of assisted reproductive technologies (ARTs),[9] a significant bid to address the global problem of infertility.[10] Within the framework of ARTs, the focus is on Pre-Implantation Genetic Diagnosis (PGD), developed in the mid-1980s as an alternative means to prenatal genetic screening,[11] and "revolved around the determination of gender as an indirect means of avoiding an X-linked disorder."[12] PGD is defined as the testing of pre-implantation stage embryos, gametes (male or female), or oocytes[13] for genetic defects. It has been developed to assist couples, whereby one or both of them, have known genetic abnormalities, which may pass down through

---

[5] Shanks (2005).

[6] Kendler and Greenspan (2006), pp. 1683–1694.

[7] Resnik and Vorhaus (2006).

[8] Stock (2003).

[9] The term "ART" is commonly used to refer to a host of techniques used to assist infertile couples achieve pregnancy through anon-coital methods of contraception. The techniques in ART involve the manipulation of gametes and was introduced together with in-vitro fertilization. Other variations of ART were later developed, including, inter alia, intracystoplasmic sperm injection (ICSI), gamete intra-fallopian transfer (GIFT), zygote intra-fallopian transfer (ZIFf). It must be noted that ARTs do not provide for a cure for infertility, and do not necessarily guarantee success for a couple to have a baby.

[10] The World Health Organization (WHO) estimates that 1 in every 4 couples in developing countries are found to be affected with infertility problems, although it cannot estimate with accuracy, a global infertility rate, due to different reproductive health denominators.

[11] Nayal (2013).

[12] Id. The United States National Library of Medicine explains what an X-linked disorder is. It is essentially a recessive, sex-linked genetic disorder attributable to the X chromosome from a male or female parent. See https://www.nlm.nih.gov/medlineplus/ency/imagepages/19097.htm. Examples of x-linked disorders include haemophilia, muscular dystrophy, colour blindness, Hunter's disease, and Lesch-Nyhan Syndrome, amongst many others.

[13] The Farlex Medical Dictionary for the Health Professions and Nursing, 2012, explains an oocyte: it is the female sex cell, which, *"when fertilized by a sperm....... is capable of developing into a new individual of the same species; during maturation, the oocyte, like the sperm, undergoes a halving of its chromosomal complement so that at its union with the male gamete, the species number of chromosomes (46 in humans) is maintained..."*.

the germ line cell[14] to their potential offspring. PGD requires in vitro fertilization (IVF),[15] embryo biopsy and using either fluorescent *in situ* hybridization (FISH)[16] or polymerase chain reaction (PCR)[17] at the single cell level. Therefore, it is a complex procedure, which requires extensive experience on the part of the medical practitioner.

If an embryo is found to be unaffected by any genetic abnormalities, it will then be transferred to the uterus for implantation. In this regard, PGD is superior to current pre-natal, post-conception diagnostic procedures such as amniocentesis[18] or chorionic villus sampling.[19] In the latter diagnostic procedures, couples would often need to grapple with the difficult decision to terminate a pregnancy if the results of the diagnostics are found to be unfavorable. PGD is therefore a more attractive option in preventing heritable genetic disease, thereby eliminating the dilemma of pregnancy termination following unfavorable prenatal diagnosis.

Because of the manner in which PGD operates, there are grave moral and ethical concerns that plague its application. This includes the destruction of prenatal life in PGD, which has been placed on equal footing with the destruction of prenatal life in abortions[20]; non-medical or non-therapeutic uses of PGD (sex selection as an example)[21]; fears about selection of embryos leading to eugenic outcomes[22]; and concerns about "designer babies" being made possible,[23] amongst others. Other rel-

---

[14] As defined by the National Human Genome Research Institute, germ line cells are "the sex cells (eggs and sperm) that are used by sexually reproducing organisms to pass on genes from generation to generation. Egg and sperm cells are called germ cells, in contrast to the other cells of the body that are called somatic cells."

[15] The United States National Library of Medicine describes IVF as the process by which a woman's egg is fertilized by a man's sperm outside the body, always in a laboratory setting.

[16] Also known as FISH. The National Human Genome Research Institute defines this as "a laboratory technique for detecting and locating a specific DNA sequence on a chromosome. The technique relies on exposing chromosomes to a small DNA sequence called a probe that has a fluorescent molecule attached to it. The probe sequence binds to its corresponding sequence on the chromosome".

[17] Also known as PCR. The National Human Genome Research Institute explains that a polymerase chain reaction is "a fast and inexpensive technique used to 'amplify' or copy small segments of DNA. Because significant amounts of a sample of DNA are necessary for molecular and genetic analyses, studies of isolated pieces of DNA are nearly impossible without PCR amplification."

[18] Amniocentesis (also referred to as amniotic fluid test or AFT) is a medical procedure used in prenatal diagnosis of chromosomal abnormalities and fetal infections, and also used for sex determination in which a small amount of amniotic fluid, which contains fetal tissues, is sampled from the amniotic sac surrounding a developing fetus, and the fetal DNA is examined for genetic abnormalities.

[19] Also known as CVS. This is another form of prenatal diagnosis to determine chromosomal or genetic disorders in the fetus. It entails sampling of the chorionic villus (placental tissue) and testing it for chromosomal abnormalities, usually with FISH or PCR.

[20] Botkin (1998), pp. 17, 20.

[21] Deeney (2013), p. 333.

[22] Scott (2006), pp. 153, 161.

[23] Fox (2010), p. 170.

evant concerns also include how resources may be adequately allocated for PGD applications,[24] and the specific PGD-related issues of germ line[25] therapy and genetic enhancement.[26] From the legal perspective, the main consideration would be whether there are appropriate laws or other forms of regulation that are able to contain PGD applications within the framework of the legal system.

As it presently stands, there is a wide divergence of opinions at the international level regarding the moral, ethical and legal applications of PGD. Hence, it becomes even more interesting and pressing when new forms of biomedical technologies, such as gene editing technologies or other similar technologies that 'disrupt' reproduction[27] are thrown into the equation. This is certainly true of the recent breakthrough genome editing tool called CRISPR/Cas9 (CRISPR)[28] which allows scientists to edit the human genome with unprecedented precision, efficiency and flexibility. The possibilities that are presented by what CRISPR may be able to achieve with PGD have therefore prompted fresh concerns.

The most recent international controversy regarding CRISPR's use at the preimplantation, embryonic level is the case of Chinese scientist, Dr. He Jiankui. At the Second Human Genome Editing Summit in Hong Kong in November 2018, Dr. He claimed to have edited the genes of twin embryos successfully using the CRISPR technology.[29] The international community, including the Chinese government, responded immediately with condemnation. Within the span of a few weeks from Dr. He's announcement at the summit, investigative materials emerged from the international discourse relating to Dr. He's apparently secret experimentation with CRISPR. The controversy still continues presently, demonstrating the urgency and timeliness of regulatory governance over the technology, and therefore, what this book aims to achieve.

With our current state of technological growth, the National Human Genome Research Institute has already raised difficult questions that need to be considered in light of PGD and prospective genetic interventions on pre-implantation embryos. For example, how can one distinguish appropriate uses of germ line gene therapy? Who is the arbiter in deciding traits that are 'normal' and those, which may constitute disability or disorder? What are the costs likely to be involved in this kind of

---

[24] Botkin (1998), p. 25.

[25] The germ-line is the reproductive cell lines in organisms. This means that any changes to DNA made to the germ line would be carried forward to future generations. In consequence, modification of the germ-line has invoked strong divided opinions, and have been prohibited in some jurisdictions.

[26] Botkin (1998), p. 25.

[27] Inhorn (2007b).

[28] Hsu et al. (2014), p. 1262. Gene editing tools are, by scientific means, not new inventions. CRISPR is the acronym for Clustered Regularly Interspaced Palindromic Repeats, which is essentially the repetition of DNA in bacteria. However, CRISPR is novel because it is fast and precise. The use of the Cas9 RNA-guided enzyme as a pair of molecular scissors to 'cut' strands of DNA, which carry certain genetic information, allows certain genetic information in the DNA to be removed; and even for new genetic information to be inserted.

[29] BBC (2019).

therapy, and will it only be available to the wealthy? Societal considerations about accepting differences need to be taken into account as well. And most crucially, what are the limits to which this kind of therapy must be subject? Should people be allowed to use germ line gene therapy to enhance basic human traits like height, eye color, other physical characteristics, intelligence, athletic ability and the like? 'Designer babies would become the new aspiration for reproduction. Can the law effectively regulate the use of this process without implicating a loss of human rights and civil liberties, in particular, towards the possible small class of elites who are able to afford the cost of these enhancement procedures? How can the law also ensure that genetic discrimination does not occur towards those who cannot afford these treatments? It is likely that questions of class segregation based on genetic alteration are likely to occur and must be addressed. In the future, sociological and cultural questions will come into play, when we consider the possible effects on society if 'synthetic' people (where whole new genetic codes are created) were to populate freely and become "artefacts of genetic engineering, literally selected from a catalogue."[30]

There is still a lack of international consensus as to the scope and breadth of acceptable pre-implantation genetic interventions, because even the scope of PGD regulation is varied. Discussions on the regulatory, legal or quasi-legal framework in the selected jurisdictions as well as relevant international instruments, serve to indicate how PGD 'regulation'[31] may be reflected upon and complement genetic interventions where necessary, but the key problem we face is a lacuna in law, even within these jurisdictions; and a markedly absent acceptability of a proper regulatory framework, from the world stage perspective.

1.2   Why Is This Work Important?

Due to the nature of PGD technology, how it is applied and the consequences of such technology,[32] it is not surprising that there are various legal and ethical issues[33] that arise in respect of its use.[34] These issues, together with issues of potential genetic interventions that may be applied to humans,[35] continue to be widely debated by geneticists, bioethicists, medical anthropologists, legal practitioners,

---

[30] Silver (1997).

[31] By 'regulation', this refers to differing legislation between states, for example, in the United States and Australia. Other jurisdictions may operate on the basis of 'soft law' (guidelines issued by a medical council or other relevant professional body) or secular pronouncements by religious councils, which raises questions as to whether these have the force of law.

[32] Risks of embryo destruction, misdiagnosis, contamination of genetic analysis tests and such still occur in PGD, as evidenced by a landmark lawsuit in Australia resulting from PGD misdiagnosis.

[33] Morrison et al. (2016), http://bmcmedethics.biomedcentral.com/articles/10.1186/s12910-016-0157-6.

[34] Botkin (1998). See also Everton (2014).

[35] Beauchamp and Childress (2011).

sociologists, science research experts, and academicians.[36] The majority of present academic literature, reproductive health practice and to a certain extent, legislation or other form of regulatory guidelines, indicate the use of PGD in a limited set of circumstances. With the expansion of the repertoire of biomedical technologies in recent years, specifically genome editing, new questions have emerged about the possibilities of pre-implantation genetic interventions. In the meantime, genome editing has simultaneously raised a whole plethora of legal, socio-legal and ethical issues that needs to be considered.[37] Although the bulk of literature suggests that genetic interventions can be very useful in the alleviation of human medical conditions *vis-à-vis* genetic somatic cell treatment in adult patients, the conversation takes on a grimmer tone when considering pre-implantation genetic interventions that may potentially alter the germ-line in human embryos.

I write this book in the anticipation of a vital conversation in the forthcoming years regarding pre-implantation genetic interventions, which I claim is starkly distinguished from "designer babies"[38] concerns. Concerns about designer babies have largely been founded on an illogical projection of dystopia and science fiction, weaving consternation through possibilities of assembly-line type manufactures of blond and blue-eyed babies with athletic prowess, unparalleled musical abilities and astounding cognitive intelligence.[39] The designer baby arguments stem from a fear of repeating eugenic outcomes of the past. Historically, eugenics earned its pejorative nature through one of the largest scale genocides in human history, fueled only by a madman's desire to populate the world with a homogeneity regarded as superior. In contemporary settings, therefore, it is easy to cry out 'eugenics' at the faintest glimpse of characteristic or trait selection, which suggests a desire for genetic or social improvements. This, I believe, is largely grounded on fear and misunderstanding.

As such, I differentiate the necessity for legal regulation of pre-implantation genetic interventions from the speculative designer baby arguments, and instead focus on an attempt at reasoned and reflected conversation about the actuality of this scientific endeavour, and how legislation and/or other regulations can respond to it, based on events such as Dr. He's recent secret experimentation. The idea is not that designer baby concerns are not valid; but that the chasm between egalitarian distribution of use versus inequalities of social structures will continue to be widened if a form of governance is not urgently undertaken. The distinction lies in this book's premise on the necessary framing of legislation that may be accepted on a global level, complemented with other forms of regulatory governance or "new governance",[40] all of which must be contemporarily negotiated (instead of its assumed 'universality') to find balance and encourage voluntary and meaningful adoption and enforcement. In doing so, the formulated criteria that run through the

---

[36] Glenn (2003), pp. 27–28.

[37] Nuffield Council on Bioethics (2018).

[38] Fox (2010).

[39] Yong (2017).

[40] Solomon (2010), p. 591.

course of this book uses philosophical, historical and cultural, and ethical and legal categorizations to create an awareness regarding the future of regulating biomedical technologies, and more specifically, pre-implantation genetic interventions.

In this regard, however, there are two main problematic scenarios that I address in this book. Firstly, the lack of an international consensus on the key issues that would form a truly "international" or global biomedical law approach that has prevented different constitutional systems from incorporating it into their jurisdictions. By this, I refer to the predominantly Western-centric discourse on biomedical technologies and interventions into the human body (as rich and insightful as these may be), that has failed to consider the rising provision of questionable services[41] by Southeast Asian biomedical markets like India, China, Malaysia and Thailand. Some of these are markets have, over the last several years, perpetuated and magnified problems of reproductive trafficking, "cycling" overseas,[42] and fertility tourism, all of which may remain unchecked unless there is appropriate legal intervention. Secondly, the necessity to cast a much wider net in terms of recognition of a multi-layered, multi-level stakeholder interest engagement with the legal and regulatory framework in the provision of biomedical processes, and with more specificity, the pre-implantation genetic interventions framework. The reality of this kind of engagement has been largely ineffective in the last several years, due to a difficulty in effectively managing purposeful stakeholder involvement[43] and the way in which legal scholarship contributes by considering these perspectives.[44]

To achieve these claims to importance, this book engages in the main existing controversy of embryo selection in PGD, and project consequential outcomes of its complementarity to genetic engineering technologies with regard to the research and related science concerning pre-implantation genetic interventions. A narrower tailoring of the issues will explore the prospective boundaries of genetic interventions as part of PGD applications in two main ways. Firstly, by referring to the role of international law in addressing or formulating a general consensus on genetic interventions; and secondly, by undertaking a detailed study of the existing legal or regulatory framework in selected jurisdictions that provide for PGD use, and whether it extends to genetic interventions, and how human rights considerations fits into the equation. The categorizations that shape the legal framework of the jurisdictions are vital in this part of the study. Any potential gaps in these frameworks are also addressed, especially in relation to the treatment of fundamental rights within such spheres. In essence, what is sought, is to advance the ideas of a

---

[41] These services which may include, for example, the practice of sex selection in PGD services, which is banned in most Western countries. In addition, until very recently, countries like India and Thailand finally enacted laws against commercial surrogacy, for foreign citizens; but this may still not adequately address the other pertinent surrogacy provisions that apply regardless of citizenship, including provisions that relate to the welfare of the surrogate child, parenthood provisions, and the like. In Malaysia, the discourse on how to regulate commercial surrogacy has only just begun this year (June 2018).

[42] Whittaker and Speier (2010), p. 363.

[43] Morrison (2017).

[44] Nalbandian (2011), pp. 9, 142.

suitable regulatory or legal framework within the field of biomedical technologies, which are capable of adaptability and evolving to meet the changing needs of PGD and biomedical genetic interventions.

The novelty of this exercise is further presented in the insight of perspectives from South-East Asia, particularly Malaysia and Thailand; insights that will reveal that notions of 'universal' legal standards and the envisioning of human dignity and personhood, may not be as 'universal' in nature presently, especially where legal standards in genetic interventions are concerned.[45] The inclusion of non-Western insight into the future of genetic interventions goes to the heart of the proposition made by this book. This proposes that the regulatory biomedical framework and approach be evaluated on the basis of an updated shared values system that should be constructively negotiated upon and which finds commonality to most, if not all, constitutional legal systems in the world. This may not achieve complete consensus, but if meaningfully constituted, the involvement of perspectives that demonstrate the social and cultural pluralism in non-Western Asian regions (often referred to as the 'Asian values') may equally be important considerations.[46]

The study carried out in this book is an exercise in balancing the conflicts between law and science, and examining the fringes of overlap with the general population of society vis-à-vis human rights. In seeking to explore a balance between the infinite perplexities of legal frameworks, with the monumental advancements of biomedical and scientific technologies, the book puts forward a theoretical recommendation for the establishment of a regulatory or legal or quasi-legal framework that may regulate biomedical technologies on a more concentrated level, particularly where pre-implantation genetic interventions are concerned. I attempted to frame the most appropriate regulatory or legal or quasi-legal framework that is fluid and adaptable enough in the field of biomedical technologies so that a similar framework may be employed on a more constitutionalized level across the globe.

In identifying clear components of commonalities that underlie international human rights instruments, and translating this into a prospective legal and ethical framework for genetic interventions in PGD regulatory frameworks, a comparative methodological outlook from the United States (US), the United Kingdom (UK), and Australia is employed. These countries were specifically chosen because the different implementations of PGD regulation within their respective legal frame-

---

[45] In Chap. 6, I introduce the 'entry points' of regulation. These would refer to points or concerns, which would prompt a particular state to regulate on technologies, depending on the priority of these concerns in the country.

[46] It must be pointed out that although some reference is made to the principles that underlie "Asian values", I do not make the claim that Asian values must take precedence when we consider the 'universality' of human rights issues. Instead, I pose that these Asian values may, in some aspects, be very useful in understanding more deeply the cultural, historical and political landscape of some non-Western jurisdictions, which may lead to a clearer picture why there may be reluctance or inability to implement fully some notions of human rights which may mean a different thing in those non-Western jurisdictions. In addition, the interpretations of notions that are accepted as 'universal', such as human dignity, in international legislative texts and conventions, are often present through different practical elements in an Asian context.

works are useful in exhibiting the primary concerns relating to the technology, and how the law can respond to these concerns. In the meantime, diverging perspectives from Southeast Asia (Malaysia and Thailand) are also examined as PGD regulation in these countries are fragmented and incomprehensive, and have difficulties being enforced. As these states embody some form of legal pluralism and where whole or partial legal authority is based on non-secular and/or religious law, this understanding contributes to a deeper value in comparing responses to PGD application in non-Western settings. Understanding the background of PGD regulation aids in greater clarity on how pre-implantation genetic interventions may also be legally treated in the jurisdictions.

The objectives determined at the outset of the development of the research plan for this book are those directed at the main legal and ethical questions arising from the science of genetic interventions in current and prospective PGD use. There is no question that given the right manner of use and egalitarian access to all sections of contemporary society, PGD is a useful tool in eliminating serious, genetic diseases described in the manner above. However, PGD has its limitations and may not be the right tool for everyone—and it must be borne in mind that PGD does not guarantee that a child will be free from any and all genetic diseases or conditions; it only guarantees freedom from a particular known genetic condition that was being screened for. The practical and lesser spoken about realities of PGD is that serious psychological consideration and trauma can arise on the part of the parents; and that it almost always involves the destruction of embryos.[47] More often than not, questions of morality come into play at this juncture. What moral questions are then engaged when the process requires the destruction of human embryos to achieve its scientific end? Is the prospective value of the process one that can be considered in cost/benefit terms, or is there a transcendent moral imperative that operates to prohibit or at least, limit its application?

In 1998, Jeffrey Botkin raised a variety of concerns about issues pertinent to PGD at that time: but the main ones concerning the pre-implantation stages included germ line gene therapy and genetic enhancement.[48] In 1998, these issues were mainly theoretical in nature, questions that would foreseeably arise in the future. Botkin reminds us of the almost 'sacred' stand about modifications to the germ line gene therapy and quotes Leroy Walters and Judy Palmer in recognizing the literature against germ line gene therapy.[49] With PGD increasing in popularity over the last decade, and with tools such as CRISPR breaking grounds in medical technology, Botkin's fears have become our present. He had stated in his article that "the possibility of introducing functional genes into an in-vitro zygote or embryo seems quite reasonable in the foreseeable future".[50] And the future has now arrived. Botkin's concerns have now become pressing issues that need to be addressed with careful consideration and a sense of urgency.

---

[47] Botkin (1998), p. 20.

[48] Ibid, p. 25.

[49] Ibid.

[50] Ibid.

PGD is no longer a new advancement in the field of ARTs, but its potential for use may not yet be fully explored. However, with this recognition comes the startling discovery that, despite its popularity trajectory in many countries around the globe, PGD as an emerging and continually growing technology is subject to different levels of regulation (and in some countries, none at all). In addition, due to the remarkable advances in science and technology, coupled with greater financial resources and an elevated academic and scholarly pursuit of technological knowledge, the emergence of genome editing tools such as CRISPR is likely to change the landscape of medical and scientific treatments. If genome editing becomes a viable possibility in PGD use, the once-taboo questions of germ line gene therapy and genetic enhancements or interventions, may now represent a dramatic change to how PGD may be marketed and offered as part of fertility treatment services. In this book, I tie in these concerns by answering the following questions:-

(i) Why is it so critical to ensure that pre-implantation genetic interventions are regulated?

(ii) What kind of regulatory or legal models are observed in the analysis of the selected jurisdictions, and what are the possible challenges or implications of regulating biomedical technologies (generally) and pre-implantation genetic interventions (specifically) on both a national and international level (taking into account an understanding of legal, ethical, sociological, religious, industry and cultural markers in the selected comparative jurisdictions)?

(iii) What is the role of international law and a 'universal' shared values system in addressing pre-implantation genetic interventions with PGD use?

(iv) What is the role of human rights discourse on a constitutional level in the selected jurisdictions, and whether it would be possible to impute a negotiated, flexible, shared values system as a means of an international "constitutionalization" of human rights protections, and the formation of a unified model of consideration for all relevant applications of biomedicine in pre-implantation genetic interventions?

1.3   The Comparative Jurisdictions

The research study in this book covers the following selected jurisdictions: the US, the UK, Australia, Malaysia and Thailand. In addition to legislation or other regulations relating to PGD in each of the selected jurisdictions, I also examine how individual national systems employ protective mechanisms of human rights generally to determine its general attitudes and priorities in how pre-implantation genetic interventions may be legally treated.

The analysis of the jurisdictions is made on a thematic basis that considers the level of legal regulation in each of the jurisdictions. In the first group, the US is selected based on the fact of its unique Federalist system, where fundamental constitutional rights are often at odds with each other, and may be fragmented in the individual states. However, there is a wealth of research and development about PGD and genetic interventions in the US, and meaningful and impactful discussions about these technologies, but also because the US has made great progress in terms of refining, creating and innovating ways of treatment for genetic conditions and diseases. In fact, the CRISPR technology referred to above was invented in 2012 by scientists at the University of California, Berkeley, and numerous reports on the CRISPR technology[51] have created simultaneous mass excitement and mass trepidation about the technology. In April 2015, scientists in the US have already announced the use of this technology to challenge Mendel's laws of inheritance,[52] which has remained the cornerstone of genetics for more than a hundred years. Although the US does not have Federal legislation relating to PGD and genetic interventions, the proactivity of the states have proved to be useful in some form of regulation, particularly with the seminal case of *Roe v Wade*[53] by the US Supreme Court which declares that 'a state can assert important interests in safeguarding health, in maintaining medical standards and in protecting potential life'.[54] This is in line with the definition of the concept of federalism under Amendment X of the United States Constitution, which, in fact, guarantees the state's reservation of powers, to enact legislation as they may deem fit.[55]

In the second group, the UK and Australia are included together, because both national legal systems, although through different means, exhibit a sophisticated degree of protective mechanisms for forms of reproductive and genetic interventions that also consider human rights elements. This does not specifically mean however, that the overall human rights protection is strongly ingrained in their respective constitutional cultures.

The UK has been selected because it has employed a national policy on PGD via the National Health Service Commissioning Board (NHSCB). The NHSCB has published a concise Clinical Commissioning Policy for PGD in April 2013, which provides guidance as to the use of PGD.[56] There is also an established oversight for PGD practices by the Human Fertilization and Embryology Authority (HFEA); it must approve and agree that a particular genetic condition is sufficiently serious before any authorized clinics are allowed to use PGD to test for that disease or condition. As such, the UK appears at this juncture to have a regulatory framework, by which to regulate PGD and possibly also genetic interventions, which will be useful for the comparative analyses in this book. In some other aspects, the UK has also

---

[51] Commentators on science and technology blog boards indicate that there are at least 20 CRISPR reports being published every week since January 2015 because the technology has surpassed its expectations and made great inroads with break neck speed.

[52] Miko (2008), p. 134.

[53] 'Roe v. Wade, 410 U.S. 113 (1973)' (*Justia Law*) https://supreme.justia.com/cases/federal/us/410/113/.

[54] Ibid, p. 154.

[55] Sholley (1951).

[56] Prepared by the NHS Commissioning Board Clinical Reference Group for Genetics, April 2013, Reference No. NHSCB/E01/P/a dated April 2013.

become the first country in the world to approve of mitochondrial donation, which, in essence, legalizes three-person-IVF combining genes from three parents.[57] Although this has been recognized as a remarkable feat, the first of its kind in the world, commentators are also wary of the new range of legal and ethical debates that open up discussions of humankind descending on the slippery slope to the legalization of "designer babies".[58]

In the meantime, Australia was selected as it is recognized as a forerunner in Asia in the field of biomedicine and biotechnology. Like the UK, Australia boasts both regulatory and administrative approaches taken by the Australian Commonwealth Government to "control and monitor research and commercial applications of gene technology."[59] On a global scale, Australia has also created an important precedent in the treatment of intellectual property rights (patent rights) pertaining to genetics/genomics and precision medicine,[60] which have the propensity to affect the interpretation of patentable rights in isolated DNAs in the future.

The third and final group comprises Malaysia and Thailand, countries that are not fundamentally skewed towards the protection of human rights generally and with relatively lax regulations on PGD and other reproductive biomedical technologies. Although some respite is offered in these jurisdictions concerning reproductive rights,[61] the more general over-arching priority appears to focus on reproductive tourism[62] and protection for the equality and emancipation of women more specifically in this context. Malaysia is chosen because of its unique position in Asia as a nation with saturation of multi-cultures, languages and religions, and its ability to make a meaningful contribution in the region; Thailand is very much similar but the fundamental commonalities between these two jurisdictions lie in the fact that PGD or ARTs too remain largely unregulated, let alone genetic interventions. For example, in Malaysia, there are no specific laws by which PGD or even ARTs and genetic interventions may be regulated. Guidelines issued by the Medical Council of Malaysia on Medical Ethics and Genetic Services[63] are merely operative guidelines, which do not have the force of law and are severely limited in scope. Thailand's 'legislation' is also similar and appears to be largely symbolic in nature and in response to international hue and cry. In both jurisdictions, because of this lacuna, there are an alarmingly large number of private fertility clinics offering PGD services,[64] amongst others, allegedly for purposes of sex selection and non-medical

---

[57] The draft regulations for mitochondrial donation have been made a UK statutory instrument. See the Human Fertilisation and Embryology (Mitochondrial Donation) Regulations 2015 No. 572.

[58] Turriziani (2014), p. 595.

[59] Polya (2008).

[60] Please see D'Arcy v Myriad Genetics Inc. [2015] HCA 35.

[61] Center for Reproductive Rights (2014). See also Umeda (2015).

[62] Stasi (2015).

[63] Guideline 010/2006 to be read in conjunction with the Code of Professional Conduct and Guidelines applicable to medical practitioners defined under the Medical Act 1971.

[64] Reaney (2004).

needs.[65] Another similarity lies in both these countries' "dualist" legal system. Hence, there may be clashes in formulating suitable laws in Malaysia that results from the need to respect Malaysia's official religion of Islam. Religious bodies such as the Department of Islamic Development and the National Fatwa Council are involved in this process. Thailand's dualist legal system is installed through a military junta. In these circumstances, potential questions of ethics and propriety and the need to address possibilities of medical negligence or malfeasance, often not covered by private insurance claims, have largely been shelved because of these differences.

Further, the necessity of the inclusion of South-East Asian perspectives into this book forms part of the negotiated requirement of drawing on an updated shared values system in moving towards a meaningfully constituted and inclusive international unitary model of governance. The position I put forth in this book is that a shared values system is not simply an acceptance of 'universality'; but instead, must be negotiated meaningfully for proposing a more inclusive international regulatory framework and approach to biomedical technologies.

1.4   Globalization and the Main Concerns

The main hypothesis of this book indicates that despite the globalization phenomenon, current regulatory measures on present and prospective PGD uses (including future use with genetic interventions) are limited in various countries and very much dependent on the type of legal framework adopted, the constitutional culture, and the manner in which human rights (and other) considerations may be taken into account. Whether these measures also have the flexibility and adaptability to regulate modern genetic interventions are also questionable. This fact is a powerful indication of the dilemmas posed by the science. There is a significant challenge to establish a working regulatory framework that will define the boundaries acceptable to different sections of society in different jurisdictions. I highlight the three main considerations of existing academic literature thus far, as follows.

1.4.1   Socio-Legal and Ethical Concerns of PGD and Genome Editing

The vast amount of scholarly work and literature relating to PGD (for human embryos) is focused on primarily the ethical and socio-legal implications of the technology. Across international boundaries, the literature relating to PGD raises very similar concerns (although it may be regulated in some jurisdictions, and not

---

[65] The first so-called designer baby in Malaysia was born in December 2004 in a private medical centre in Kuala Lumpur.

in others). Early objections have been raised by Jürgen Habermas, who draws parallels between eugenic practices and morality, ethics and our pre-supposed sense of autonomy as human being, which could be eroded by PGD and stem cell research. Some other socio-legal, ethical concerns were raised during the late 1990s and early 2000s period, evidenced by the work of Jeffrey Botkin and John Robertson. Botkin identifies ethical issues in PGD that relate to the destruction of prenatal life,[66] the limitation of PGD uses for serious conditions only,[67] how resources to provide PGD services may be allocated,[68] and finally, germ-line gene therapy and genetic enhancements.[69] In the meantime, Robertson focused on the medical versus non-medical dilemma of PGD use, touting possibilities where PGD may be used for gender selection,[70] perfect pitch,[71] and even other non-medical traits.[72] Despite the passage of almost two decades, scholars continue to debate on the socio-legal and ethical controversies that result from PGD. In this period of time, some countries have come to recognize the beneficial propensity of what it offers, and therefore framed the appropriate legislative efforts to govern the provision of its uses. Indeed, Judit Sándor states that perhaps a misconception about PGD is that it has been conflated into a theoretical possibility of "perfecting humankind.",[73] although in actuality it is used to screen for very severe illnesses and diseases. However, I theorize that the main problem about PGD, to which all these socio-legal and ethical concerns are connected, is the underlying fact that it enables a process of selection (of traits, genetic conditions, etc.) in pre-implantation human embryos. In truth, I believe that it is this process of selection that is therefore disturbing to a large majority of scholars.[74]

I further identify that contemporary literature on PGD or other forms of pre-natal screening and testing technologies are much more concerned with this enablement of selection: whether it is for sex selection,[75] the reduction of inheritable cognitive or mental disabilities,[76] or for a matching process to an existing child with a genetic condition (the savior sibling purpose),[77] or more objectionably, for non-medical, non-therapeutic traits (the enhancement debates).[78] This selection process of human embryos is further 'complicated' by also needing to recognize the privacy, autonomy and individual liberties of prospective parents or mothers who choose to engage

---

[66] Botkin (1998), p. 20.
[67] Ibid, p. 21.
[68] Ibid, p. 25.
[69] Ibid.
[70] Robertson (2003), pp. 213, 214.
[71] Ibid, p. 215.
[72] Ibid, p. 216.
[73] Sándor (2015), p. 357.
[74] Brock (2009).
[75] Deeney (2013).
[76] Basas (2014), p. 1035.
[77] Gross (2003), p. R541.
[78] Savulescu (2007a).

in that process of selection. This is particularly acute in jurisdictions that employ a much broader mechanism of protection for a right to privacy, which correctly includes the right for a person to decide on their bodily integrity. When these conundrums are put together for consideration, it is very clear why there are varying degrees of disagreements regarding PGD indications.

The emergence of much more sophisticated genome editing tools in 2011, however, has invited a new storm of concerns, both socio-legal and ethical. The emergence of CRISPR revitalized conversations and discourse about the future of humankind, creating a great divide in opinions. The same is no less disconcerting even within the scientific and medical fields. In fact, Dr. Jennifer Doudna, the leading biochemist who pioneered the discovery of CRISPR/Cas9 is acutely aware of the possibilities that may be offered by the technology; stating the importance for ethical discussions to keep up with the technology.[79] Possibly one of the most controversial issues that CRISPR has raised is the renewed possibilities of creating "designer babies"[80] and the genetic enhancement debates, where the forefront of proponents include renowned libertarian philosophers such as John Harris and Julian Savulescu, and the main supporter of liberal 'eugenics' such as Nicholas Agar. Therefore, the debates about germ-line gene therapy, editing or other forms of interventions, have resurfaced with a vengeance.

Although I do not believe that the creation of designer babies is feasibly and scientifically possible at this juncture,[81] the concerns, even from the viewpoint as a not-so-distant spectator, are valid and reasonable. However, the most enthralling aspects of these renewed concerns are likely to be the genetic enhancement debates (that, once fully considered, implicates the designer baby concerns as part of its discourse). In this realm, besides the obvious issues about risks and safety of the technologies per se, the discourse has also been focused on the dramatic elements that attach to the characteristics and pointers of humanity, both from the ethical and legal perspective.

Michael Sandel made his views known on "medically unnecessary genetic engineering",[82] reinforcing his stand from his work, *The Case Against Perfection*.[83] Citing how the moral fabric of humankind may be eroded through the trend of hyper-parenting, the rush for genetic mastery is a "troubling overall trend."[84] In his work, *The Wisdom of Repugnance*,[85] Leon Kass argues that the principles of reproductive freedom has the effect of venturing down the slippery slope of "producing

---

[79] Lincoln (2018).

[80] Fox (2010).

[81] Scientific reports regarding the creation of designer babies in the mold of a supermarket-type selection of traits and characteristics have indicated that although this is theoretically possible, the multitude of genes within the human organism that interact with each other and influences certain traits or characteristics is not so easily isolated.

[82] Brookings (2001).

[83] Sandel (2004), p. 51.

[84] Brookings (2001).

[85] Kass (1997), p. 17.

children whose entire genetic makeup will be the product of parental eugenic planning and choice."[86] Francis Fukuyama's work is abundant with dire predictions of the future of humankind, drawing on the dystopian fictional fantasies of George Orwell's *1984* and Aldous Huxley's *A Brave New World*. Although there is some truth in the visions he has projected, it is easy to overlook that his main concern is how technologies shape our future from a political viewpoint: "human nature shapes and constrains the possible kinds of political regimes, so a technology powerful enough to reshape what we are will possibly malign consequences for liberal democracy and the nature of politics itself."[87] In essence, Fukuyama's premise hinges on the effects of genetic engineering on the political functioning of democracies, because "political institutions rest on a notion of equality that is rooted in our shared biological heritage........constrains each of us in certain ways and imposes a relatively uniform set of abilities and talents across the board."[88]

On the other end of the spectrum, John Harris in his work, *Enhancing Evolution: The Ethical Case for Making Better People*,[89] puts forward his hypothesis in two parts, given the possibilities of human enhancement through genetic engineering. First, that "human enhancement is a good thing and our genetic heritage is much in need of improvement"[90]; and secondly, because human enhancement is more likely to confer benefits and to avoid harm, then there is therefore a moral obligation for parents to enhance their children or future offspring. Although I concur with Harris' first point, I would be more inclined to his second contention (regarding the moral obligation of parents) if it is developed within a flexibly permutable framework of reasonable regulations, limitations or other regulatory constraints. Savulescu also puts forward the following provocative position:

> I want to argue that far from being merely permissible, we have a moral obligation to enhance ourselves and our children. Indeed, we have the same kind of obligation as we have to treat and prevent disease. Not only can we enhance, we should enhance.[91]

In addition to this, Savulescu has also written extensively on reproductive ethics, and coined the principle of "procreative beneficence",[92] touting reasons why parents should enhance their children.[93]

In temperance of these opposing ends of the spectrum, I. Glenn Cohen presents a more middle-ground view regarding enhancements, going into clear divisions between the various types of enhancements and rejecting policies of a 'one-size-fits-all' nature. Instead, Cohen suggests that "decision making has to be done at the

---

[86] Ibid, p. 24.

[87] Fukuyama (2003), p. 7.

[88] Cook (2002).

[89] Harris (2010).

[90] Ibid, p. 8.

[91] Savulescu (2009), p. 417.

[92] Savulescu (2001), p. 413.

[93] Savulescu (2007b), p. 284.

level of particular enhancements, or in some cases, categories of enhancements."[94] The need to properly define the borders of the kind of enhancements referred to is also crucial, because genetic enhancements are not homogenous and have different implications.[95]

Taken from the perspective of considering genetic enhancements for adult human persons is enough of a dilemmatic concern. As adults, human beings have been exposed to variables of normative bias, conditioning through social environments and education, life experiences, and the encounters with a gamut of physical, emotional and psychological emotions and feelings. However, as adult persons, decisions made are presumed to have been arrived at through the awareness of the self. Individual autonomy is presumed. The outcomes of genetic engineering concerns for adult persons therefore, are no less concerning, but have verily come to be accepted as part of the process of adult autonomy. Having established this premise, then we must ask the question: what about the life of future, prospective children? If adults may be able to partake in the benefits of genetic engineering, should future children not be accorded that equivalency; especially if these benefits could be reaped from as early as the embryonic stage?

The answer truly is not as clear cut. Bearing all these concerns in mind, and as is evidenced by the underlying premise of this book, my claim in respect of enhancements for future children is to attribute these choices or decisions in respect of individual (parental) autonomy, but also recognizing that this autonomy cannot be completely unfettered, unabated and unregulated. If we choose to recognize that the desire to enhance is distinct from the obligation to enhance, the implications on how the law can and may respond to such distinctions can be refined more adequately. As such, this is a most timely endeavor as we enter the age of genetic engineering, and may have the ability to alter human life at the pre-implantation level in reproductive technologies.

1.4.2 Differences in Legal/Regulatory Frameworks & Attitudes

In the adoption of a suitable legal regulatory framework in a particular jurisdiction, variables such as political will, economic growth and sustainability, religious considerations, societal values and industry-driven attitudes, amongst others, are relevant factors of how useful a specific regulatory measure in a particular country will prove to be. The prioritization of these variables as important national interests, will affect the development and success of any foregoing regulatory measures.

In addressing questions of PGD and genetic interventions, it is helpful to study the types of models that would be suitable to attain the objectives sought here. For example, a comprehensive legal framework will likely encompass equally comprehensive approaches in handling, processing and licensing PGD and genetic inter-

---

[94] Cohen (2014), pp. 645, 686.

[95] Ibid, pp. 646–648.

ventions; with strong emphasis on dialogues with new organizations and a veritable system of sanctions and punishments for violations of the newly minted legislation. If law is the tool by which to manage PGD and genetic interventions, it is important to bear in mind that it should "avoid over and premature regulation, which happens when the law jumps in too quickly to the latest scientific advances without leaving sufficient time for reflection on the ethical and social implications of a new technology".[96] This is the *raison d'etre* of the model of foresightedness, which may not always be the most effective framework because of the over-zealous implications that may ensue. However, the opposing extreme of foresightedness is a deep reluctance on the part of legislators, and even the judiciary, to extend their knowledge to matters of science.[97] The French model of regulatory framework, for example, is based on principles of re-examination and revision of the provisions of a particular law on a periodic basis; this demonstrates some form of flexibility to adapt to changing conditions in technology, societal attitudes and the interpretive role of law in scientific matters.

In conceiving a regulatory state, Cass Sunstein offers some useful guidance[98] in his book covering regulatory policy in the United States. In essence, Sunstein is a proponent of government regulation and recognition of rights in regulation, for the purposes of promoting the health, welfare, safety and autonomy of citizens, and emphasizes that the key functions of such regulation must be borne in mind. This translates to the *purpose* by which a regulatory state is to be conceived. On this basis, this book borrows from some of Sunstein's considerations of the essential factors to take into account when proposing a regulatory framework. Although Sunstein's treatises focus on a defense of regulatory statutes and a very involved role of judicial interpretation in regulatory policy, this may provide us with an understanding of the relevant approaches that may be looked at when conceiving a general consensus or threshold for law's entrance into this scientific, technological research domain, allowing us to conduct a balancing test of risks versus benefits.

Legal and ethical questions regarding PGD and genetic interventions have certainly been raised across the world. However, it is critical to note that even though different countries and regions have raised similar issues, they have also raised contrasting views. The selected jurisdictions, in fact, give credence to these contrasting views. These jurisdictions have attempted to impose (and some, with a measure of relative success) legal boundaries relating to human genetic engineering[99] specifically. Legislation and/or some type of regulatory framework with regards to PGD and genetic interventions are less consistent, however. For example, in the US, the vociferous feature of the US Constitution as a source by which to derive inspiration, may be one of the reasons why regulation of PGD and genetic interventions at the federal level is markedly absent. The oft-described *laissez-faire* approach of the

---

[96] Sándor (2012), pp. 1142–1161.

[97] Ibid, p. 1160.

[98] Sunstein (1993).

[99] Annas et al. (2002), pp. 151–178.

federal government to reproductive issues has left a gap in this growing area of reproductive rights. Bratislav Stankovic wrote:

> The dearth of federal public policy guidance in reproductive matters extends to a lack of professional PGD self regulation. No national regulations are officially sanctioned, although opinions and recommendations of reproductive medicine agencies have been published. Heterogeneity of clinical liability exists on a national level resulting from clinics in different states performing discrete steps of the IVFPGD procedures. In such a heterogeneous environment of plausible deniability, a regulatory gap exists. This gap is partially filled by a voluntary mandate against the performance of procedures that are morally repugnant.[100]

Michael Sandel[101] launched a philosophical interpretation about human desire for materiality of enhancements, which, in his opinion, is objectionable.[102] Accounts of philosophy and public policy are necessary discussions in techno scientific research of this nature.[103] The width and breadth of literature, however, reveals that there is a gap in regulating PGD in the US, although authors often espouse public policy arguments in this area. Numerous academic opinions have called for the need to regulate PGD and genetic interventions, whether at the federal or state level. The challenges faced at this juncture, though, hinge upon the fundamental American values of due process and equal protection,[104] subjecting any kind of legislation to strict requirements of narrow construction that must be rationally related to a legitimate government interest.[105] At present, the governing position in the US for PGD is of 'soft law' nature, the 'unspoken' agreement that no germ line cell therapies be permitted; and although there appears to be a large consensus agreeing that the use of PGD to create 'designer babies' be prohibited, there is also a minority that embraces the freedom of genetic enhancements. With the passage of time and as genetic research continue to break new ground, however, there may be a greater need for a more effective and enforceable legal framework to tackle rising issues of greater complexity.

The literature regarding the role of law in the UK reveals a sophistication of a seemingly stable and adaptive legal framework,[106] which is an attractive means by which to emulate best practices. In addition to making general provisions with regards to human genetic engineering, the Human Fertilisation and Embryology Act

---

[100] Stankovic (2005), p. 3.

[101] Sandel (2004).

[102] Michael Sandel believed in the purity of human talent which, in his opinion, does not need genetic enhancements.

[103] Essentially, these areas of medical enhancements involving human subjects, may fall within the broad heading of eugenics, which according to Frederick Osborn, involves "social philosophy".

[104] Amendment XIV of the Constitution which states: "No state shall make or enforce any law which shall abridge the privileges or immunities of citizens of the United States nor shall any state deprive any person of life, liberty, or property, without due process of law; nor dent to any person within its jurisdiction the equal protection of the laws."

[105] In US constitutional jurisprudence, there is a host of Supreme Court decisions emphasizing the unwavering narrow tailoring of a matter that is compelling to state interest. See Stone et al. (2013).

[106] The Human Fertilisation and Embryology Act 1990 administered by the Human Fertilisation and Embryology Authority.

(HFEA Act) has, in the past decade, allowed for the legality of sex selection only when there is substantial risk concerning gender-related diseases being passed on; and at the same time, prohibiting such selection for family balancing reasons.[107] From the selection of embryos for its genes, to the oft-debated concept of the savior sibling selection,[108] and most recently, the legalization of mitochondrial DNA donation, it is clear that the UK has made substantial progress in dealing with PGD-related issues and genetic interventions. An oversight body in the form of the HFEA under the HFEA Act as the arbiter of conditions that would necessitate PGD interventions, serves as a useful system of supervision and avoids arbitrary, economic considerations on the part of medical professionals in providing PGD services. Although the regulation of genetic interventions in the UK is limited at the moment, there is a sensitive recognition of the numerous ethical concerns that arise in dealing with human subjects.[109] The literature review also indicates a non- statutory body of regulations that provides ethical guidelines for research and use of genetic interventionist technologies on human subjects.[110] However, despite the progressive nature of the UK's regulatory framework, many critics are of the view that these progressions have intensified the "slippery slope" debate of the use of PGD, with great terror that, for example, 'designer babies' could be a realization of the near future. Critics also question the ability of the law to grapple with issues of disability, non-discrimination and the rights of future children, amongst others, which largely remain unaddressed by legislation.

The prominence of literature in Australia, however, paints a different picture as to the treatment of PGD and genetic interventions. Unlike the US, Australia, which is also a legal system highly dependent on its early roots in federalism, has managed to achieve greater polarity and clarity in regulating PGD and genetic interventions. A survey of Australian literature indicates a relatively stable and continually growing legislative effort (both at the Commonwealth and state level, together with national ethical guidelines) towards the treatment of reproductive rights (including PGD) and genetic interventions.[111] At the state level, Victoria, Western Australia, South Australia and New South Wales have specifically legislated on ARTs[112] (which include PGD). Although it is true that Australian federalism operates in a somewhat similar manner as the US, the literature review indicates that there is a consistent application of PGD guidelines in the different territories of the Australian

---

[107] Francioni (2007).

[108] Upon the recommendation of the Department of Health, the Human Fertilisation and Embryology Act was amended in 2008 to permit the creation of saviour siblings *"to enable the identification of a tissue match for an older sibling suffering from a life-threating illness, where umbilical cord blood is to be used in treatment"*.

[109] Doherty and Sutton (1997).

[110] UK Medicines for Human Use (Clinical Trials) Regulations 2004.

[111] Polya (2008).

[112] These include: The Assisted Reproductive Treatment Act 2008 in Victoria; The Western Australian Human Reproductive Technology Act 1991 in Western Australia; The Assisted Reproductive Treatment Act 1988 in South Australia; and The Assisted Reproductive Technology Act 2007 in New South Wales.

Commonwealth.[113] Although these guidelines do not strictly have the force of law, the consequences of non-compliance have a significant effect on the national accreditation system of clinics that offer ARTs.[114] Notwithstanding these positive achievements, Helen Szoke has criticized the Australian regime for being "…a rich tapestry of diversity in terms of regulatory structure, or a patchwork of regulatory stitching lacking cohesion and order."[115] Michelle Taylor-Sands[116] further quotes Isabel Karpin and Belinda Bennett in recognizing that some aspects of child welfare, for instance, are not clearly enunciated in the Australian regulatory framework. The Australian system also does not adequately frame the use of PGD in savior sibling selection.[117] These authors continue to emphasize a need for greater clarity in PGD regulations, and better protection of the welfare of the child in such processes.

Despite small disparities on the ethical issues raised in the selected jurisdictions, most of these ethical issues are similar in many dimensions. One, they are all concerned with the potential health, mental and human rights implications that may arise due to PGD and processes of genetic interventions. Two, they focus on the long-term impacts of the process on society and future generations. Three, they are concerned with the social and ethical reviews or controls that should be placed on these processes. Four, fundamental human rights issues cannot be separated from the operation of these processes. Since science and technology is universal, there is a very high possibility that the impact of genetic interventions has been or will be seen in nearly all parts of the world, even if in the slightest way.[118]

Some literature that expands on PGD discussion in the Middle East could also be useful in the context of recognizing social, cultural and religious aspects in the formulation of suitable legislation in this area of 'technoscience' for Muslim communities.[119] Although the Middle Eastern jurisdiction does not form part of this work, some of these perspectives may provide a suitable ground for comparison with Malaysia, for example, which is also a Muslim country. Indeed, Marcia Inhorn has recognized that the Muslim way of life in practising ARTs and PGD is significantly different from that of Western practices,[120] due to compliance with the Islamic

---

[113] See "The Ethical Guidelines on the Use of Assisted Reproductive Technology in Clinical Practice and Research", issued by the National Health and Medical Research Council of the Australian Government in June 2007.

[114] Sands (2013), p. 24.

[115] Szoke et al. (2006), p. 187.

[116] Polya (2008).

[117] A saviour sibling is "created" by selecting an embryo that, when born, will be able to provide stem cells or healthy tissue to an older sibling suffering from a serious medical condition, that may be treated by the donation from the saviour sibling.

[118] Smiley (2005).

[119] This refers to communities whose regulation of daily public and private life is based on a legal system governed by Islam.

[120] Inhorn (2007a).

Syariah[121] law with prominent emphasis on religious *fatwas*[122] issued by the head of the religious councils, or recognized religious experts in each respective country.[123] In addition to these religious tenets, other factors such as social, cultural, economic and political considerations have impacted tremendously on the way ARTs and PGD is viewed in these Muslim states.

As a starting point, what has been emphasized in these societies is a 'right' to reproductive health[124] following the deliberations around issues of infertility and a form of social need by many childless Muslim couples in the region. For example, the famed religious Al-Azhar University in Egypt issued the first *fatwa* on ARTs in 1980 and this particular fatwa has been deemed to be the foremost authority in this area and is widely accepted throughout the Sunni Muslim world.[125] However, despite the authoritative nature of the *fatwa* and national guidelines, there is a lack of legal regulation in PGD in Egypt. Inhorn, and indeed, many other commentators, question the extent to which those in the medical profession in the Muslim world accept these *fatwas*. Similar to the position within the UK, questions therefore arise as to the legal effect of these guidelines and *fatwas*, and this is not an easy question to address due to the varied interpretation of the *fatwas* in different Islamic countries. The same is true of the interpretation of *fatwas* in a multi-racial, multi-cultural and multi-religious nation like Malaysia.

1.4.3 An International Regulatory Benchmark?

In seeking to address a possible framework of international regulations, what I have found has demonstrated a material lack of consensus on how PGD and genetic interventions may be governed. There are some existing frameworks for governing genetic interventions and also some other instruments regulating reproductive health and rights, which may loosely deal with other instrumental issues of

---

[121] Shari'a law is a body of religious instructions that have a quasi-'legal' framework, regulating various aspects of daily private and public life of observant Muslims living in a system based on Islam. The sources of Islamic Shari'a law include the Quran, Sunna (traditions) and Hadith (statements) of Prophet Mohammed, as well as opinions of Islamic scholars and fatwas issued by mufti or an Islamic council in a particular state or nation.

[122] The Islamic Supreme Council of America defines a fatwa as "an Islamic legal pronouncement issued by an expert in religious law (mufti), pertaining to a specific issue, usually at the request of an individual or judge to resolve an issue where Islamic jurisprudence (fiqh) is unclear."

[123] This is also largely dependent on the treatment of religious morality depending on the sects of Islam, comprising Sunni Islam and Shi'ite Islam.

[124] In Islam, it is believed that the holy Quran encourages marriage, the creation of a family and reproduction. It is not surprising, therefore, that due to increasing rates of infertility among Muslim couples, other means of enjoying reproductive health are sought by these childless couples. Since the Quran also expressly prohibits legal adoption and the donation of sperm or ovum by a third party to a childless Muslim couple, traditional Islamic views on reproductive technologies that would encourage conception, pregnancy and children, began to change.

[125] Inhorn (2007a), p. 190. Please see also: Serour (2005), pp. 185–190.

reproduction itself. However, the problems of these 'regulations' are often related to issues of consent and ratification in individual nations who have subscribed to them. As ideal as these instruments set out to be, it must be equivocally recognized that there is often a challenge in subjecting a 'violation' of these instruments to proper sanctions. The Americas and the European Union have demonstrated some legislative capabilities but are limited in applicability on a jurisdictional basis. Any attempts at this kind of framework are not present in South-East Asia.

This raises questions on the role of international law in addressing or framing consensus on genetic interventions in PGD use; and in the process, recognizing possible variables that need to be taken into account in subscribing to an international regulatory framework. The relevant problems associated with this kind of internationalized effort will, without doubt, present a wide-ranging plethora of considerations that must be sensitive to territorial sovereignty and political peculiarities.

In hoping to ascribe to an international regulatory framework, consensus as the foundational point of acceptability is a hurdle to deal with. Consensus, for various reasons, is likely the most challenging notion to achieve on the international front. Indeed, in many instances, there can be international gridlock when states attempt to negotiate matters of international affairs.[126] The concept of consensus, therefore, may be impeded because of differing national interests and multi-layered, multi-factorial international outlooks. Combining the interplay between state consent and international negotiation in these instances, the role that is played by international law, and particularly, the manner in which its laws are being made, have become much more diverse than before. Alan Boyle's and Christine Chinkin's work, *The Making of International Law* is perhaps a suitable reference in understanding how international law is no longer confined to processes of treaties and conventions at the state level. Instead, international law is made "in a large number of fora, including multilateral processes, tribunals, and the organs of international organizations."[127]

Bearing this in mind, this book also looks to questions of universality, or common values that have traditionally been regarded as universal in nature, within the scope of international biomedical laws. In the same manner that international law is negotiated between states in the very first place, I also suggest that common values should likewise be subject to a process of negotiation in contemporary settings, particularly where countries in the global south may be concerned.

Due to the globalization phenomenon, and in an effort to bridge the gap between the divergence of attitudes in understanding the impact of biomedical technologies on pre-implantation genetic interventions and its instrumentalities, the concluding hypothesis in this book is that human rights as a key approach, plays a fundamental role in shaping the awareness, growth, development and fine tuning of a regulatory framework in a constitutional environment, because it reveals values that are capable of transcending international boundaries, regardless of geographical location, constitutional pluralism, cultural and social determinism, and historical, legal,

---

[126] Chatham House (2015), https://www.chathamhouse.org/london-conference-2015/background-papers/elusive-consensus.

[127] Pronto (2008), pp. 601, 602.

social and economic climates. Flowing from this is the remarkable differences, but also similarities, in various countries that adopt a variance of observable regulatory, legal or quasi-legal models of regulation. These models may prioritize, for instance, religion over constitutional law or fundamental rights. Other countries may consider industry drives and economic development and sustainability as the main factors in its regulatory regimes, signifying that there may be prevalently similar industries in different countries that adopt this model. Ultimately, a study of the components of these models of laws provide utility and elements that may serve as a flexible and adaptable blueprint to augment the basis of a regulatory framework on an international level in respect of biomedical pre-implantation genetic interventions.

References

Annas G, Andrews L, Isasi R (2002) Protecting the endangered human: toward an international treaty prohibiting cloning and inheritable alterations. Am J Law Med 28:151–178

Basas CG (2014) What's bad about wellness? What the disability rights perspective offers about the limitations of wellness. J Health Polit Policy Law 39:1035

BBC News (21 January 2019) China Turns on "Gene Editing" Scientist. https://www.bbc.com/news/world-asia-46943593

Beauchamp T, Childress J (2011) Principles of biomedical ethics. Oxford University Press

Botkin J (1998) Ethical issues and practical problems in preimplantation genetic diagnosis. J Law Med Ethics 26:17

Brock DW (2009) Is selection of children wrong? In: Human enhancement. Oxford University Press

Brookings (30 November 2001) Event summary: a debate on the ethics of genetic engineering. Brookings. https://www.brookings.edu/opinions/event-summary-a-debate-on-the-ethics-of-genetic-engineering/

Center for Reproductive Rights (20 February 2014) Malaysia. Center for Reproductive Rights. https://www.reproductiverights.org/our-regions/asia/malaysia

Chatham House (2015) The elusive consensus in international affairs. Chatham House. https://www.chathamhouse.org//node/17691

Cohen IG (2014) What (if anything) is wrong with human enhancement? What (if anything) is right with it? Tulsa Law Rev 49:645

Cook A (17 July 2002) Our post-human future: consequences of the biotechnology revolution by Francis Fukuyama. PopMatters. https://www.popmatters.com/our-posthuman-future-2496243108.html

Deeney MS (2013) Bioethical considerations of preimplantation genetic diagnosis for sex selection. Wash Univ Jurisprud Rev 5:333

Doherty P, Sutton A (1997) Man-made man: ethical and legal issues in Genetics. Four Courts Press

Everton K (2014) Walking the edge with controversial use of Pre-Implantation Genetic Diagnosis (PGD): opinions and attitudes of genetic counsellors. University of South Carolina Scholar Commons

Fox D (2010) Retracing liberalism and remaking nature: designer children, research embryos, and featherless chickens. Bioethics 24:170

Francioni F (2007) Biotechnologies and international human rights. Hart Publishing

Fukuyama F (2003) Our posthuman future: consequences of the biotechnology revolution. Farrar, Straus and Giroux

Glenn LM (2003) Crossing species boundaries. Am J Bioethics 3:27–28

Gross M (2003) Dawn of the saviour sibling. Curr Biol 13:R541

Harris J (2010) Enhancing evolution: the ethical case for making better people. Princeton University Press

Heinemann T, Honnefelder L (2003) Principles of ethical decision making regarding embryonic stem cell research in Germany. Bioethics 16:530

Hsu P, Lander E, Zhang F (2014) Development and applications of CRISPR-Cas9 for genome engineering. Cell 157:1262

Inhorn M (2007a) Reproductive disruptions and assisted reproductive technologies in the Muslim world. In: Reproductive disruptions: gender, technology and biopolitics in the new millennium

Inhorn M (2007b) Reproductive disruptions: gender, technology and biopolitics in the new millennium, vol 11. Berghan Books

Kass L (1997) The wisdom of repugnance. New Republic:17

Kendler K, Greenspan R (2006) The nature of genetic influences on behaviour: lessons from simpler organisms. Am J Psychiatry 163:1683–1694

Lewis R (2017) Human genetics: concepts and applications, 12th edn. McGraw-Hill Publishing Company

Lincoln R (16 May 2018) CRISPR pioneer Jennifer Doudna explains gene-editing technology in Prather lectures. Harvard Gazette. https://news.harvard.edu/gazette/story/2018/05/crispr-pioneer-jennifer-doudna-explains-gene-editing-technology-in-prather-lectures/

Miko I (2008) Gregor Mendel and the principles of inheritance. Nat Educ:134

Morrison M (2017) "A good collaboration is based on unique contributions from each side": assessing the dynamics of collaboration in stem cell science. Life Sci Soc Policy 13

Morrison M, Dickenson D, Lee SS-J (2016) Introduction to the article collection "Translation in healthcare: ethical, legal, and social implications". BMC Med Ethics 17

Nalbandian E (2011) Sociological jurisprudence: Roscoe Pound's discussion on legal interests and jural postulates. Mizan Law Rev 5:9

National Human Genome Research Institute. (National Human Genome Research Institute (NHGRI)) https://www.genome.gov/

Nayal MB (2013) Preimplantation genetic diagnosis. http://emedicine.medscape.com/article/273415-overview

Nuffield Council on Bioethics (2018) Genome editing and human reproduction: social and ethical Issues. Nuffield Council on Bioethics

Polya R (2008) Chronology of Genetic Engineering Regulation in Australia: 1953–2008. https://www.aph.gov.au/About_Parliament/Parliamentary_Departments/Parliamentary_Library/pubs/BN/0809/ChronGeneticEngineeringA

Pronto AN (2008) Some thoughts on the making of international law. Eur J Int Law 19:601

Reaney P (25 September 2004) In search of baby perfect. Center for Genetics and Society. http://www.geneticsandsociety.org/article.php?id=1469

Resnik DB, Vorhaus DB (2006) Genetic modification and genetic determinism. Philos Ethics Humanit Med 1

Robertson JA (2003) Extending preimplantation genetic diagnosis: medical and non-medical uses. J Med Ethics 29:213

Roe v. Wade, 410 U.S. 113 (1973) (Justia Law) https://supreme.justia.com/cases/federal/us/410/113/

Sandel M (2004) The case against perfection. Atl Mon 293:51

Sándor J (2012) Bioethics and basic rights: persons, humans and boundaries of life. In: Rosenfeld M, Sajos A (eds) The Oxford handbook of comparative constitutional law. Oxford University Press, pp 1142–1161

Sándor J (2015) The ethical and legal analysis of embryo preimplantation testing policies in Europe. In: Sills ES (ed) Screening the single euploid embryo. Springer International Publishing

Sands MT (2013) Saviour siblings: a relational approach to the welfare of the child in selective reproduction. Routledge

Savulescu J (2001) Procreative beneficence: why we should select the best children. Bioethics 15:413

Savulescu J (2007a) Genetic interventions and the ethics of enhancement of human beings. In: The Oxford handbook of bioethics. Oxford University Press

Savulescu J (2007b) In defence of procreative beneficence. J Med Ethics 33:284

Savulescu J (2009) Genetic interventions and the ethics of enhancement of human beings. Read Philos Technol:417

Scott R (2006) Choosing between possible lives: legal and ethical issues in preimplantation genetic diagnosis. Oxf J Leg Stud 26:153

Serour GI (2005) Religious perspectives of ethical issues in ART: Islamic perspectives of ethical issues in ART. Middle East Fertil Soc J 10:185–190

Shanks P (2005) Human genetic engineering: a guide for activists, skeptics, and the very perplexed. Nation Books

Sholley JB (1951) Constitution of the United States of America. In: Cases on constitutional law. Bobbs-Merrill

Silver L (1997) Remaking Eden: how genetic engineering and cloning will transform the American family. Avon Books Inc

Smiley S (2005) Genetic modification: study guide (exploring the issues). Independence Educational Publishers

Solomon JM (2010) New governance, preemptive self-regulation and the blurring of boundaries in regulatory theory and practice. Wis Law Rev:591

Stankovic B (2005) "It's a Designer Baby!": Opinions on regulation of preimplantation genetic diagnosis. UCLA J Law Technol:3

Stasi A (2015) Maternal surrogacy and reproductive tourism in Thailand: a call for legal enforcement. Ubonratchathani University 17

Stock G (2003) Redesigning humans: choosing our genes, changing our future. Mariner Books

Stone G et al (2013) (Chapter 7) Constitutional law, 7th edn. Wolters Kluwer

Sunstein C (1993) After the rights revolution, reconceiving the regulatory state. Harvard University Press

Szoke H, Neame L, Johnson L (2006) Old technologies and new challenges: assisted reproduction and its regulation. In: Freckelton I, Petersen K (eds) Disputes & dilemmas in health law. Federation Press

Turriziani J (2014) Designer babies: the need for regulation on the quest for perfection. http://scholarship.shu.edu/cgi/viewcontent.cgi?article=1595&context=student_scholarship.   Law School Student Scholarship 595

Umeda S (6 April 2015) Thailand: new surrogacy law I global legal monitor. Library of Congress. http://www.loc.gov/law/foreign-news/article/thailand-new-surrogacy-law/

Whittaker A, Speier A (2010) "Cycling Overseas": care, commodification, and stratification in cross-border reproductive travel. Med Anthropol 29:363

Yong E (2 August 2017) The designer baby era is not upon us. The Atlantic. https://www.theatlantic.com/science/archive/2017/08/us-scientists-edit-human-embryos-with-crisprand-thats-okay/535668/

Chapter 2
The Legacy of Eugenics in Contemporary Law

Abstract This chapter introduces the philosophical foundations of eugenics as a starting point, because this supports the reasoning that proposes a form or regulatory or governance framework for pre-implantation genetic interventions. Through a historical exploration of the laws of human inheritability of conditions, and the rise of national eugenic policies, the premise made here is that a wholesale free-for-all use of emerging biomedical technologies, particularly where those technologies involve possibilities to intervene into the human genome, may be interpreted to result in eugenic consequences though a process of selection, and also impacts the operability of contemporary laws. Even if the principle of autonomy is respected, as it is in the case of a new form of "liberal eugenics", I provide three main reasons why this concept is flawed, and why a more meaningful capitulation of the effects of genetic interventions particularly in the scope of human reproduction must be very carefully evaluated. Instead, I advance the call for a reinterpretation of eugenics in light of embryo selection in biomedical and reproductive technologies; founded upon limits that do not encroach on another individual's rights and liberties.

This book begins in this chapter by addressing the philosophical foundations of eugenics as a starting point. The modern understanding of eugenics is admittedly markedly different from the discourse of yester-years; the reason this begins with an explanation about eugenics is to impose a challenge to existing mindsets that pre-implantation genetic interventions are unequivocally eugenic in nature. This is crucial in the theoretical discussion on genetic interventions in the reproductive health and technological landscape, because the general motivation that spurs the human race towards advancement, life expectancy, and quality of life involve elements of 'selection',[1] which is a key feature of eugenics.

---

[1] This understanding of 'selection' lies at the heart of this book, and the discussion why pre-implantation genetic interventions may be supported or opposed. The embryo selection phenomenon (which is a necessary part of PGD) is one of the key components in various polarizing debates.

© Springer Nature Switzerland AG 2019
P. L. Lau, *Comparative Legal Frameworks for Pre-Implantation Embryonic Genetic Interventions*, https://doi.org/10.1007/978-3-030-22308-3_2

In the basest of terms, the tendency for human beings to exercise full personality and autonomy through the selection process, whether it be for a life partner, a career and even choices of food—all strive towards a common goal to ensure a meaningful lifespan, the inurement of our fit into an ever changing society. The importance of this social aspect is valuable as an underlying and overarching logic which supports the reasoning that proposes a form of regulatory or governance framework for genetic interventions and Pre-Implantation Genetic Diagnosis (PGD) in this book. (For ease of reference, this will hereinafter be referred to as pre-implantation genetic interventions). In essence, the natural proclivities of humankind in these social settings, in essence, can carry forward into the use of technologies; an unfettered use of, and access to, these technologies may result in consequences that smack of 'bad' eugenics through a process of selection even if it is based on individual autonomy.

Section 2.1 traces the historical and developmental roots of eugenics in its infancy, emphasizing that there is no single pinpointed theory of eugenics. Instead, this section observes that there is a collection of societal perceptions of eugenics, the narrow formulations of what is ideal in accordance with acceptable norms of selective breeding to exhibit the best human traits among a general population. An exploration of the goals of several eugenics programs throughout the course of history will be examined briefly; and Sect. 2.2 further propagates the discourse on these past paradigms and their legacy on contemporary legal discourse in bioethics, highlighting some of the darker moments of humankind's history of eugenics through practices of racial genocide, forced sterilizations and other similar experimentation processes. This elucidation covers examples of the early eugenics movements in the United States, and the horrors of the Shoah (Holocaust) during World War II. To lend comparative analysis from an Asian perspective, the eugenics movement in Japan and Singapore from the late 1970s to 1990s period is also outlined.

The recognition, however, is that the reality of present-day disguised eugenics hints at an evolved, more libertarian, modern, and post-metaphysical form now commonly known as 'liberal eugenics'.[2] In an attempt to encompass a more complete and well-rounded understanding of how eugenics influences genetic interventions and reproduction specifically *vis-à-vis* this evolved form of liberal eugenics, Sect. 2.3 appropriates part of the philosophical and post-metaphysical discourse by Jürgen Habermas, raised in his essay called *The Future of Human Nature*[3]; a discussion which has been touted to be one of the most important contributions to both national (in Germany) and the international controversies on scientific and medical advances in biomedical research.[4] More crucially, a discussion on the very garrison of the concept of human nature and what it means, is central to the role that it plays in any form of arguments or debates in relation to the specificity of human beings and their development.[5] The concept of human nature additionally lends some

---

[2] Agar (1998), p. 137.

[3] Habermas (2003), https://philpapers.org/rec/HABTFO-2.

[4] Rorty (2003).

[5] Groll and Lott (2015), p. 623.

thought to the development of legal personhood in contemporary laws. However, it has not escaped the attention that the Habermasian theory on human nature, and how it affects personhood, has been developed from a continental European perspective. Besides the Habermasian rhetoric as a response to the horrors of Nazi Germany in World War II, it is also reminiscent of the fear of post-metaphysical, transhumanism.[6] This concern is not present in the Asian philosophical dimensions, as will be briefly demonstrated; the section will demonstrate a difference in how human nature is viewed in Asian philosophies, but also point to convergences with Habermasian theory.

The perforation from the concept of human nature subsequently flows into the more controversial concept of 'liberal eugenics'; controversial because the outcome of intended actions is shifted from the state, to parents as the main actors.[7] In greater particularity, the concerning beneficiaries often involve parents, an intended offspring/children and the governing sphere of governance or the state. I use Agar's concept of liberal eugenics to demonstrate that a shift in balancing of the needs of the incumbent beneficiaries, contributes to meaningful dialogue on the interpretation of reproductive rights and its later derivate translations into the realm of privacy, autonomy, personhood and human dignity. I put forward three arguments that I believe support the notion that autonomy in liberal eugenics is flawed; through a spectrum of considerations that dispels the true notion of autonomy. It must be stressed however that this does not mean that I reject any form of selection, which may, whether correctly or not, be categorized as eugenics. Instead, my claim is premised on the manner in which autonomy is presented as an integral part of the selection process.

The intended and purposeful deliberation at the end of this chapter, in Sect. 2.4 therefore aspires to put forward a call to consider the impact of the legacy of eugenics (whether defiled or otherwise) on relevant, contemporary laws, and new or emerging medical and scientific technologies. What is also sought is to present the beginnings of how legal approaches and sensitivities have contributed to the development of legal frameworks in the different jurisdictions that will later be examined in this book. Through our understood knowledge thus far, the ingrained, underlying motivations for the improvement of overall health and avoidance of debilitating diseases sometimes overlap into the realm of the improvement of future progeny. What is also needed is a fundamental reinterpretation of what amounts to eugenics. A selection of traits or characteristics alone may erroneously be classified as eugenics, when its main aim is to eliminate genetic or inheritable diseases or conditions. A distinct differentiation must therefore be made, with focus on rationalized purposes that is consistent with legal traditions and human rights, as opposed to simply mere outcomes or effects of the selection process that may or may not be viewed negatively.

---

[6] Thomas (2017).
[7] Handyside (2010), p. 978.

2.1   Relevance of Eugenics in the Contemporary Genetic Debates

A question that begs to be asked at this juncture is whether the roots of eugenics may be traced to a single normative theory of society, a social philosophy of sorts. What would be helpful, however, would be to locate the commonality of most eugenics programs, which is its goal. Identifying the goal of the program is pertinent to identifying mindsets in eugenic practices. What can be discovered is that the basic premise of this goal is simply that human beings are in a position to, and should, manipulate and control their breeding/reproductive systems in ways that can influence the future offspring to manifest the best attributes of this program. Whether to eliminate undesirable characteristics on a societal level, such as the early migrants in North America,[8] to the horrors of the Holocaust, where an estimated eleven million lives (both Jewish and non-Jewish) were exterminated by Adolf Hitler's Nazi regime,[9] to Singapore's population policy in the 1980s,[10] and China's national one-child policy that stressed upon the health of the child, and bill on Maternal and Infant Health Care of 1995,[11] in the Far East; what is demonstrably vivid with these examples is the misconceived and nominal idea that the 'purity' of society reigns paramount in an existing community. The operational workings of modern day 'eugenics' has been disguised to the point of policy justification, and is far removed from the historical vestiges of the horrors of the past; but there can be no detraction from the fact that despite its attempts at concealment, some modern healthcare policies may still hint of a eugenics-based logic.

The quandary in any type of eugenics movement is that there can be no real, workable mechanism in which to promote the legitimacy of such practices without thus involving the use of force or sanctions against sections of the population who have been identified as possessing undesirable characteristics, and as such, carry inferior genetic material (which may be passed through the natural, generational

---

[8] The beginning of the revolutionary eugenics-based programmes in North America was initially targeted against Asian migrants, specifically the Chinese. This resulted in the Chinese Exclusion Acts of 1882 and 1902.

[9] During World War II, Hitler's Nazi regime carried out what is now known as one of the deadliest genocides in history, termed "The Final Solution to the Jewish Question", where the regime began a series of broad acts of oppression, violence and murders for the extermination of the Jewish people specifically, but also included ethnic Poles, Soviet citizens and prisoners of war, other Slavs, Romanis, communists, homosexuals, Jehovah's Witnesses and the mentally and physically disabled.

[10] Singapore's initial eugenics-based program was introduced in 1984 and had the goal of increasing fertility of university-educated women and the provision of major subsidies for voluntary sterilization of poor and uneducated parents, in a bid to dramatically increase population growth in tandem with its phenomenal socio-economic growth. This initial eugenics program has since been modified.

[11] The key provisions of the Bill called for the "termination of pregnancy if the fetus is suffering from a genetic disease of a serious nature or the fetus has any other defects of a serious nature."

processes). In addition, whether or not the general human condition may be improved with the intervention of modern genetic technologies that have the ability (theoretically) to eradicate some genetic abnormalities is still being determined. Even if it would be possible, does this amount to a form of eugenics born of a depraved need for perfection, or does it speak to something necessary for the full living and being of human life? The transformative power of science has had a tremendous impact on the future of human existence, raising an equal depth of staggering moral implications.

The disrepute of early eugenics movements in the course of history has left its indelible mark on how we view the progress of scientific technologies in today's contemporary society. Francis Galton's project on the human trait heredity[12] started by posing to us interesting questions about the biological future of humankind, igniting a spurious race towards the utopian idea of improving the quality of human beings; the birth of the term 'eugenics'.[13] The fear imposed upon modern societal conception of moral wrong or right, treads upon the fatal possibilities of repeat performances of the Americans' early twentieth century eugenicists-led regulatory framework of the amelioration of 'degenerate' persons that they believed threatened the good of society.[14] Even more significantly, the *Nationalsozialist* eugenics campaign, which has impacted individuals worldwide in some way or other, is not far removed from our memory, that view the tyranny of governments through the manipulation, commodification and abuses of, and in the name of science.[15]

There are lessons that we have learnt about the role early eugenics campaigns played, and made their lasting effects felt to date. The persistent state of eugenics, however, is not in a fulcrum of stasis, and as new philosophies emerge, they may serve to provide some insight into what has now been described as a new form of eugenics, often referred to as 'liberal eugenics'. Described as the apparently much more effective, compassionate and a more socially acceptable form of eugenics or enhancement science, its aim seeks to improve human life and provide us all with the best qualities that we may be able to afford. This is a re-energized movement that has taken the important lessons from a painful, historical past, and may transform our understanding about the role of science, genetics and how we can use this new knowledge in a positive manner to influence the plans that we choose to make in life through alleviation of the human condition.

Part of the flaws in contemporary understanding is to give weight to eugenics as a flawed mechanism in choosing to select embryos, especially where the reasons for selection are non-medical in nature. Judit Sándor states that this is a problematic approach because the contemporary discourse is such that "a classification for eugenics does not necessarily follow a medical vs. non-medical distinction."[16] In

---

[12] Kango-Singh (2010).

[13] Selgelid (2014), p. 3.

[14] Kango-Singh (2010).

[15] Chesterton (2000).

[16] Sándor (2015), p. 355.

fact, owing to the old meanings ascribed to eugenics, a modern interpretation of 'eugenics' *vis-à-vis* selection processes is also misunderstood, when, in essence, "it encompasses more individual choice rather than the expectations by society."[17] Understanding the historical accounts of eugenics helps us put in context its connection with selection in pre-implantation genetic interventions. This understanding contributes to the development of legal approaches that characterizes the selected jurisdictions in this book. Notwithstanding these lessons to be learnt, this new form of 'liberal eugenics' is also not without its particular stressors.

2.2   Historical Aspects and Paradigms of Eugenics Movements

2.2.1   Galton's Theory of Heredity as the Birth of Early Eugenics

In what may only be described as a fateful turn of events that has led contemporary bioethicists to attribute to Francis Galton the birth of "eugenics", the latter's theory of heredity and talent is posited along the lines that we, as human beings, should be in a position to take charge of our own evolutionary progress. Often known also as the cousin of Charles Darwin,[18] and the founding father of genetics, Galton's purely scientific research into improving the quality of human breeding and population has, unfortunately, been misinterpreted as value judgments of sycophancy, one of the earliest forms of minority oppression and racism, and a vindicator of mass murder of populations that were deemed to be less than "suitable races or strains of blood."[19]

Historical accounts, however, indicate that Galton's main point of contention was simply that human beings "should attempt to exert control over organic evolution in the same way as we exert control over the physical world and to direct it into channels of our own choosing."[20] Although it is not disputed that Galtonian ideals stressed upon a form of 'positive' eugenics (by cultivating larger representation in society of persons whose genetic traits were deemed to be valuable for the flourishing of said society) and 'negative' eugenics (by the elimination, incarceration or removal of the socially undesirable sections of society with negative traits such as diseases, illnesses, and cognitive and physical disabilities),[21] this theoretical classification provides, not only a congruent motive into Galton's research, but also into

---

[17] Ibid.

[18] Galton (2002), p. 78.

[19] Kevles (1999), p. 435.

[20] Kango-Singh (2010), p. 81.

[21] *Encyclopedia of Bioethics* (3rd edn, vol. 2, Thomson Gale 2004), p. 848.

the flawed conception of value judgments used to denote desirable and non-desirable traits of human beings within Victorian England society.

Galton proposed some fundamental scales of measurement on the estimation of Victorian England values, both traits that were deemed desirable and non-desirable, with their corresponding 'worth' virtues, which would consequently be entered into a form of eugenics register.[22] He propounded for a various array of social incentives that would encourage positive eugenics, as such extensive genealogical mapping of family records, and a bonus points-system for 'pedigree' family lineage in competitive civil service examinations and employment opportunities[23]; the more drastic measures that included segregation and incarceration were instituted against the "feeble-minded, habitual criminals and the insane",[24] often prevented by way of forced sterilization or simply, complete segregation, to discourage and/or prevent the birth of children. However, these suggestions never broached the level of atrocities in implementing 'good breeding' that was later carried on by the eugenicists in North America and Germany.

The spillover effect of a kind of thinking similar to Galtonian ideas was overshadowed in many respects by the impetus of Mendelian laws of inheritance. Nevertheless, it is unfortunate that he is often remembered not as the father of genetics, as many had attributed to him before the arrival of Mendel; but as the father of eugenics, which negatively continues to denote the nasty vestiges of human dignity and violations of humanity. In all instances, however, it would be wise to note that Galton's form of eugenics never did, in any manner whatsoever, condone the use of coercion or force as a method of widespread societal compliance; and many supporters of the eugenics program[25] at that time also did not condone the kind of state intervention that could easily be subject to grave abuse and place itself within the reach of an elitist, authoritarian government, as we will observe in later movements of eugenics-based programs.

2.2.2   Mendel's Laws of Inheritance and Their Relationship with Early Eugenics

It is, without doubt, that Gregor Mendel's experimental work in the eighteenth century, *Experiments in Plant Hybridization*,[26] has contributed to our understanding of gene expression this very day, and which has spurred countless research projects

---

[22] Kango-Singh (2010), p. 84.

[23] Ibid 85.

[24] Ibid.

[25] Some prominent supporters of the American eugenics programme included renowned biologist, Charles Davenport, psychologist Henry Goddard, and lawyer and conservationist Madison Grant, amongst others. The American eugenics programme also received immense funding from foundations such as the Carnegie Institution and the Rockefeller Foundation.

[26] Mendel, J.G. (1866). "Versuche über Pflanzenhybriden", *Verhandlungen des naturforschenden Vereines in Brünn*, Bd. IV für das Jahr, 1865, *Abhandlungen*: 3–47, [1]. For the English translation,

and studies into the human genome and its various instrumentalities.[27] Although Mendel's insights remained largely unnoticed until 1900, his work quickly became solidified as one of the most classic leading paradigms into human genetic studies, influencing modern human genetics, as we know it today.[28] What makes his work uniquely important is the fact that his experimentations on pea plants revealed fascinating insights into the fundamental laws of inheritance in human gene expression, and completely changed the scientific landscape in our understanding of the human genome and its complicated sequencing patterns.

What is now known collectively as Mendel's Laws of Inheritance (observable from his pea plant experimentations) can be summarized in the following manner: first, because inherited human traits are defined by a pair of genes, randomly separated by parental genes into the germ-line cells, the resulting children from the parental genes often inherit one genetic allele from each parent when the germ-line cells unite in fertilization. This is known as the Law of Segregation.[29] Secondly, the Law of Independent Assortment refers to genes bearing different human traits being sorted separately from one another, resulting in the non-dependence of one trait on the inheritance of another.[30] And thirdly, the Law of Dominance coined by Mendel indicates that a gene that carries alternate organism forms will be capable of expressing what is dominant in that form.[31]

Why is Mendel's work fundamentally important to human genetics? What began as a form of statistical evidence provable by the results of Mendel's experiments of artificial insemination of new color variants into pea plants, has formed the base pair sequence of deoxyribonucleic acid (what we now know as DNA). And, through the advent of the Human Genome Project,[32] we now know that DNA is the repository of information for protein synthesis and for life in all its forms. It is through Mendel's work that the evolving triggers of how we begin to understand the inherent divisibility of complex human disorders, both physiological and physical, play into the realm of the transmission of human genetic traits that become the key subject of grave concern into the evolution of eugenics campaigns.

Although scholarly opinions have indicated that Mendel's work was intended as an alternative supplement to Darwin's theory of evolution, the magnanimous nature of Mendel's findings became a catalyst that shifted the public attention to the inheritance of traits, or heredity bearing fruition from genes, as the source of any form of undesirable characteristics in human population. At this juncture, what is also

---

see: *Druery, C.T.; Bateson, William (1901). "Experiments in plant hybridization" Journal of the Royal Horticultural Society 26: 1–32.*

[27] This led to the Human Genome Project, indisputably the largest, internationally-collaborative global scientific and biological project spanning a period of thirteen years, with the aim of completely mapping all the genes of the human genome.

[28] Heinemann and Honnefelder (2003), p. 530.

[29] 'Mendel as the Father of Genetics:: DNA from the Beginning' http://www.dnaftb.org/1/bio.html.

[30] Ibid.

[31] Ibid.

[32] Botkin (1998), p. 17.

observed is that the intentions or motivations of Mendel, very much like Galton, did not anticipate the reactions that were provoked once their respective theories became more accessible and understood by the general scientific population at large.

It was largely due to the work of both Darwin and Galton that influenced the eugenics movement in North America later, the theoretical underpinnings of which became the source of inspiration for a 1911 Treatise[33] that targeted specific groups of the population and outlined methods in which the eugenics project to create a superior population, and even "Better Babies",[34] could be systematically carried out in the country.[35]

2.2.3 Beyond Historical Conceptions: A Violation of Rights in Eugenics Practice

2.2.3.1 Forced Sterilization as a Form of the Early Eugenics Movements in the United States (US)

Beyond the evolution of understanding how the transmission of human genetic diseases occur, following Mendel's laws of inheritance, the possibilities that have been evoked have unfortunately led to revolutionary eugenics movements in the United States (US) in the late nineteenth century, beginning with the Chinese Exclusions Acts of 1882 and 1902,[36] (fueled by conceptions that immigrants were of less than desirable characteristics, and marriages and/or free movement of these immigrants amongst the general population, would surely taint the products of society that resulted), scattered with various forced sterilization laws in individual states, such as Indiana, Pennsylvania, and California.[37] Incidentally, although the Third Reich's

---

[33] This report was called the "Preliminary Report of the Committee of the Eugenic Section of the American Breeders' Association to Study and Report on the Best Practical Means for Cutting Off the Defective Germ-Plasm in the Human Population".

[34] The "Better Babies" project in the United States was the idea and culmination of Mary deGormo, who developed the first "Scientific Baby Contest" in Louisiana, complete with grading sheets designed together with a paediatrician, and the traits that were viewed favourably in the contest included physical measurements and measurements of intelligence. The contribution that such contests made to societal development at that point of time was seen as being a form of "social efficiency" movement that advocated and encouraged certain "standardized" aspects of ideal American life.

[35] Some of the methods described in this 1911 Treatise report offer a startlingly disturbing glance into the visceral ideals of what is viewed as the perfect human person. It also included several visceral descriptions on how the objectives described in the report could be carried out, including suggestions for the euthanization of 'imbeciles', the 'feeble-minded' and any other members of the population that were deemed to have defective characteristics; and also the creation of gas chambers to eliminate these people.

[36] Galton (2002), p. 91.

[37] Ibid 92.

eugenics movement of a 'master race' has remained prominently in the historical context, what is lesser known is that this idea originated from the United States several decades prior to the Nazi regime's race to create an Aryan[38] human race. Indeed, in the United States, the early eugenics movement began with the establishment of the American Eugenics Society,[39] which, included in its mission, solid actions not only to segregate, but to "cleanse" the vilified characteristics purportedly possessed by primarily immigrants, deemed to be of lesser blood due to race, disability, 'feeble-mindedness' or any other characteristics inundated with the perception of what amounts to an 'imbecile'.[40]

Forced sterilization as a method of discouraging the growth of the 'undesirable' population was the choice of the day. At the height of the eugenics movement in the US, forced sterilization laws were adopted in more than 30 American states, resulting in the forced sterilization of at least 60,000 disabled individuals (the mentally-challenged or those deemed to belong to undesirable sections of the population).[41] Decades later, it appears that various individual states have come forward to accept responsibility for these violations of the past. In 2003, the state of California issued a public apology to the victims of the state's forced sterilization programs.[42] 2010 saw the state of North Carolina taking positive actions to compensate victims of the eugenics movement by approving a compensation fund and setting up the Office of Justice for Sterilization Victims in the state to assist and compensate victims of forced sterilizations of the state's eugenics program.[43] The state of Virginia also followed suit in compensation to victims in 2015[44]; and despite all these recent developments, we must remember that it was only in the late 1970s where there was a complete eradication of these forced sterilization programs in the US.[45]

---

[38] In this context, the Aryan race envisaged was one of German descent, heavily influenced the theories of German social Darwinists of the nineteenth century. Social Darwinists attributed both positive and negative stereotypes of ethnic group appearance, behaviour and culture as unchangeable and rooted in biological inheritance, immutable through time and immune to changes in environment, intellectual development or socialization. Therefore, for Hitler's Nazi regime, the assimilation of a member of one race into another culture or ethnic group was impossible because the original inherited traits could not change, they could only degenerate through race mixing.

[39] The American Eugenics Society was established in 1926 for the purpose of promoting awareness of the eugenics programme to the American public. See R. Gur-Arie, American Eugenics Society, The Embryo Project Encyclopaedia, 22 November 2014, https://embryo.asu.edu/pages/american-eugenics-society-1926-1972.

[40] Ibid.

[41] Lutz Kaelber, 'Eugenics: Compulsory Sterilization in 50 American States' https://www.uvm.edu/~lkaelber/eugenics/.

[42] Ingram (2003).

[43] North Carolina Administration, 'NC DOA: Welcome to the Office of Justice for Sterilization Victims'   https://ncadmin.nc.gov/about-doa/special-programs/welcome-office-justice-sterilization-victims.

[44] Portnoy (2015).

[45] Lutz Kaelber, 'Eugenics: Compulsory Sterilization in 50 American States' https://www.uvm.edu/~lkaelber/eugenics/.

Forced sterilization is an important consideration as part of the eugenics movement for the purposes of this book, mainly because of its relationship with a person's fundamental right of "life, liberty, or property".[46] By this extension in modern constitutional jurisprudence, the interpretation of such fundamental right has also been extended to cover penumbral rights that stem from the Constitution, including the right to privacy, for example. The right to privacy can encompass a variety of issues relating to bodily integrity, control over one's individuality and autonomy, choices made within the realm of private and personal life, and the like.

Some prominent case law has also emerged from the perspective of the role of eugenics with reproduction. The forced sterilization movement in the US cannot be complete without a discussion of the case of *Buck v Bell*.[47] This was a landmark decision by the Supreme Court that clearly demonstrated the mindset of American society during that period; Justice Oliver Wendell Holmes' most famous statement disseminated from this case, in which he stated "it is better for all the world; three generations of imbeciles are enough".[48] The case concerned a challenge to the constitutional validity of the state of Virginia's forced sterilization statute that targeted the intellectually-challenged population and was centered on an 18-year old girl named Carrie Buck who had been institutionalized at the Virginia State Colony for Epileptics and Feeble Minded. It was determined by the superintendent of the colony that Carrie represented a "genetic threat" to society, having been descended from a non-desirable societal lineage, her mother having been determined to be of feeble mind, with a history of prostitution and immorality, and Carrie herself deemed to be "incorrigible" and eventually also giving birth to an illegitimate child, where her pregnancy had been widely regarded as caused by some kind of "immorality" on her part.[49]

The overwhelming majority of the Supreme Court determined the validity of the Virginia sterilization statute in order to further the purpose of the "protection and health of the state",[50] stipulating that this was a legitimate cause for state regulation and therefore, did not violate the due process clause in the Fourteenth Amendment of the US Constitution. Tragic though the consequences may be in this case, it is often also seen as cementing the position of the Supreme Court in endorsing the operation of negative eugenics. It is also crucial to note that after the decision by the Supreme Court here, this effectively laid the foundation, not only for Virginia to widely carry out forced sterilization on the mentally-disabled, or mentally-deficient; but to also encourage the fall-out effect that served as a validation that the other

---

[46] Sholley (1951). Please see: the Fourteenth Amendment of the Constitution.

[47] 'Buck v. Bell, 274 U.S. 200 (1927)' (*Justia Law*) https://supreme.justia.com/cases/federal/us/274/200/case.html.

[48] Ibid.

[49] A turn of events much later revealed that Carrie had been raped by her adoptive mother's nephew, and her family committed Carrie to the institution in the hopes of concealing the rape and resulting shame.

[50] Acts of Assembly, Chapter 394, Virginia SB281, Eugenical Sterilization Act of 3/20/1924, p. 569.

American states could also adopt and carry out similar forced sterilization procedures within their jurisdiction. In a larger scheme of things, the decision became the catalyst of a eugenics-based justification in creating the ideal, Anglo-Saxon American race of people. Virginia, incidentally, remained one of the last states in the US to finally outlaw its forced sterilization laws in 1970s.[51] Although some states have made public apologies and issued monetary compensation to the surviving victims of the forced sterilization/eugenics programs of the past, we must view this with some form of trepidation, because the Supreme Court has not, to date, expressly overruled its decision in *Buck v Bell.*

It is this particular trajectory, in the light of rapid developments in techno-science, and the completion of the Human Genome Project[52] where the advancement of reproductive technologies give way to a higher echelon of reproductive freedom, that raises an almost "spine-chilling"[53] concern about the decision not being overruled. From a constitutional perspective, the question arises as to the validity of such decision and whether reliance may still be had on its ratio at some point in the future. If we surmise, from the doctrine of precedent (*stare decisis*), as is the tradition of the court system in the US, this would appear to be so. Therefore, by this implication, this would mean that because *Buck v Bell* had not been expressly overruled by the Supreme Court in its later decisions, this judgment specifically, purely from a technical perspective, would still hold some weight in consideration in future cases. Hence, one may only simply speculate as to why *Buck v Bell* has not been explicitly overruled.

From a practical perspective, however, the open condemnation of eugenics-based laws and/or forced sterilization in general, as well as the wealth of case law before the various courts in the US being brought and decided on as constructive basis for the due process and equal protection clause of the Fourteenth Amendment, the decision of *Buck v Bell* may very well be eviscerated by these factors, and even some later decisions of the Supreme Court, for example, the case of *Skinner v Oklahoma.*[54] The reality of constitutional adjudication and practice in the US court system does not dictate the necessity of an express overruling, by virtue of the relevant implications that may be drawn from the later decisions, which had the effect of overruling the previous decision.[55] If we take this to be an acceptable feature of the US

---

[51] Virginia Eugenics, https://www.uvm.edu/~lkaelber/eugenics/VA/VA.html.

[52] The Human Genome Project has been, to date, the largest-scale, international collaborative effort in genetics research, whose goal was to map out all the entire genome of a human being. The results of the project has enabled us to now understand the development and function of a human being by reference to the human genome. At the same time, because of the propensity and gravity of this newfound genetic knowledge, care must be taken to take account of any ethical, legal and social implications that results from the use or possible abuse of this vast pool of knowledge.

[53] Cohen (2016).

[54] 'Skinner v. Oklahoma Ex Rel. Williamson, 316 U.S. 535 (1942)' (*Justia Law*) https://supreme.justia.com/cases/federal/us/316/535/case.html.

[55] From the US constitutional perspective, Miranda v United States 384 U.S. 436 (1966) may be a suitable example as later decisions that continue to cite Miranda have also gone further by endors-

constitutional system, then the grave concern of a return to *Buck v Bell* is severely diminished.

An additional reason as to why *Buck v Bell* was perhaps not expressly overruled was also the hugely-political and social realities of the time in connection with how the benefits of eugenics could have been viewed as a betterment of the quintessential "upper-class, white" kind of 'desirable' society. The judgment written by Justice Oliver Wendell Holmes, decorated war-veteran and known as one of the most prolific, outspoken and eloquent Supreme Court Justices, delivered a scathing blow that glorified the "white-ness" of the upper class elite. The judgment also signified the elite's failure to recognize their own misunderstanding about science, and its relationship with social factors. It is remarked[56] that Justice Holmes claimed that *Buck v Bell* was "one decision that I wrote gave me pleasure, establishing the constitutionality of law permitting the sterilization of imbeciles." Further rhetoric indicated that Holmes himself was a supporter of the eugenics policy in the United States, and by his judgment in *Buck v Bell*, was able to embody his own opinions into law.

However, the change of attitude towards individuals with mental deficiencies have changed over time; the American movement on civil rights and women's rights are steps taken in the positive manner to pace the way for legislators and policy makers alike, to begin thinking about the fundamental rights attributed to members of society who were disabled, whether physically or mentally, and these rights include the reproductive agenda, linking closely with the fundamental right of privacy that has continued to pervade American constitutional jurisprudence over the last fifty years.

In fact, the importance of the realm of privacy has been largely felt beginning in the 1940s, beginning with the seminal case of *Skinner v Oklahoma*,[57] one of the crucial decisions by the United States Supreme Court, which recognized that forced sterilization amounted to a violation of the accused person's fundamental right to pro-creation, that was provided for under the equal protection clause of the Fourteenth Amendment of the United States Constitution (instead of relying on the constitutional liberty of the due process clause of the same Amendment). The reasoning in *Skinner v Oklahoma* was instrumental in creating a dialogue in the United States about how a recognition of a person's right to privacy was a fundamental element of personal autonomy that the state should not unnecessarily interfere with,

---

ing more specificity in the availability of proper legal defense, and good faith loopholes for the police. This is not to say that Miranda is irrelevant, but simply means that its stronghold has now become weakened in light of these later decisions that have broadened the scope of Miranda's initial holding. The Supreme Court also famously "overruled" itself in Plessy v Ferguson, 163 U.S. 537 (1896), by the issuance of its judgment in Brown v Board of Education of Topeka, 347 U.S.483 (1954) concerning the issue of racial segregation in schools. No express words were given in these examples as to how the court made a formal declaration of overruling its own previous decisions, but the culminated effect of the later decisions do indicate that a de facto overrule has taken place.

[56] Burrus (2011).

[57] 'Skinner v. Oklahoma Ex Rel. Williamson, 316 U.S. 535 (1942)' (*Justia Law*) https://supreme.justia.com/cases/federal/us/316/535/case.html.

later leading to landmark cases in reproduction privacy such as *Griswold v Connecticut*[58] and *Roe v Wade*.[59]

The facts of *Skinner v Oklahoma* concerned the constitutionality of Oklahoma's Habitual Criminal Sterilization Act of 1935 which imposed the punishment of forced sterilization on individuals who had committed three or more "crimes of moral turpitude". In the enforcement of these sterilization laws, also existed the framework of prejudice and discrimination in the definition of what amounted to "crimes of moral turpitude", since the law and its instrumental enforcement provided a form of immunity, for instance, against white-collar crimes, or embezzlement crimes (which, translated into the realities of American society then, signified the prejudice in categorizing individuals of a certain race, upbringing, income group and other such minorities as being those prone to commit these "crimes of moral turpitude"). However, the judgment of the Supreme Court revealed much more than met the eye. In essence, what can be gleaned is that *Skinner v Oklahoma* gave birth to an unconventional strand of constitutional law, known as the doctrine of equal protection of fundamental interests; and the main tenant of this doctrine is that judges would calibrate the aggressiveness of scrutiny or review of particular legislation depending on the identity or target group at which such legislation was directed, determined on a tiered standard of review in US constitutional law.

A close-up review of *Skinner v Oklahoma* would compel us to question as to why this doctrine of equal protection was employed, instead of reliance by the court on due process on a procedural level of the law, since the facts indicated that Skinner was never given an opportunity to address or refute why he should not be compulsorily sterilized. The default operation of the Habitual Criminal Sterilization Act of 1935 meant an automatic and blanket application of forced sterilization over individuals who had committed three or more of these particularly described crimes, and hence, prevented them from the due process of the law. Instead, Justice Douglas, who delivered the opinion of the court, communicated its reasoning that the legislation dealt with one of the civil rights of man,[60] relying on the 'right' of an individual to pro-create which had been violated by the Oklahoma statute because it deprived some individuals, and not others, of their fundamental right of procreation.

There is a variety of conjectural speculation as to why the court may have chosen to rely on the equal protection clause of the Fourteenth Amendment instead of the due process of the law requirements. For one, it is possible that the decision operated as a stop-gap measure because the court did not want to then court that declare a free-standing fundamental right[61]; and secondly, a judicial reliance on equal protection showed a respect for the political process by giving the legislature another

---

[58] 'Griswold v. Connecticut, 381 U.S. 479 (1965)' (*Justia Law*) https://supreme.justia.com/cases/federal/us/381/479/case.html.

[59] 'Roe v. Wade, 410 U.S. 113 (1973)' (*Justia Law*) https://supreme.justia.com/cases/federal/us/410/113/.

[60] He states at p. 316, that "marriage and procreation are fundamental to the very existence and survival of the race."

[61] Franklin (2015).

option to possibly 'rectify' the circumstances.[62] The almost-deferent attitude to the legislature here presumes that there is no meaningful way in which the legislature would commit to passing a forced sterilization law of that nature. But more importantly, *Skinner v Oklahoma* demonstrated "a fear about the invidious and potentially genocidal manner in which government control over reproductive matters might be exercised if the choice shifted from the individual to the state." The sanctity of reproduction and procreation in general employs the choice of social roles and concepts of the self. If a state were to be permitted to deny such a choice in an organized society, it would also be tantamount to denying an individual a presumptive right to be treated as a person. Hence, the reasoning in Skinner goes beyond skin deep, although critics have stated that it lacked a "secure constitutional foundation"[63] for calling into question the validity of a piece of unfavorable and ugly legislation.

It is very interesting to note that the way in which eugenics had been perceived at the times when important court cases involving sterilization were brought, ultimately impacted on how the court ruled. *Buck v Bell* was brought before the court at the height of the eugenics movement and the law had been upheld, to no one's surprise and chagrin. Conversely, in *Skinner v Oklahoma*, it was brought to the court at a very low point in time in American eugenics about the time of the atrocities committed by Hitler in Germany during 1942, and unsurprisingly, in global solidarity and condemnation against Hitler's actions, the forced sterilization law was condemned. Hence, both these cases are paramount in the discussion on how forced sterilization has had an impact in understanding reproductive rights and its underlying connotations pertaining to discrimination, autonomy and liberty.

2.2.3.2 Eugenic Policies in Nazi Germany

At the same time, the early eugenics movement in Germany began, even prior to Hitler's Nazi regime, the latter of which will continue to live on in history books as one of the most atrocious occurrence of genocide of humankind. At this juncture, however, the preoccupation with the purity and good breeding characteristics of an Aryan human race was of paramount concern to a group of Germans (apparently heavily influenced by Galtonian ideals of human breeding), who made it their mission to found a Teutonic colony.[64] Historical accounts indicate to us an obsession with the purity of the Nordic race and a great fear about the contamination of the race, thereby fueling general eugenic ideas as part of German Fascism and the developing Nazi ideology in the later part of the century.[65] At this juncture, it must be highlighted that this part of the book will not delve into the accounts of Nazi Germany, the literature and study of which is plentiful and widely available for

---

[62] Ibid.

[63] Stone et al. (2005), p. 773.

[64] Galton (2002), p. 95.

[65] Ibid 96.

further reading. What this book intends to highlight, however, by recounting the Third Reich's historical perspectives, is the role of medical professionals in Nazi Germany and how blind indoctrination, the promise of power and career advancement, opportunisms of money and a skewered idealism of perfection of the human race; has influenced and contributed to the role eugenics (and emerging genetics, during that time) has played.[66]

Although some research has emerged indicating that eugenics had not been one of the goals of Hitler's *Nationalsozialist* policies,[67] eugenics' indivisible link to the Third Reich cannot be denied, whatever its original motivations had been. This school of thought puts forward the claim that factors such as sensationalism, misinformation, dissemination of ideologies by Marxists, and factions who were opposed to the eugenics movement, amongst others, had contributed to the pejorative conception of Nazi abomination with eugenics. Notwithstanding, the authors of the study[68] do contend that there has been guilt by association by the Third Reich, but that the *Nationalsozialist* did not specifically address eugenics as their main advocating policy and guiding force. The paper presents an alternative view on understanding the lines of argument often attributed to the Nazi eugenics movement, refuting that race and eugenics had never been part of the original intention, and instead, posits that the persecution began with a deep-rooted hatred of the European Jewry due to an economic accumulation of national sources and finances. This may be a plausible conjecture, given that some ideas have surfaced and pointed to Hitler's "deep and genuine ignorance of biology and genetics."[69] Nevertheless, the evolution of the prejudicial feelings towards European Jewry had escalated to a point beyond the macabre and grotesque, that we cannot downplay, in any manner, the role of the *Nationalsozialist* in perpetuating and contributing to eugenics, one of the greatest demonstrations of humanity at its worse.

One of the most abhorrent and disquieting facts about the *Nationalsozialist* policies, in addition to the mass murders and euthanizations, and forced sterilizations, all in the name of medical science and a conviction in social Darwinism,[70] as one pushing for the 'survival of the fittest', was the fact that medical professionals such as doctors and neuroscientists, were complicit in carrying out these odious acts.[71] It has only been in recent years that the role of bioethics and science was more fully considered vis-à-vis the role played these Nazi doctors in the name of medicine and science.[72] These accounts now serve as illuminating fodder to reflect on what had

---

[66] Kater (1987), p. 31.

[67] Saetz et al. (1985).

[68] Ibid.

[69] Galton (2002), p. 97.

[70] Hediger (2016), p. 5.

[71] Zeidman (2011), p. 696.

[72] Kater (1987). In this article, Kater highlights the tremendous challenge in charting the historical perspectives of the Nazi doctors, due particularly to two reasons: one, a lack of resources, documents and files; and secondly, the reluctance of German historians to "probe the more recent past of their professional, because they have been unwilling to come to grips with the moral and ethical problems posed by the perversion of medicine in the Third Reich."

gone 'wrong' and how our more evolved understanding of diversification and human rights has taken lessons from these past atrocities. For example, one of the most notorious doctors at Auschwitz was Josef Mengele,[73] a SS officer and doctor who was responsible for selecting the people to be killed in the gas chambers, and to conduct deadly and inhumane experiments on those detained at Auschwitz. A television documentary series titled Science and the Swastika[74] offers a view into the human experimentation and eugenics-based program of the Nazis, raising critical questions that we must also consider the role of medical professionals. This section aims to do just this fact, examining the key pointers of Nazi Germany and its influence of our understanding of a spectrum of issues, for example, relating to human dignity, and its impact on contemporary bioethics and laws.

After the Nuremberg trials and the end of Hitler's 'reign' of terror, what was even more shocking was that many of these medical professionals, (who had not only been responsible for carrying out the eugenic policies, but also served as consultants to the Third Reich in the formulation of the first German sterilization laws in 1933), were never prosecuted for their role in the atrocities. In addition to the grave crimes levied upon countless innocent lives, the justification put forth by these doctors was skewed to fit the complicated circumstances of that dark time. Upholding the main tenets of the Hippocratic Oath to "do no harm or injustice" had been conveniently displaced, giving way to the much more desirous allure of power, financial gain and opportunistic, nihilistic considerations. Indeed, interesting perspectives have pointed to there being a symbiotic relationship between these medical professionals and the Nazis.[75] The Doctors' Trials, as it later become known as part of the Nuremberg Trials, subsequently culminated in The Nuremberg Code of 1931[76]; still considered to be one of the most important documents in medical and research ethics, and possibly give rise to the more modern international instruments such as the Declaration of Helsinki.[77]

Historical accounts from witnesses to these unspeakable horrors describe an escalation of acts, starting out with the forced sterilization of those judged to be possessed of congenital feeble-mindedness; human experimentation and research into the "idiocy" of members of the lower strata of society; segregation into concentration camps and severe ill-treatment, and finally, the mass euthanizations, calculated by the doctors to incur much lower costs than the prevailing sterilization programs. The mass genocide perpetrated by the Nazis remains till today, one of the most atrocious and violent accounts of crimes against humanity. In light of these terrible events, and how it has changed the landscape of humanity in the course of its short history, it is therefore incumbent upon future generations that the same not

---

[73] 'Josef Mengele, Angel of Death' http://www.auschwitz.dk/mengele.htm.

[74] IMDb, *Science and the Swastika* http://www.imdb.com/title/tt0808104/.

[75] Zeidman (2011).

[76] Shuster (1997), p. 1436.

[77] World Medical Association (2018).

be repeated under any circumstances whatsoever. On this basis, any forms of policies or movements that hint at eugenics undertow are therefore understandably condemned.

However, as will be demonstrated further in this chapter, it is found that despite its pejorative history, eugenics can still continue to find its way into contemporary societies through a plethora of advancements made possible through science and technology.

2.2.3.3 State-Supported Eugenics Movements in Asia

A sketch of the Asian paradigm provides a useful comparative insight into the length and breadth of eugenics programs post-World War II, notwithstanding the impactful atrocities perpetuated in the war. Discussions in this area cannot be complete without the mention of Japan, China and Singapore, all of which, at the relevant points of time, are (and still remain) Asian countries that have integrated successfully into the international sphere insofar as financial, developmental, and economic structures are concerned.[78] A brief account of the eugenic policies in these countries contributes to the comparative perspectives of eugenics that is not confined to the Western sphere. Indeed, Yoko Matsubara quotes Frank Dikötter, who emphasizes that this examination is important under contemporary biomedical discussions.[79] He attributes this to how "specific cultures is dependent on the local construction of medical knowledge, and that the strictly genetic determinism or the hard Mendelian discourse familiar from studies of Britain and Germany are not necessarily prototypical of eugenics."[80]

Notwithstanding the monumental, incremental state progression that has transformed the landscape of these developing countries, scholarly research has revealed that the non-Western history of eugenics is closely linked to colonialism and nationalism in these countries: for example, Japan's sterilization laws in 1948[81]; China's eugenics and forced sterilizations laws as part of Chinese political reforms,[82] and Singapore's population policies under the ministerial helm of Lee Kuan Yew[83] during the 1980s.[84] The discussion of these policies in Japan, China and Singapore

---

[78] World Bank (2018).

[79] Matsubara (1998), p. 187.

[80] Ibid.

[81] Matsubara (1998).

[82] Nature (1998), p. 707.

[83] Often touted to be the founding father of Singapore, Lee Kuan Yew was Singapore's first Prime Minister for over 30 years. Under his leadership, Singapore transformed from a third world country to a first world one in an immensely short period of time, gaining a name for itself in the international arena for high-quality education, economic expansion, and financial stability, amongst others, and continues to make its presence felt on the world stage. Lee Kuan Yew passed away in March 2015, and was succeeded by his son, Lee Hsien Loong.

[84] Chan (1985), p. 707.

show that the eugenics reach traverses cultures and international boundaries; demonstrating state-supported mindfulness in a misinformed preservation of 'breeding' in its citizenry population, but historically and culturally evolved from the Western discourse of eugenics.

The remnants of Japan's post-war eugenic policies have recently made the international headlines. Victims of forced sterilizations under Japan's post-war Eugenics Protection Law of 1948[85] have come forward in early 2018 to battle with the government in Tokyo over the past forced sterilization procedures to which they have been subjected, and to seek compensation from the government.[86] This 'battle' is still presently being waged in the courts in Tokyo. Historical accounts have indicated that the 1948 eugenics law, as a predecessor to the earlier 1940 National Eugenic Law[87] was motivated by strong feelings of Japanese nationalism, what has been termed *shin'nipponjin*.[88] It was believed that the *shin'nipponjin* or New Japanese possessed the ideal characteristics to carry forward Japanese lineage to expand the empire. A fascinating body of literature on the history of Japanese eugenics indicates the importance placed on factors such as *junketsu* (pure blood) and *kenzen* (wholesome)[89]; the fear of diseases such as mental illness, leprosy, intellectual disabilities, alcoholism, and other such social ills,[90] were attributed to *iden* or heredity. Bearing the traits similar to the Nazi eugenic policies, the imperial Japanese government, and subsequently, the post-war government, took legislative steps to ensure the purity of the Japanese race and popularized the discourse into daily lives, and official national policies.[91]

Statistics have indicated that at least 25,000 people were forcibly sterilized under the Japanese eugenics law,[92] due to reasons such as intellectual disabilities. The eugenics law also encompassed "maternal health and life"[93]; and therefore, to give birth to an "unfit offspring" or *furyo na shison*[94] was unacceptable. In addition, the eugenics law also covered matters relating to abortions, birth control and drugs. In 1995, a study by the Japanese Ministry of Health and Welfare showed that 16,520 involuntary sterilizations had been performed.[95] All these indicators reasonably tell us that the privacy, autonomy and individual liberties of the victims were violated. But it is also surprising and interesting that progressive supporters of sexuality, reproduction, and human dignity in Japanese scholarship, simultaneously empha-

---

[85] Robertson (2010).

[86] Watts (2018) http://www.atimes.com/article/eugenics-case-highlights-dark-chapter-japanese-history/.

[87] Robertson (2010), p. 1.

[88] Ibid 2.

[89] Ibid.

[90] Matsubara (1998), p. 189.

[91] Robertson (2010), p. 3.

[92] Watts (2018).

[93] Robertson (2010), p. 14.

[94] Ibid.

[95] Ibid 15.

sized the strength of autonomy and liberties through a counter-veiling measure: that the eugenics movement and legalization of euthanasia as a valid means of improving society, respects individual autonomy and freedom by removing the "burdens" to society because of their less than desirable characteristics.[96]

In Singapore, the 'eugenics' policy was less direct and less aggressive, in that the governmental policy chose to target population control, as opposed to an outright eradication vis-à-vis express sterilization or eugenics laws. Under the premiership of the late Prime Minister Lee Kuan Yew, the country's national policies on the key issue of education and particularly Singapore's rise as a competitive nation was given due to "the importance of the quality of human material".[97] It has been noted that Lee Kuan Yew's social policies in Singapore focused on his belief that Singaporean society would flourish through the qualities of its citizens[98]: referring particularly to the "no more than 5 percent" group of elites in the country.[99] In popular contemporary Singaporean and Malaysian culture, this elitist mind-set has often been humorously mocked; the term kiasu (literally translated to mean 'afraid to lose out')[100] in colloquial Hokkien was coined as a means of describing the perceived mentality of Singaporean culture.

As Singapore began its rise on the world stage as a formidable economic and infrastructural force, it was also faced with a fertility problem. As iterated above, Singapore chose to address its fertility and population issues vis-à-vis a social policy in the 1980s. Studies within the field of fertility and population by Singaporean scholars, particularly, have shown that the government began to encourage female university graduates to marry partners of equal 'quality'.[101] This was exercised through the government's "Graduate Mother Scheme."[102] The government also introduced incentives in a national procreation package, amending its previous two-child policy and encouraging university graduates particularly to have more children if they could afford to do so.[103] The practice of sterilization, unsurprisingly, also featured as part of the government's population policies. This was targeted at those who were poor or were of low education, incentivizing women or families who underwent voluntary sterilization if their maximum household income did not exceed a particular threshold.[104]

These measures on the surface level do not appear to be overly problematic. However, there are two points to note why these policies amount to a soft form of

---

[96] Hovhannisyan (2018), p. 28.

[97] Chan (1985), p. 707.

[98] Michael D Barr, 'Lee Kuan Yew: Race, Culture and Genes' 18.

[99] Singapore Democratic Party 'Eugenics in Singapore' http://yoursdp.org//news/eugenics_in_singapore/2008-11-09-558.

[100] Ganesan (2016).

[101] Chan (1985).

[102] Yap (2003), pp. 643, 652.

[103] Ibid 644.

[104] Ibid 652.

positive eugenics. First, the policies must be poised against the background of Lee Kuan Yew's personal philosophy on the betterment of Singaporean society; the emphasis on marriage and procreation of university graduates, with a variation of incentives to these persons. This appears to have been propagated to the exclusion of other members of Singaporean society. Secondly, the policies tread into the realm of family, family planning and personal relationships, which in confluence with egalitarianism, should always remain outside the purview of government control. By putting forward these policies, the Singaporean government sought to gently penetrate the family domain of its individuals; in contemporary jurisprudence, this would surely be viewed as a violation of privacy rights.

China is another country, which is not immune to the allegations of its eugenics past. Often touted to be one of the next superpower equal to its Western counterparts, China's phenomenal economic and technological growth has made it one of the fastest growing powerful nations in the world presently.[105] Recent events indicated in international headlines show the might of China's force in economy, science and technology. Within the context of this book, China has shown itself to be unafraid to undertake endeavors that its Western counterparts may have a problem with. For example, it was the first to experiment with gene editing on non-viable human embryos in 2015,[106] creating a hue and cry in the international bioethical community. China has also used gene editing on adult human patients thus far.[107] And in early 2018, a Chinese laboratory successfully cloned monkeys[108] using the same techniques to clone Dolly the Sheep.[109] China has also had very little ethical qualms about the use of PGD, focusing instead on its health benefits, as opposed to its ethical concerns.[110] And, of course, the recent international debacle of Dr. He Jiankui's alleged secret CRISPR project involving the successful birth of twin girls, must be mentioned here as well. All these are simply contemporary examples of the lengths that China is willing to pursue in the name of health improvement.

These measures, however, comes as little surprise if we view the historical accounts of China's stand on eugenics as part of its political reforms process. The paramount concern in Chinese society was public health following several national crises after its defeat in the Opium War in 1840.[111] The campaigns for political reforms began, touting issues of disability (foot binding) and hygiene and sanitation; in the course of combatting this phenomenon, the Chinese formulated what was termed a "degeneration discourse",[112] and women were primarily 'revered' as

---

[105] Tarrant-Cornish (2017).
[106] Cyranoski and Reardon (2015).
[107] Foley (2018).
[108] Briggs (2018).
[109] Gabbatis (2018).
[110] Cook (2017).
[111] Yuehtsen (2010), p. 260.
[112] Ibid 261.

the "breeders of the nation",[113] with policies of forced sterilizations that targeted persons with mental and physical deficiencies. In totality, the Chinese approach to social health and hygiene and sanitation incorporated eugenic elements[114] as a manner to improve Chinese society and strengthen the country after the crises it had faced. Today, China has shown its willingness to be on the forefront of cutting edge science and technological experimentations; demonstrating that its eugenics past had never really seen an end.[115]

2.3   The Theory of 'Liberal' Eugenics: Autonomy and Freedom from State Intervention

2.3.1   Liberal Eugenics: A Recognition of Autonomy?

In Habermas' work titled *The Future of Human Nature*,[116] he employs specific arguments in opposition to genetic interventions, embryo research and PGD. These arguments are seminal because of their co-relation to a post-metaphysical contribution to the history of eugenics and how it impacts on all human personhood. His line of reasoning is also consistent with Germany's conservatively staunch stance on all forms of contemporary medicine and bioethics that could, in any manner, be inextricably linked to, or be the result of, eugenics as we understand it to be (namely, in its pejorative form), or even a liberal form of eugenics that he has envisaged in his essays. Although Habermas' discourse is from a continental point of view, this is important because he raises some fundamental questions about the decisions made by parents (very prevalent in Anglo-American bioethics discussions) and the burning issue of morality, ethics and our pre-supposed sense of autonomy as human beings. It may be surprising to learn that these fundamental questions about parental decision-making also have the same prevalence in many Asian historical and cultural discourse.[117] It is believed that these considerations may pave the way in which we are conditioned to understand the true nature of our corporeal selves; whether from the Habermasian perspective or the Eastern philosophical perspective; and it is

---

[113] Ibid.

[114] Ibid 262.

[115] Pellissier (2015).

[116] Habermas (2003).

[117] The emphasis of Asian values, on filial piety and the role of parents in a family are ingrained in many Eastern cultures. It is often accepted without question that parents are the main arbiters of determining the best interest of their children. As a product of such an environment myself, it is now interesting to see that these similar values described by Habermas, converge in the determination of our sense of individual autonomy.

these elements that Habermas views as essential to the discussion on both positive[118] and negative[119] eugenics.

Before delving into Habermas' dialogue in his book, it is important to bear in mind at this juncture that the crucial underlying consideration of his discourse is formulated upon the interest of either the parent or the child, whether adequately represented or not, by the variation of legal and/or regulatory frameworks that he perceives, can be individualistic in nature to achieve consistency with the autonomy of the person. Arne Johan Vetlesen provides a concise insight on the breadth of Habermas' work: that the latter's discourse should not be viewed as a cure-all frame of answers to the questions that plague us about the morality of human nature, but should be taken as an interesting and important philosophical understanding of how we can seek to formulate general principles to guide legislators and policy makers in the way forward for dealing with genetic interventions, PGD and human embryonic stem cell research.[120] This is an important point that can be applicable in all discourse, regardless of historical, cultural, social, philosophical, and religious outlooks.

In the meantime, the crux of Habermas' argument launches a soft blitzkrieg against the specificity of morality and human personhood in what he views as problematic issues in genetic engineering, embryo research and PGD. Habermas' main argument delves into human personhood and moral agency as part of that personhood, which presupposes that human beings possess a self-assessing mechanism of their image as being free, autonomous, and in control of decisions made in their lives as self-legislating beings; and primarily because of this state of affairs, also requiring the treatment of other moral agencies as a putative equal. He refers to this as the 'species ethics'[121] that validates the individual ethical questions and considerations of who we are *qua* human, and the values that are intrinsically linked in the histories and connection of our lives. The existence of species ethics can be taken to mean that we are birthed with a form of inherent morality, a sense of innate 'goodness' attributed to human beings by virtue of Nature's processes: the vital convergence of social interactions, cultural assimilations, parental influence and upbringing, religious beliefs, worldly exposure, experiential teaching and learning, and such other factors that contribute to the personhood development and growth of an individual.[122]

In Habermas' view, this is a universal concept of morality that is possessed by all human beings from the moment of conscious interactions with the external world,

---

[118] Positive eugenics refers to the enhancement procedures that take place to heighten or amplify an individual's genetic make-up, also referred to as a form of enhancement or non-therapeutic treatment in human genetic engineering technologies.

[119] Negative eugenics serve a therapeutic purpose (traditionally), for instance, to eradicate diseases or abnormalities in an individual's genetic makeup by removing the genes that cause a particular problem in the individual.

[120] Vetlesen (2005), p. 232.

[121] Habermas (2003), p. 40.

[122] Ibid 58.

which begins at the time of birth. This perception of self-realization advances the opinion that human beings are capable of distinguishing between the 'made' (nature) and the 'grown' (genetic interventions) based on their inherent moral agency which is formed by the relevant influences mentioned above.[123] The influence of interventions from technological advances that have the ability to manipulate the genome therefore violates this "realm of [potential] symmetry"[124] between human nature and the equality of the self-understanding species. Habermas further explains that the intervention of genetic technologies intrude upon, and can amount to an arbitrary determination of an individual's make-up, specifically where parents make conscious decisions in 'choosing' characteristics that they feel would be in the best interest of their children. This is the kind of positive eugenics that Habermas speaks vehemently against.[125] Although Habermas does not dive into the commodification of the human body, he teases with his claims that the operation of positive eugenics in this manner by parents, in effect, amounts to an objectification of the embryo or intended future child. It is in this claim of parental decision-making either to enhance, strengthen, improve, or provide some form of genetic advantage to a future child so that they would be able to lead 'better lives' that Habermas takes issue with. His theoretical aspect in this regard hinges upon these concepts as arbitrary characteristics fueled by the preferences of a force external to the intended child, or future individual—which, in turn, distorts a particular implementation of growth in the natural sense.

The understanding of human nature in Asian or Eastern philosophy is, however, much more complex. Where the prominence of Western philosophical thought grew from the likes of Plato and Aristotle, and the Western understanding of human nature is more or less pervasive,[126] the diversity of Eastern philosophy, encompassing Buddhism, Taoism, Hinduism and Confucianism, amongst others, do not have a "strict equivalent to either the word itself or the philosophical problem."[127] The varied *raison d'etre* of each of these Eastern philosophies do not simply operate as philosophies, but in some instances, because of cultural engrailment, are also treated as religious guidance. This is true of Taoism, for example. Taoism was the official state religion in ancient China, but in contemporary settings, has been seen as one of the most important Eastern philosophies. The Tao Te Ching, widely hailed as both a philosophical and spiritual text, explains the connections between human and nature as part of the order of natural things.[128] In another example, the essence of Buddhist and Hindu philosophies are entrenched in the knowledge of the self, which is a vital part of discovering the true reality of human nature.[129] The human nature

---

[123] Ibid 40.

[124] Chesterton (2000).

[125] Habermas (2003), p. 32.

[126] Barnhart (1997), pp. 417, 422.

[127] Ibid.

[128] Deppe (2010).

[129] Dhammanada (2002), p. 188.

of the self in Buddhist philosophy views the component of freedom of moral choice as an integral dimension in the Theory of Karma.[130] Viewed as anti-metaphysical in nature, Buddhist philosophy as an example, focuses on the Four Noble Truths,[131] as a way to alleviate suffering of the human condition. Although this may not be impacted in the same manner as Western scientism, for example, in formulating the essence of human personhood, and has been subject to criticism that its normative existence does little in "formulating actual ethical standards and norms",[132] these views are significantly associative in pointing towards a different understanding of human nature. On this basis, the interpretation about how the fears of eugenics that is linked to human nature and human personhood is also different.

In Habermas' argument, he presupposes a kind of sentient perceptiveness that exists the moment human beings are born and are capable of communicative efforts with their surroundings. Although I do not intend to elaborate on this specific point, it should be mention that this may be regarded as a form of disengagement if we do not delve into the conception of pre-natal life as part of the discourse on human nature). There is a wealth of literature over the past century, on when life begins and at which point a fertilized egg possesses awareness and a self of being. This book does not seek to engage in this discourse at this point of time, but the removal of this from the equation of life cannot provide us with a complete picture on Habermas' claim of personhood of the individual. In addition, the pre-supposition of a form of innate morality also means that there is pre-supposition or belief in a form of divine power, an omnipresent force; which accounts for the nature and existence of human life. Again, engaging in this discussion of a divine being as the force of creation of all life forms on Earth, will necessitate inquiries into theological discussions and evolutionary perspectives which cannot be contemplated by the scope and breadth of this book. For these reasons, and harnessing an evaluation from an empirical perspective, I do not view the possibility of the strength of this argument as a clear opposition to the view of one's self-image as a distortion of "nature". Rather, I pose that the clear conception of a human being and its surrounding-influenced normative values also come from a place of nurturing development and influences which does not always necessarily involve any kind of 'morality' growth. I do, however, agree with Habermas' suggestions that in the journey towards the impending technological future, there is a need to ensure that care is taken in the formulation of 'acceptable' limits (another point of contention which will be addressed further below), with grave distinctions drawn between therapeutic (negative) and non-therapeutic (positive) genetic interventions in reproductive technologies.

Habermas addresses specific issues in PGD as the gateway that leads to possible genetic manipulations in the generation of future offspring. In the meantime, the

---

[130] Ibid 80. The Theory of Karma is posited as the law of cause and effect; that every action or inaction contributes to a specific outcome.

[131] Ibid 98.

[132] Barnhart (1997), p. 421.

recently emerged genome-editing tool, called CRISPR/Cas-9 (CRISPR),[133] allows scientists to edit genomes with unprecedented precision, efficiency and flexibility. As such, the field of possible genetic interventions in PGD processes, and with the advent and potential of CRISPR,[134] raises some very serious questions about possibilities that are unprecedented or may not have been foreseen, and with this, the role of law in medical research and the differences of its treatment of PGD and its progressively haptic use in various countries. In addition, the much-feared idea of germ-line therapy[135] could no longer simply be just a matter of fiction. Botkin's fears have become our present. He had stated in his article that "the possibility of introducing functional genes into an in-vitro zygote or embryo seems quite reasonable in the foreseeable future."[136] And the future has arrived; Habermas' concerns about the profound possibilities of PGD in the new wave of liberal eugenics, may quite well be reasonably founded[137] and we may well proceed to err with caution,[138]

---

[133] Hsu et al. (2014), p. 1262.

[134] It is recognized that CRISPR is still an evolving technology, although it has been preliminarily successful in gene therapy treatments to correct sickle-cell mutation in human cells. In January 2015, researchers in China reported that they had created genetically modified monkeys using CRISPR, raising alarm bells that it is theoretically possible to alter a person's genome before birth if the changes were made to the germ cells of a prospective parent.

[135] The idea of germ line therapy is still a highly controversial topic of discussion among geneticists, bioethicists and members of the medical profession. It involves therapy that targets the germ cells (reproductive cells), either removing or enhancing changes in the DNA, which will then allow for the "correction" of disease-causing gene variants that are certain to be passed down from generation to generation. In effect, germ line therapy manipulates and changes the DNA of basic instruction in a person's body. Current gene therapy does not involve germ lines, and only targets specific somatic cells for treatment.

[136] Botkin (1998), p. 20.

[137] Some promising advances using CRISPR has been indicated recently. For example, genetic mutations (for favism and thalassemia) in early embryos have been successful corrected by scientists in China. See https://www.newscientist.com/article/2123973-first-results-of-crispr-gene-editing-of-normal-embryos-released/. As recently as August 2017, a team of scientists from the United States and Korea demonstrated promising results using CRISPR to alter genetic mutations relating to a heart condition, in early stage human embryos. See https://www.sciencedaily.com/releases/2017/08/170802142844.htm.

[138] For instance, in 2016, it was reported that Duke University researchers had successfully (to a certain extent) edited the genes that mutated into Duchenne muscular dystrophy in mice, see: https://gizmodo.com/first-successful-gene-editing-in-live-mammals-brings-us-1750908059 accessed 2 October 2017. Further promise is shown in research experiments conducted in 2017, including gene editing of non-human primates' embryos (rhesus) by Michigan State University; see https://www.sciencedaily.com/releases/2017/05/170501112525.htm accessed 2 October 2017; the elimination of the HIV DNA in animal models by researchers in Temple University and University of Pittsburgh, see: https://www.sciencealert.com/a-new-gene-editing-technique-has-eliminated-acute-hiv-infection-in-living-animals accessed 2 October 2017; and also a release of results relating to gene editing of normal human embryos (early embryos) by researchers in China, see: https://www.newscientist.com/article/2123973-first-results-of-crispr-gene-editing-of-normal-embryos-released/.

although certain scholars believe that the human race is still far behind in terms of the dystopian life[139] envisaged in the 1997 Hollywood film, Gattaca.[140]

Habermas' opposition to the new liberal eugenics specifically focuses on the parents as the main actors in the determination of employing not only tools of genetic engineering, but also PGD, in the 'creation' of their future or intended child. Habermas recognizes that this form of liberal eugenics excludes the intervention of the state, and modern reproductive biologists such as Alan Handyside, take the view that the parents become the instruments of autonomy in deciding for themselves the key characteristics or aspects of genetic interventions that they wish to participate in, thereby preserving the liberty and personal autonomy in such decisions.[141] This new line of liberalism in eugenics has emerged chiefly as an interesting topic of debate among modern philosophers who are engaged in contemporary dialogues about the morality and ethics of engaging in human genetic engineering technologies, specifically where such uses are non-therapeutic in nature. An examination of the literature in this arena reveals that liberal eugenicists' thought-processes seek to provide a form of reoriented justification about the tacit use of eugenics on modern society, and the tremendous benefits that may be gained from this practice so long as it is well-modulated and reasoned.

One of the more prominent proponents of liberal eugenics is Nicholas Agar; and arguably, his conceptual view on liberal eugenics[142] is an interesting foremost attempt to explain the role of genetic engineering technologies as a justification for individual choices, autonomy and state neutrality. A proponent of scientific and technological advances in improving the quality of human life, Agar's seemingly well balanced theory hovers between a conservative bioethics view,[143] and the more radical trans-humanist ideals.[144] Agar's main contention for the benefits that can be

---

[139] Yong (2017).

[140] Some interesting philosophical viewpoints regarding the premise of the film can be seen here: http://www.philfilms.utm.edu/1/gattaca.htm.

[141] Handyside (2010).

[142] Agar (1998).

[143] Prominent bioconservatives include Leon Kass and Francis Fukuyama; bioconservatists view a post-human state (a state whereby genetic interventions have been embraced completely by individuals) as being degrading, dehumanizing and an affront to human dignity. Often, bioconservatist views are rooted in some form of religious or crypto-religious sentiments that generally condemns the mastery over human nature through the use of genetic engineering technologies. The gist of this condemnation hinges strongly upon the concept of human dignity, which bioconservatists view as an important element of the recognition of individual personhood.

[144] Transhumanists embrace the advent of genetic interventions with open arms, believing in the wide use and dissemination of genetic technologies to the public, based purely on each individual's desires, intention and choice to engage in the use of such technology, whether for enhancement or other therapeutic reasons. The general view held by transhumanists is that the advances of science and technology have made it possible for human nature to be improved, strengthened and enhanced; but transhumanists also strongly campaign for a strong framework and recognition of human rights and individual choices, believing in the evolution and systematic revamp of a social system that metes out understanding and compassion. Julian Savulescu, for example, a prominent bioethicist and philosopher, even provokes conservative styles of thinking about genetic interventions, by

reaped from therapeutic genetic treatments and engineering tools focus on what he terms a 'life plan', which is an individual's projected plan for his or her life in the future determined by many variable factors such as values, upbringing and background, abilities and skills, and the like.

In essence, although Agar is a believer in genetic interventions, and views that there is no fundamental difference between genetic modifications made upon the human genome, and environmental modifications in the concept of child-rearing, in decisions taken by parents, (for instance, in endeavoring to provide a high quality education, bio-organics nutrition and superior healthcare, and extra-curricular activities to nurture or enhance talents or skills in children,[145]) he is also cautious to adopt a sensible stance in the recognition that new genetics medicine and technology must be approached with a respect for the future life plans of individuals.[146] Although there is great autonomy placed on the parents of a potential genetically-engineered child or individual, thereby excluding the intervention of the state, and shifting a value-driven burden onto parental choices, Agar is simultaneously cognizant of the fact that a set of very dominant values exhibited on the part of parents, for instance, can also contribute to a subset of paternalism evocative of the old pejorative eugenics movement. In this regard, there is a need to take into account the very thin dividing line between the impositions of one's personal values (the parents) in accordance with their view of what amounts to an idealist state of affairs, versus a true, good demarcation of the best interest of future offspring.

The formulation of 'life plan' is, by no means, something completely novel. However, what is interesting about Agar's perspective in this domain is his contention that genetic interventions are beyond the reach of a life plan, and a systematic modification of that life plan cannot necessarily be curtailed, for a variety of reasons.[147] In this regard, Agar rejects what is the commonly accepted moral importance in conventions that shape human beings.[148] First, the position of liberal eugenics draws no distinction between the improvements of human beings, whether through their environments (for example, likening environmental influence to things like "experimental vitamin enriched diets or hothouse schooling"[149]) or whether through genomic modification.[150] Secondly, liberal eugenicists reject the notion of making a distinction between "therapeutic goods of genetic engineering"[151,152] and

---

stating that parents would be under an obligation, in such times, to genetically enhance their offspring in order to give them the very best quality and enjoyment of their lives.

[145] Agar (1998), p. 139.

[146] Ibid 138.

[147] Ibid.

[148] Agar (1998), p. 139.

[149] Ibid.

[150] Ibid.

[151] Ibid 141.

[152] The "therapeutic goods of genetic engineering" encompass treatments to 'normalize' an individual's health, specifically focused on the treatment of diseases.

"eugenic goods".[153,154] Agar refers to the work of Philip Kitcher, who repels the accounts of our understanding of the function of disease from the social constructivist and objectivist biological perspectives.[155] Kitcher further formulates an account of how parents should seek the improvement of their children,[156] which should be concerned with "well-being, welfare or quality of life."[157] It is also interesting, at this juncture, that Agar makes no excuses for his reference to this kind of scientific intervention as a form of 'eugenics'[158]; indeed, the clear recognition of this often negatively viewed term, has been modified in the context of Agar's arguments that focus on the strength and tailoring of individual improvements, as opposed to a state-created, mold of societal acceptability in the manner of the old eugenics movements. The morality assessment, according to Agar, will always be a grave one; but he verily states that "a eugenics program appropriately sensitive to the range of potential life plans of future persons will not seek to enhance capacities with any one life plan in mind."[159]

It must be made abundantly clear, that this book is not a wholesale opposition to contemporary interpretations of what it means to engage in eugenics, neither is it an opposition to decision-making processes exercised by parents for their offspring. The demonstration that is sought here is that the term, in itself, is one loaded with evocations of humanity's behavior at its worst in the understanding of our historical accounts. Additionally, the hesitation to use the word 'eugenics' is an understandable one, but the fallacy of humankind is to presume that we do not make eugenics-based decisions in our daily lives. The historical orbit of the theory of eugenics that we now know of has been systematically characterized by an over-zealous commitment to the (perhaps misconceived) betterment of human citizens in a particularized society in a state. Notwithstanding the absence of a unified, single theory about the social philosophy of eugenics, historical accounts have demonstrated to us that the saturation of eugenics' components have influenced a variety of state-sanctioned eugenics practices. The general drive to enhance the improvement of society, throughout the course of eugenics' history, has been marred by forced sterilization programs, mass murders and euthanizations, all of which acutely violate an individual's fundamental right of "life, liberty, or property",[160] and viewed through the lens of modern constitutional jurisprudence, is even more appalling because of the fact that these were state-sanctioned. This is ironic, because it is often accepted that the role, purpose and responsibilities of government or state, is, first and foremost,

---

[153] Agar (1998), p. 141.

[154] It is recognized that the main aims of "eugenic goods" is targeted at individual enhancement.

[155] Agar (1998), p. 141.

[156] Ibid 142.

[157] Ibid.

[158] Agar (1998).

[159] Ibid 138.

[160] Stone et al. (2005).

to protect its citizens.[161] But governments, state or federal, are often suffused with volatility, and we are no strangers to the reality of cronyism, mismanagement, corruption and tainted ideological or political agendas; and governments, like any changeable artefact in the corporeal world, can be overthrown or simply, fail in its duties and responsibilities.

On this basis, it is facile to accept that a shift in the paradigm of decision-making, a gift of autonomy to individuals in the body of citizenry, is a most attractive ideal, even if the underlying connotations swim the turbid waters of eugenics. From as early as the late 1930s, attempts were made within the European context (with specific reference to Germany), to found a more liberating form of developing eugenics philosophy. Scholarly denouement put forward strides to recognize eugenics selection in a more "natural and voluntary process",[162] by propagating a "freedom of parenthood",[163] which would consequently be in line with "the concepts of individual liberty and of non-interference by government".[164]

Hence, the contemporary movement of liberal eugenics, in itself, is premised on the fact that should technological advancements progress to the point of safety and availability, then parents should be at liberty to use at their disposal, the full spectrum of these technologies for the purposes of enhancement of their future offspring. The allure of liberal eugenics pivots on the centrality of this choice: the shift in autonomy from state to individual, and the freedom from state interference in its subsequent exercise by individuals. As a firm supporter of scientific and technological developments seeking to improve the quality of human life, Agar contends for the benefits that may be reaped from genetic treatments and engineering tools.[165] Agar would be quick to argue that, should we focus on the veritable sustenance and orientation of a variety of "life plans",[166] the 'new' eugenics foothold *vis-à-vis* tools of genetic engineering technology, is capable of presenting adequate constrains built into the exercise of autonomy (in this regard, bearing upon the parents of the future offspring), which will not interfere into this varied projected plan of the offspring's future, and will not be capable of directing the offspring only into the direction of one life plan.[167] This supports the view that "the family is the level of implementation",[168] and because the target of 'new' eugenics is robustly rooted at the DNA level (earmarked in the scientific, chemical or structural discrepancies in DNA), and not within societal structure in the manner of the old eugenics,[169] the voluntariness and flexibility of liberal eugenics is a far cry from its ominous ances-

---

[161] Heyman (1991), p. 507.
[162] Osborn (1937), pp. 389, 395.
[163] Ibid 391.
[164] Ibid 395.
[165] Agar (1998).
[166] Ibid 143.
[167] Ibid 141.
[168] Wiesenthal and Wiener (1999), pp. 383, 385.
[169] Ibid.

tor. If we implement this reasoning within the context of Asian culture as a way of comparison, then the emphasis on the family level is a strong factor which influences the use of the technologies, since family is viewed as a paramount component in Asian values too.

Enthusiastic defenders of liberal eugenics further enunciate that the removal of the 'offending' element of past eugenics movements (that is, the coercive edict issued by the state) would result in varied, diverse options in the choices to be exercised by parents for their future offspring. The old-fashioned eugenics reeked of extreme disparity, and was guilty of its somewhat successful efforts to "produce citizens out of a single centrally designed mold".[170] The implication of this "single centrally designed mold", as Michael Sandel correctly identifies, is that "its burdens fell disproportionately on the weak and the poor, who were unjustly sterilized and segregated".[171] In the past, it is clear that the marginalization of these specific groups; of the 'feeble-minded' or "imbeciles",[172] the disabled, the diseased, the lepers, for example, have carried forward in the recesses of our darkest memories. The fear of a return to these dark times, therefore, is a reasonably founded one.

On this basis, political philosophers have put forward arguments that if the abhorrent aspects of state-sponsored eugenics are removed from the equation, and the outcomes of the tools of genetic engineering are evenly and fairly distributed to the general population at large, then the 'eugenics' as we know it is no longer in existence; and therefore, becomes unobjectionable.[173] Given this reasoning, then the practice of liberal or modern eugenics based on the fundamental tenets of individuality, and liberty, should be differentiated and accepted. Gyngell and Douglas further illuminate by reference to Robert Nozick's proposal[174] for a "genetic supermarket"[175] whereby "it involves no centralized decision fixing the future human type(s)".[176] The "genetic supermarket" would be the Walmart of modern genomic societies; and potential parents would be the sole arbiters in determining the 'products' (characteristics, attributes, and such) that they intended to buy. Despite these justifications, this new form of liberalism also evokes a form of squeamishness from the less convinced, for reasons that primarily evoke the ethics of humanity and human nature,[177] *vis-à-vis* the attribution of choice and responsibility to parents, instead of a rumble through the lottery of chance and the cosmos. The contention presented in this book recognizes this paradigm shift, but also questions whether the transposition of choice and autonomy, is truly an enlightened one,

---

[170] Agar (1998).

[171] Sandel (2004), pp. 51, 60.

[172] Burrus (2011).

[173] Buchanan et al. (2001).

[174] Nozick (1974).

[175] Gyngell and Douglas (2015), pp. 241, 242.

[176] Nozick (1974).

[177] Habermas (2003).

which should allow unfettered access and a complete freedom cognizant of a "right" bestowed upon parents in respect of their offspring.

With advancements made in medical technologies and research environments, a modification of the exercise of autonomy should be encouraged to commensurate with the changing dynamics of informed consent in clinical health settings as well. The proponents of the liberal eugenics coalition appears to have struck a chord insofar as the exercise of autonomy is concerned in the decision-making process to reap the benefits of gene editing or engineering tools. At first blush, it is possible to find that the equivalency of choice and autonomy afforded to parents of future offspring is more acceptable on a moral level in comparison to eugenics movements of the past. However, the heart of the matter is intractable insofar as "[making] children into products of deliberate design"[178] is concerned. This book posits that the legality of the concept of autonomy in this instance continues to falter in the light of the more discursive issues that provoke the framework of human rights considerations. In particular, the coupling of parental autonomy in liberal eugenics (for non-medical, non-therapeutic purposes), together with one of the most contentious debates in the field of bioethics, that of gene enhancements and the accompanying germ-line modification concerns, is a surefire, explosive combination that will continue to provoke the fiery discourse on interventions into the human genome, particularly where the exercise of choice through the options proffered by gene editing technologies, are not exercised by the intended beneficiary or recipient of the technology—the future offspring. Without the intention of disrespect for the freedom of parenthood and choices made by parents for and on behalf of their children, I posit that choice and autonomy within the contemplation of liberal eugenics is merely a cloak, and does little to dispel the salient legal and ethical debates that continue to mar the landscape of controversial uses of new and emerging technologies.

Even to the untrained neophyte, it may be inescapably logical that non-therapeutic enhancement treatments for future offspring cannot be a bad thing. John Harris states that we should view enhancement with positivity because the very meaning of enhancement is to make things better, and therefore, he questions how an enhancement could be viewed as something negative.[179] Defenders of genetic enhancements, such Julian Savulescu[180] believes in a moral requirement on the part of parents to 'enhance' their children, *vis-à-vis* a principle he calls "procreative beneficience".[181] This principle, in essence, puts parents at the forefront of choice-making, to ensure that they "select the child, of the possible children they could have, who is expected to have the best life, or at least as good a life as the others, based on the relevant, available information."[182] Some measurable contentions have been levied against how one would ascertain the 'goodness' of a human life, citing

---

[178] Sandel (2004), p. 60.
[179] Harris (2010).
[180] Savulescu (2001), p. 413.
[181] Ibid 413.
[182] Ibid.

distinctions that may be drawn between medical treatment and genetic (medical) enhancement. Savulescu[183] and Harris,[184] however, refute this claim and both deem that this distinction is not morally significant.[185] In fact, the provision of reasoning in this context rests on the supposition that to enhance is to therefore increase the general well-being and welfare of an individual, and by this invocation, treatment is "a special case of increasing the well-being and/or functioning of those with diminished amounts of such things"[186] and is therefore a "subset" of enhancement. Harris provides provocative arguments about how we already accept and widely practice medical and non-medical enhancements in the context of modern life, such as vaccinations, special schools, food supplements, music lessons, and the like,[187] all of which pursue the bid to increase the abilities of children. If these practices in child-rearing are commonplace, and in fact, viewed as universally acceptable, then we should question why genetic enhancements through a similar process of selection of the 'best' attributes, is morally problematic. (This presupposes, above the other legal and ethical concerns, that safety and risks are no longer objectionable issues, in the same manner that medical or other therapeutic treatments are also considered safe enough).

On the intersection between the enhancement debates, with the new eugenics, Agar argues that the freedom and autonomy of choice and decisions imparted upon parents provides a deeper respect for reproductive liberty and human rights,[188] certainly a far cry from the old eugenics programs, which provoked extremely vehement objections because of "the use of state power in the pursuit of eugenic aims."[189] The defenders of contemporary eugenics enhancements further pronounce that the development and continuous evolution of human rights protection in modernized democratic societies has been implanted deeply enough to ensure that the past atrocities of the old eugenics movements are not repeated.[190] The fundamental protections and freedoms accorded by international and regional human rights instruments such as the Universal Declaration of Human Rights (UDHR), and the Convention for the Protection of Human Rights and Fundamental Freedoms (the European Convention on Human Rights, or ECHR in short), amongst other key human rights instruments, accord contemporary societies with a much more stringent and higher level of human rights recognition and protection than they had been imbued with in our past histories.[191]

---

[183] Savulescu (2007).

[184] Harris (2010).

[185] Selgelid (2014), p. 9.

[186] Savulescu (2009), p. 417.

[187] Harris (2010).

[188] Agar (2006), p. 4.

[189] Selgelid (2014), p. 8.

[190] Ibid.

[191] Ibid.

The dexterity of this discourse, I pose, is truly due to a concern regarding the cultural fears of transhumanism. In the European context (and Western context more generally), the imagination of the horrors of the World Wars, the advancement of technologies, and the increased representation of individual self in dominant facets of life, are likely to be factors that contribute to why, and how, debates and laws have taken shape into their existing forms. In our contemporary digital technological age, it is not difficult to see how technologies have become so embedded in our daily lives, that it is possible to think that we may, as a humankind, someday become products of technologies ourselves. In the trans-humanist framework, human beings are highly encouraged to "transcend their current natural state and limitations through the use of technology".[192] Since technologies are now available to harness both therapeutic and non-therapeutic applications to alleviate human conditions, transhumanism is seen as the logical extension of this alleviation, tethering into the realm beyond treating, and instead, focused on enhancing. Raymond Kurzweil wrote of 'The Singularity' in 2005,[193] a phenomenon that he believes will merge humans and machines through the exponential growth of technologies; and the achievement of artificial general intelligence (AGI)[194] by machines which would surpass human intelligence. Since it is debatable that transhumanism found its roots within the discourse on human nature, or humanism, particularly within Western conceptual philosophies,[195] this allows us to make sense of Habermasian theory on the future of humankind where biomedical technologies, such as genetic engineering, is available.

This kind of fear is rarely present in many Asian societies. As illustrated in Sect. 2.2 of this chapter, the Asian perception of technological advancements is rarely rooted in a similar metaphysical perspective. Instead, the historical and cultural vantage of many Asian societies, particularly those that have been colonized by Western powers in the past, seeks to emerge through key indicators that power national economic and growth and development. Although some convergence is observed, *vis-à-vis* the parental autonomy theory argued by Habermas and the cultural values of family that bind many Asian communities, the differences in how the future of humankind is viewed is significant. It is also likely that the transhumanist concerns may be too far removed from the historical, cultural and societal connections to Asian communities. Countries like Japan and China have made tremendous progress in all forms of technologies simply because of the view that technological advancements seek to liberate and make lives better for individuals. A further exam-

---

[192] Thomas (2017).

[193] Kurzweil (2014) http://link.springer.com/10.1057/9781137349088_26.

[194] AGI is not to be confused with Artificial Intelligence (AI), although it may be considered a subset of AI. AI enables technologies like computers to operate intelligently, to mimic human-like behavior. AGI, however, is differentiated by the fact that computers in themselves would be able to perform actual, intellectual tasks that human beings can do, and not simply mimic in the manner of AI. The element of independence, therefore, is what differentiates AI from AGI. Current technologies, however, do not demonstrate that AGI is close to being achieved just yet.

[195] Fred Baumann, '*Humanism and Transhumanism*' 17.

ple is the manner in which South Korea has normalized the enhancement of the female form and image through cosmetic surgery,[196] because its external idealization of the 'perfect' beauty contributes to job prospects, advancements in career, and even social integration in communities. In this regard, it may be startling to discover that the Asian interpretation of modification technologies that may have 'eugenic' outcomes, is seen as a positive thing and simply an evolution of the human condition.

2.3.2   A Re-interpretation of Autonomy in Liberal Eugenics

Notwithstanding the permeability of the human rights discourse, the practical realities are often much harder to reconcile. Within the discourse of disability rights, for example, the narrative still remains highly polarized. Some disability advocates criticize the selecting-out or negative selection of hereditable disability traits, or the avoidance of bearing an offspring that may have serious disabilities. Conversely, these advocates state that disability should be treated as "just another manifestation of human diversity"[197] because disability is a "mere difference".[198] On the other hand, the "medical model of disability"[199] takes an opposite stand by stating that disabilities such as dwarfism, deafness, or mental disabilities, for example, should be corrected if the opportunities to do so were to be made available to parents, because to consciously make a procreative choice to have a child with these disabilities, would be wrong. Since the promulgation of the liberality of eugenics rests on parental autonomy, which encompasses these decisions to either select-out, or maintain these disability characteristics, the implications become tangled in a web of irreconcilable debate. Because of this, and particularly in cases of mental disabilities, the implicational throwback of familiarity to old eugenics in its aim to eliminate "the feeble-minded" or "imbeciles"[200] becomes very acute. Indeed, in Savulescu's and Kahane's attempt to consider a "welfarist approach" to disability,[201] they recognize that "conceptions of disability that associate it with deviation from the normal are entrenched in the public discourse, medicine and law"[202] and therefore, a dramatic "conceptual revolution"[203] would be necessitated.

Within the context of the principles of equality and distributive justice, and with the recognition that inequalities do exist within the stratified layers of society, the

---

[196] Leem (2017), p. 657.

[197] Bognar (2016), p. 46.

[198] Basas (2014), p. 1035.

[199] Savulescu and Kahane (2011), p. 45.

[200] Burrus (2011).

[201] Savulescu and Kahane (2011), p. 45.

[202] Ibid 50.

[203] Ibid 51.

intensification of these inequalities may possibly be exacerbated.[204] Selgelid states that "given that ordinary medical technology is not equally available to all, there is no reason to believe that enhancement oriented technologies would be either."[205] Whether there is truth to the trepidation that enhanced individuals would, on a practical level, be able to pose advantageous over others, is still a subject matter that is being debated. Singer provides some clarity by stating that "many of the advantages people will seek to ensure for their children will be advantageous for them only in comparative, not absolute terms",[206] and he emphasizes the need to differentiate between an "intrinsic good" and "positional good", both of which essentially involve value judgments and the necessity to consider the benefits it may bestow on a social level.[207] It is also concurrently irrefutable that inequalities that exist within societies should be a matter of social reconstruction by states, but this does not mean that we should also downplay the existing problems of inequalities: "the extent of inequality is a key consideration"[208] in the determination of arguments that seek to justify the restriction on autonomy and liberty in the use of these technologies.

Posed against the background highlighted above, I attempt to demonstrate that the root of the exercise of autonomy and choice may be a delusive constituent within the liberal eugenics framework, and in fact, is not as clear-cut as it may appear to be. Through the over-arching framework of the state working its 'invisible' hand (which I do not view as a necessarily evil occurrence), I posit that first, parental autonomy in making decisions relating to genetic enhancement of their future offspring, cannot be completely value-free; secondly, I echo the sentiments of Wiesenthel & Wiener[209] that put forward the illusory and false sense of security in autonomous power as an extension of societal structures; and thirdly, I refer to Foucault's discourse on the existence of power relations in every human interaction, even between parents and children, "subject to negotiation, each individual having his place in the hierarchy, no matter how flexible it would be",[210] as well "bio"-power and politicization of the human body by subjugation through social and covertly-political controls.[211]

First, following Harris' justifications[212] for the enhancement of future offspring, and by virtue of existing mechanisms of improvement in which parents already do engage for their children, I now examine the highly criticized but tongue-in-cheek portrayal of 'Asian'-style parenting in Amy Chua's *Battle Hymn of the Tiger*

---

[204] Reiss and Straughan (1996).

[205] Selgelid (2014), p. 9.

[206] Singer (2009), p. 282.

[207] Ibid 288.

[208] Selgelid (2014), p. 11.

[209] Wiesenthal and Wiener (1999).

[210] Foucault (1963).

[211] Foucault (1977).

[212] Harris (2010).

Mother.[213] In this biographical (and satirical) account of Chua's authoritarian parenting style, often referred to within parenting pedagogy as 'helicopter-parenting', I draw a preliminary hypothesis that "heavily managed, high-pressure child rearing"[214] and the "trend towards hyper-parenting"[215] does not drastically differ from concerted parental decisions taken to implement genetic enhancement technologies on their offspring. The beneficiaries of Chua's strict parenting style are her highly accomplished, Ivy-leagued daughters, Sophie and Lulu; and the biography describes Chua's hard-liner methods in bringing up her children, pushing them through enormous amounts of study and music practice on a daily basis so that they could (and did) achieve the success that only many could dream about. However, Chua also received intense criticism on this, and many termed her parental methods as "abuse".[216] In *Harvard Girl*,[217] written in Chinese and hailed in the People's Republic of China to be the foremost parenting 'manual', special focus was placed on early education and how parents could raise successful children through a strict, methodical lifestyle and to be accepted into top-tier universities in the United States.[218] These examples demonstrate that the reality of child-rearing is not only saturated and particularized as part of cultural determinism, but also truly begs the question of autonomy on the part of the children or future offspring, and their parents, and whether it may be parental ideologies instead that have been imposed on these children. There is no argument that the successes of these children are largely attributable to their parents and the manner in which they were raised. There are modern examples where parents send their children to exclusive private education schools, or music lessons to hone the abilities of their children at playing the violin or piano, and dance lessons in the hopes of raising the next prima ballerina, and even providing nutritional supplements like gingko biloba vitamins to boost focus and memory. More disturbingly, studies have shown the increase of young adults' misuse of drugs like Adderall or Ritalin, in universities, due to intense pressure to do well in their studies, with some scholars calling for these drugs to be available to them without a diagnosis of Attention Deficit Hyperactivity Disorder (ADHD),[219] and as legitimized neuro-enhancement tools within a legalized framework.

Within the field of non-medical, non-therapeutic genetic enhancements, barring the difference on a molecular level, I pose that there is no significant difference between using gene enhancement technologies for the purposes of enhancing a future offspring's cognitive and intellectual abilities, versus parenting styles that

---

[213] Chua (2011).

[214] Sandel (2004).

[215] Ibid.

[216] Cochrane (2014).

[217] Weihua and Xinwu (2000).

[218] The subject of Harvard Girl, Liu Yiting, was not only accepted into Harvard to study applied mathematics and economics, but also received competitive offers from Columbia, Yale and Wellesley.

[219] Flanigan (2013), p. 325.

serve to nudge their offspring in education, music, and other fields that may be regarded by parents as 'good'. In both instances, the desire and intentions of parents to provide the 'best' for their children is not disputed, although the motives for such desires and intentions will invariably be very subjective; but both instances also demonstrate that the heralding light of choice and autonomy is a flawed one. Agar's contention that genetic engineering cannot alter the future life plans of offspring[220] and therefore, maintains the functionality of autonomy, cannot be truly sustainable, because, as in the case of 'normal' child-rearing, a majority of parents do deliberately and concertedly push their children towards a definitive life plan; there is no guarantee that the decisions made in respect of genetic engineering will not echo the same sentiments. It is, of course, arguable that the expansion of a child's life plan, following the initial directedness of parental decisions, may evolve into an unassociated and distinct future life plan; but it should also be recognized that the subsequent trajectory of this later life plan may have been fundamentally affected and heavily influenced by the initial directedness of parental decisions; in a similar manner that decisions by parents to genetically enhance their children may be canvassed. The 'autonomy' therefore exercised by parents, in this manner, is value-laden, burdened with societal expectations of a 'good life', and in some instances, may be a projection of the parents' internal unconscious ideologies and desires of their definitions of a 'good life' for themselves.

Secondly, the concept of autonomy in liberal eugenics and gene enhancement also takes a slippery fall when we examine the structural architecture of equality, citizenry and individual relationships within the function of society. Wiesenthal and Wiener[221] posit that the new eugenics do not particularly lead to true empowerment and true autonomy. Quoting Freeman Dyson,[222] who raised the provocation question whether scientific advances do truly empower individuals, Wiesenthal and Wiener state "whether true or false empowerment exists if whether scientific advances or technology has provided more freedom of choice to the individual, or whether it has enabled the forces of social control to better direct, supervise, impose, or enforce its will and decisions upon the populace."[223] They further state that true empowerment "reduces societal control over the individual by shifting power from the government… to the individual."[224] The paradox, however, in this shift of power, is that individual choices made are often inextricably linked to some variation of societal control with links to communities; an informal mechanism of social control and a possible watered-down version of cultural and societal hegemony. With the appearance of conferring greater choice upon individuals, namely parents, how are ethical values then to be determined within the scope of communities? Parents are already choosing endowments that may lead their children down the path of "socially

---

[220] Agar (2006).

[221] Wiesenthal and Wiener (1999), p. 390.

[222] Dyson (1997), p. 46.

[223] Wiesenthal and Wiener (1999), p. 390.

[224] Ibid.

defined success",[225] and in this process, the empowerment in genetic decision-making has the effect of transforming genetic structures into a consumer by-product.[226] When consumerism and the outcomes of gene enhancement becomes entangled, so too does autonomy and decision-making; the truth of the matter lies in the fact that true empowerment only lies with those who are wealthy and may have access to these enhancement services, and therefore, reminiscent as a "symbol of conspicuous consumption."[227]

Thirdly and finally, I turn to Foucault's discourse on power relations,[228] and draw analogous parallelism to the relationship between state and citizens/individuals. More particularly, this parallelism also exists between parent and child, and shows that autonomy may be illusive in liberal eugenics and gene enhancement. Within the milieu of the natural course of human interactions, Foucault has rightly emphasized the existence of power relations in *omnium gatherum*. The expansive breadth of his work also reveals a fascination with sciences and technology: this I view to be a fully-functioning theme on the means to identify the politicization of individuality within the domination framework, referring to this as the "political economy of the body" or "political investment of the body",[229] of "power",[230] and of "knowledge"[231] where he states "we should admit rather that power produces knowledge.....; that power and knowledge directly imply one another; that there is no power relation without the correlative constitution of a field of knowledge, nor any knowledge that does not presuppose and constitute at the same time power relations"; and "the body".[232] With the backdrop of genetic modification of human beings, genetically modified organisms, embryonic stem cell research, the proliferation of bio-data and bio-banking, and other such advancements that have come to polarize our philosophical, legal and ethical discourses, this is consistent with Foucault's moniker of these issues: "biopolitics" or "biopower".[233] If we do agree with the alignment of Foucault's theories on the existence of omniscient power relations that exist at every level of human interaction, then the underlying reason for the legitimacy of autonomy may come into question through the influence of these power relations.

I pose that an analogy can be drawn from visiting the scope of power relations between citizens and state, considering the interplay between law and legal regulation, with 'architectural regulation'.[234] Hence, I pose that the architectural frame-

---

[225] Ibid 391.

[226] Ibid 392.

[227] Ibid.

[228] Foucault (1963).

[229] Foucault (1977), pp. 25–30.

[230] Ibid.

[231] Ibid 27.

[232] Ibid 25.

[233] Foucault (1976).

[234] Lessig (2006).

work of regulating societies[235] can sometimes be as powerful as the textual rules of the law and legislation. The operation of social controls that regulates norms of conduct *vis-à-vis* the law (which controls and enforces sanctions on behaviors after the fact or action) is supplemented through the *fait accompli* in the construct of architectural regulation, an incidental method that facilitates the present conditions of action, social settings and resources available to individuals through constraints on some behaviors, which then make other behaviors possible. (This can be distinguished with laws that deal with consequences, actions or behaviors <u>after</u> the fact,[236] whereas architectural regulation presents an <u>immediate</u> barrier to certain consequences, actions or behaviors.) This can change the manner in which the nature of rules are presented and enforced because architectural regulation has the capacity to "design out" individual decisions and actions.[237] In this manner, the exercise of individual liberty and autonomy, may be disfigured *vis-à-vis* the creation of social norms (established through architectural constraints) that seemingly give rise to that liberty and autonomy. Consequently, I view that this has the capacity to preserve the "politicization" of human bodies[238] in the manner put forward by Foucault, and hence, amounts to an 'invisible hand' that continues to be exerted by the state. In the very same manner, the dynamics of these power relations within the context of a family is also prevalent, and the beacon light of equality sought between relations is inevitably unbalanced in favor of the party with a stronger positioning, made apparent through the fragmentations of power, knowledge and control. This is not to say, however that this 'invisible hand' is invariably negative; in many ways, I predicate the necessity of some measure of state involvement because the discrete nature of liberal eugenics and gene enhancement technologies goes beyond the frontiers of human life and must be regulated. The conjecture made here is simply that the exercise of autonomy really is not fully autonomous, and cannot be fully appreciable in the manner described of liberal eugenics.

2.4   Eugenic Legacies in Contemporary Genetic Debates

This chapter invites a reflection of a democratic diversity of opinions that leads to the exercise of the autonomy of selection in particularized ways. I surmise that the rallied efforts by liberal eugenicists in framing autonomy are still in deficit because it is very difficult to articulate and implement. If we are faced with possibilities to choose between an enhanced versus 'normal' or disabled future offspring, we are likely to find ourselves at the crossroads of moral, ethical and in some instances, religious dilemmas. Instead of presenting the concept of autonomy in liberal

---

[235] Tien (2005), p. 23.

[236] Lessig (2006), p. 237.

[237] Tien (2005), p. 3.

[238] Foucault (1976).

eugenics as one reminiscent of the full spectrum of ease and individual liberty, I put forward that democratic innovations like these, no matter how advantageous they may be to humankind, must be balanced against a social movement of a non-radical nature in the interest of enlightened medical discourse.

The legacies left in the wake of past negative eugenic practices has provided a wealth of lessons and considerations for us to take into account into contemporary genetic debates. The foremost of these lessons, from the legal and socio-legal perspective, is that the consideration of eugenics as a legacy contributes to a contextual environment in contemporary genetic debates to understand three components of the regulatory environment in biomedical technologies. Firstly, *why* are some biomedical processes completely forbidden; for example, any form of biomedical interventions that clone human beings?[239] Secondly, *what* are some of the identifiable biomedical processes that are prohibited or forbidden? This contributes to a delineation of acceptability and compromise, with those practices that are strictly off-limits, because eugenics is 'objectively' deemed to be negative.[240] And thirdly, *how* does the law regulate and enforce the prohibitions and permissibility of biomedical processes and practices?[241] All these are the culmination of our understood notions about the concept of eugenics that has led to the state of biomedical regulation in its current form, and the emphasis on genetic germ line modifications as to the backdoor to unacceptable eugenic practices.

The key is not to allow ourselves to be deceived into a belief of full and complete idea of what autonomy or choices in biomedical interventions present. It is more meaningful and useful to understand the shortcomings of how autonomy operates in liberal eugenics, and to take advantage of these shortcomings to begin questioning how we may strike a balance. The 'invisible hand' continues to wave its influence over all aspects of contemporary societies, but the challenge then would be to strike a composite balance between a justifiable restriction of autonomy in reproductive liberties and technologies, and the simultaneous deference to the protection of fundamental human rights.

Although there is warranted and reasonable concern that eugenic practices of the past could possibly be repeated in a free-for-all, non-regulated state of biomedical technologies, the intervention through international human rights laws is likely to ensure that this will not happen. In addition to this, the fact that the words 'eugenics' is used within the framework of a selection process of human embryos does not necessarily mean that the same vision of past practices is presented. In fact, considered from the ethical dimensions of clinical treatment and practice, the most compelling reason for contemporary embryo selection, especially within PGD services, is to ensure that future embryos to be implanted are free of severe genetic illnesses

---

[239] United Nations, 'In Opening Debate On Human Cloning Ban, Some Speakers Urge Outright Prohibition, Others Favour Partial Ban To Allow For Medical Advances | Meetings Coverage and Press Releases' https://www.un.org/press/en/2002/l2995.doc.htm.

[240] Christiansen (2017) (*sciencenordic.com*).

[241] Mehlman (1999), p. 671.

that may affect them later in their lives. The possibilities of a designer baby future has been presented simply as a worst case scenario, culminating from a lack of understanding of the true nature and workings of the technologies,[242] and much more importantly, the main purpose why PGD is engaged in the first place.[243] The emphasis on autonomy and choice is the main consideration in contemporary genetic debates of this nature. It is likely that eugenics has been misinterpreted even within the purview of embryo selection processes. Despite this recognition, though, we must also be mindful that autonomy should be founded upon limits that do not encroach into a violation of another individual's rights and liberties.

Regarding the philosophical reasoning behind the eugenics fear, the comparative dimensions of East and West helps to provide a baseline for our understanding in the rest of this book. If we accept these differences, then the evolution of legal approaches and frameworks in the selected jurisdictions does not come as a big surprise. Instead, this provides greater clarity in how regulation has been shaped; for example, looking towards the manner PGD has been governed in the different jurisdictions, or how human rights considerations impact on the constitutional frameworks of the different jurisdictions. Hence, what is suggested at this point is to emphatically determine, (and not merely from the medical versus non-medical aspects) as to what may amount to eugenics. In terms of PGD, or pre-implantation genetic interventions, selecting an embryo, or facilitating the editing of that embryo's genes (where it would be made possible) should not simply be categorized as eugenics. In avoiding a possible future harm to a future child, such as genetic or inheritable diseases, this is consistent with the improvement of human life in the same way medical treatments are accorded. Sándor states most succinctly that we assume the infallibility of medical criteria in determining what is eugenics, and what is not.[244]

What is therefore needed, is an enlargement of criteria, encompassing rationalized views of multiple stakeholders, to determining what eugenics is in the context of emerging biomedical and reproductive technologies. This would enable a more enlightened understanding of embryo selection in PGD, the avoidance of diseases in the future, and the reasons why we would even consider genetic interventions in the first place. The comparative philosophical reasoning that is seen in this chapter attempts to lay the foundation for deeper sensitivity to differing legal cultures, shaped by particularized historical and cultural thinking. Ultimately, I believe that this foundation will provide a critical basis for understanding why calls for a newly negotiated set of common and universal values that may ideally be applicable in all legal frameworks, is timely and necessary.

---

[242] Belluck (2017).

[243] Sándor (2015), p. 357.

[244] Ibid 355.

References

Agar N (1998) Liberal eugenics. Public Aff Q 12:137

Agar N (2006) The debate over liberal eugenics. Hast Cent Rep 36:4

Barnhart MG (1997) Ideas of nature in an Asian context. Philos East West 47:417

Barr MD, Lee Kuan Yew: race, culture and genes. 18

Basas CG (2014) What's bad about wellness? What the disability rights perspective offers about the limitations of wellness. J Health Polit Policy Law 39:1035

Baumann F, Humanism and Transhumanism. 17

Belluck P (4 August 2017) Gene editing for "designer babies"? Highly unlikely, scientists say. The New York Times. https://www.nytimes.com/2017/08/04/science/gene-editing-embryos-designer-babies.html

Bognar G (2016) Is disability mere difference? J Med Ethics 42:46

Botkin JR (1998) Ethical issues and practical problems in preimplantation genetic diagnosis. J Law Med Ethics 26:17

Briggs H (24 January 2018) First monkey clones created in the lab. BBC News. http://www.bbc.com/news/health-42809445

Buchanan A et al (2001) From chance to choice: genetics and justice. Cambridge University Press

Buck v. Bell, 274 U.S. 200 (1927) (Justia Law). https://supreme.justia.com/cases/federal/us/274/200/case.html

Burrus T (23 June 2011) One generation of Oliver Wendell Holmes, Jr. Is Enough. Cato Institute. https://www.cato.org/blog/one-generation-oliver-wendell-holmes-jr-enough

Chan CK (1985) Eugenics on the rise: a report from Singapore. Int J Health Serv 15:707

Chesterton GK (2000) Eugenics and other evils: an argument against the scientifically organized state. Inkling Books

Christiansen K (14 November 2017) Genome editing: are we opening a back door to eugenics? Science Nordic. http://sciencenordic.com/genome-editing-are-we-opening-back-door-eugenics

Chua A (2011) Battle hymn of the tiger mother. Penguin Group

Cochrane K (7 February 2014) The truth about the tiger mother's family. The Guardian. http://www.theguardian.com/lifeandstyle/2014/feb/07/truth-about-tiger-mothers-family-amy-chua

Cohen A (2016) Imbeciles, The Supreme Court, American Eugenics and the Sterilization of Carrie Buck. Penguin Press

Cook M (19 August 2017) China rushes into embryo selection. BioEdge. https://www.bioedge.org/bioethics/china-rushes-into-embryo-selection/12399

Cyranoski D, Reardon S (2015) Chinese scientists genetically modify human embryos. Nature News. http://www.nature.com/news/chinese-scientists-genetically-modify-human-embryos-1.17378

Deppe C (2010) Tao Te Ching: a window to the Tao through the words of Lao Tzu. Fertile Valley Publishing. https://terebess.hu/english/tao/Deppe.pdf

Dhammanada KS (2002) What Buddhists believe, 4th edn. Buddhist Missionary Society Malaysia

Dyson F (1997) Can science be ethical? N Y Rev Books 44:46

Encyclopedia of Bioethics, vol 2, 3rd edn (Thomson Gale 2004)

Flanigan J (2013) Adderall for all: a defense of pediatric neuroenhancement. HEC Forum 25:325

Foley KE (23 January 2018) Chinese scientists already used Crispr gene editing on 86 human patients. Quartz. https://qz.com/1185488/chinese-scientists-used-crispr-gene-editing-on-86-human-patients/

Foucault M (1963) Naissance de La Clinique Une Archéologie Du Regard Médical. Presses Universitaires de France

Foucault M (1976) The history of sexuality volume I: an introduction. Pantheon Books

Foucault M (1977) Discipline and punish: the birth of the prison. Vintage Books, Random House

Franklin DL (29 June 2015) How the 1942 case of a one-footed chicken thief laid the foundation for marriage equality. Slate Magazine. http://www.slate.com/articles/news_and_politics/jurisprudence/2015/06/gay_marriage_supreme_court_ruling_how_skinner_v_oklahoma_laid_the_foundation.html

Gabbatis J (14 February 2018) Dolly the sheep: 15 years after her death, cloning still has the power to shock. The Independent. https://www.independent.co.uk/news/science/dolly-the-sheep-cloning-15-years-death-future-humans-monkeys-what-next-a8208896.html

Galton D (2002) Eugenics. The future of human life in the 21st century. Abacus

Ganesan JS (7 October 2016) A short history of the word "Kiasu". Esquire Singapore. https://
www.esq.sg/lifestyle/culture/news/A-Short-History-Of-Kiasu

Griswold v. Connecticut, 381 U.S. 479 (1965) (*Justia Law*). https://supreme.justia.com/cases/fed-
eral/us/381/479/case.html

Groll D, Lott M (2015) Is there a role for "human nature" in debates about human enhancement?
Philosophy 90:623

Gyngell C, Douglas T (2015) Stocking the genetic supermarket: reproductive genetic technologies
and collective action problems: stocking the genetic supermarket. Bioethics 29:241

Habermas J (2003) The future of human nature. Polity Press

Handyside A (2010) Let parents decide. Nature 464:978

Harris J (2010) Enhancing evolution: the ethical case for making better people. Princeton
University Press

Hediger R (2016) Becoming with animals: sympoiesis and the ecology of meaning in London and
Hemingway. Stud Am Nat 11:5

Heinemann T, Honnefelder L (2003) Principles of ethical decision making regarding embryonic
stem cell research in Germany. Bioethics 16:530

Heyman SJ (1991) First duty of government: protection, liberty and the fourteenth amendment.
Duke Law J 41:507

Hovhannisyan A (2018) Ōta Tenrei's defense of birth control, eugenics and euthanasia. Contemp
Jpn 30:28

Hsu P, Lander E, Zhang F (2014) Development and applications of CRISPR-Cas9 for genome
engineering. Cell 157:1262

IMDb, *Science and the Swastika*. http://www.imdb.com/title/tt0808104/

Ingram C (12 March 2003) State issues apology for policy of sterilization. Los Angeles Times.
http://articles.latimes.com/2003/mar/12/local/me-sterile12

Kaelber L, Eugenics: compulsory sterilization in 50 American states. https://www.uvm.
edu/~lkaelber/eugenics/

Kango-Singh M (2010) In: Speicher M, Antonarakis SE, Motulsky AG (eds) Vogel and Motulsky's
human genetics-problems and approaches. BioMed Central. https://humgenomics.biomedcen-
tral.com/articles/10.1186/1479-7364-5-1-73

Kater MH (1987) The burden of the past: problems of a modern historiography of physicians and
medicine in Nazi Germany. Ger Stud Rev 10:31

Kevles DJ (1999) Eugenics and human rights. BMJ: Br Med J 319:435

Kurzweil R (2014) The singularity is near. In: Sandler RL (ed) Ethics and emerging technologies.
Palgrave Macmillan

Leem SY (2017) Gangnam-style plastic surgery: the science of westernized beauty in South Korea.
Med Anthropol 36:657

Lessig L (2006) Code: Version 2.0, 2nd edn. Basic Books

Matsubara Y (1998) The enactment of Japan's sterilization laws in the 1940s: a prelude to postwar
eugenic policy. Historia Scientiarum 8:187

Mehlman MJ (1999) How will we regulate genetic enhancement. Wake Forest Law Rev 34:671

Mendel as the Father of Genetics:: DNA from the Beginning. http://www.dnaftb.org/1/bio.html

Nature (1998) China's "eugenics" law still disturbing despite relabelling. Nature 394:707

North Carolina Administration, NC DOA: Welcome to the Office of Justice for Sterilization Victims.
https://ncadmin.nc.gov/about-doa/special-programs/welcome-office-justice-sterilization-victims

Nozick R (1974) Anarchy, State, and Utopia. Basic Books

Osborn F (1937) Development of a eugenic philosophy. Am Sociol Rev 2:389

Pellissier H (22 June 2015) Do you fear eugenics? China does not, and that's a problem - interview
with Chad White. Institute for Emerging Technologies and Ethics. https://ieet.org/index.php/
IEET2/more/pellissier20150622

Portnoy J (27 February 2015) Va. General Assembly agrees to compensate eugenics victims.
The Washington Post. https://www.washingtonpost.com/local/virginia-politics/va-general-
assembly-agrees-to-compensate-eugenics-victims/2015/02/27/b2b7b0ec-be9e-11e4-bdfa-
b8e8f594e6ee_story.html?noredirect=on&utm_term=.129bf4f66cb1

Reiss MJ, Straughan R (1996) Improving nature? The science and ethics of genetic engineering. Cambridge University Press

Robertson J (2010) Eugenics in Japan: Sanguinous repair. In: Bashford A, Levine P (eds) The Oxford handbook of the history of eugenics

Roe v. Wade, 410 U.S. 113 (1973) (*Justia Law*). https://supreme.justia.com/cases/federal/us/410/113/

Rorty MV (2003) The future of human nature. Notre Dame Philos Rev

Saetz SB, Court MV, Henshaw, MW (1985) Eugenics and the third Reich. Eugen Bull

Sandel M (2004) The case against perfection. Atl Mon 293:51

Sándor J (2015) The ethical and legal analysis of embryo preimplantation testing policies in Europe. In: Scott Sills E (ed) Screening the single euploid embryo. Springer International Publishing

Savulescu J (2001) Procreative beneficence: why we should select the best children. Bioethics 15:413

Savulescu J (2007) Genetic interventions and the ethics of enhancement of human beings. In: The Oxford handbook of bioethics. Oxford University Press

Savulescu J (2009) Genetic interventions and the ethics of enhancement of human beings. Read Philos Technol:417

Savulescu J, Kahane G (2011) Disability: a welfarist approach. Clin Ethics 6:45

Selgelid MJ (2014) Modern eugenics and human enhancement. Med Healthcare Philos 17:3

Sholley JB (1951) Constitution of the United States of America. In: Cases on constitutional law. Bobbs-Merrill

Shuster E (1997) Fifty years later: the significance of the Nuremberg code. N Engl J Med 337:1436

Singapore Democratic Party, Eugenics in Singapore. http://yoursdp.org//news/eugenics_in_singapore/2008-11-09-558

Singer P (2009) Parental choice and human improvement. In: Human enhancement. Oxford University Press

Skinner v. Oklahoma Ex Rel. Williamson, 316 U.S. 535 (1942) (*Justia Law*). https://supreme.justia.com/cases/federal/us/316/535/case.html

Stone G et al (2005) Constitutional law, 5th edn. Aspen Publishers

Tarrant-Cornish T (26 December 2017) Richest country in the World: China to overtake the US as most powerful economy. Express. https://www.express.co.uk/news/world/896869/China-economy-US-richest-country-world-Donald-Trump-trade-GDP-research

Thomas A (31 July 2017) Super-intelligence and eternal life: transhumanism's faithful follow it blindly into a future for the elite. The Conversation. http://theconversation.com/super-intelligence-and-eternal-life-transhumanisms-faithful-follow-it-blindly-into-a-future-for-the-elite-78538

Tien L (2005) Architectural regulation and the evolution of social norms. Yale J Law Technol 7:23

United Nations, In opening debate on human cloning ban, some speakers urge outright prohibition, others favour partial ban to allow for medical advances | Meetings Coverage and Press Releases. https://www.un.org/press/en/2002/l2995.doc.htm

Vetlesen AJ (2005) The future of human nature. Scand J Disabil Res 7:232

Watts G (31 January 2018) "Eugenics" case highlights dark chapter in Japanese history. Asia Times. https://www.asiatimes.com/2018/01/article/eugenics-case-highlights-dark-chapter-japanese-history/

Weihua L, Xinwu Z (2000) Harvard Girl Liu Yiting: a character training record. Writers Publishing House

Wiesenthal DL, Wiener NI (1999) Ethical questions in the age of the new eugenics. Sci Eng Ethics 5:383

World Bank (2018) Global economy to edge up to 3.1 percent in 2018 but future potential growth a concern. World Bank. http://www.worldbank.org/en/news/press-release/2018/01/09/global-economy-to-edge-up-to-3-1-percent-in-2018-but-future-potential-growth-a-concern

World Medical Association (2018) WMA declaration of Helsinki - ethical principles for medical research involving human subjects. https://www.wma.net/policies-post/wma-declaration-of-helsinki-ethical-principles-for-medical-research-involving-human-subjects/

Yap MT (2003) Fertility and population policy: the Singapore experience. J Popul Soc Secur (Popul) 1(Suppl):643

Yong E (2 August 2017) The designer baby era is not upon us. The Atlantic. https://www.theatlantic.com/science/archive/2017/08/us-scientists-edit-human-embryos-with-crisprand-thats-okay/535668/

Yuehtsen JC (2010) Eugenics in China and Hong Kong: nationalism and colonialism, 1890s–1940s. In: Bashford A, Levine P (eds) The Oxford handbook of the history of eugenics

Zeidman LA (2011) Neuroscience in Nazi Europe Part I: eugenics, human experimentation, and mass murder. Can J Neurol Sci/Journal Canadien des Sciences Neurologiques 38:696

Chapter 3
The Legal and Ethical Debates in Embryo Selection

Abstract This chapter calls for a reflection on the legal and ethical aspects of embryo selection in reproduction, tracing the beginnings of the global problems of infertility and the brief advent of ARTs and other forms of prenatal testing technologies into the reproductive landscape. I surmise in this chapter that the quest for bearing a child free of diseases has led to the concept of perfection, and why reproductive technologies like PGS and PGD have gained increasing importance. In highlighting the regulatory or legal pronouncements that govern PGS and PGD in the selected jurisdictions for the book, I identify the main debates that dominate the discourse of embryo selection and potential genetic interventions that have, and shall continue to shape the framework of regulation in the various jurisdictions. These debates have been categorized into the political and socio-legal or legal debates; the religious and cultural debates; and the ethical and philosophical debates.

This chapter begins with outlining the paradigm shift of the reproductive health and rights agenda, culminating from the International Conference on Population and Development (ICPD) held in Cairo, Egypt, in 1994. Following the previous chapter's main premise regarding the need to redefine the parameters of eugenic selection, and how contemporary notions of individual autonomy impact the selection process, this chapter expands into the reasonings that explain embryo selection phenomena. The recognition of the pressing global infertility up-trend makes it valuable for this book to delve into the question of accessibility to effective assisted reproductive technologies (ARTs) that may alleviate the problems of primary or secondary infertility,[1] whether these be provided for by individual state components or private healthcare or medical facilities.[2] In Sect. 3.1, I trace the key reasons for the global infertility up-trend; these are based on extensive demographic studies

---

[1] The discourse on global infertility is a vital component in understanding how pre-implantation genetic screening technologies made their foray into the realm of reproduction. Not only does infertility impact on matters of population and economic growth, it is also the main motivation behind reproductive technologies like PGD and PGS.

[2] ICPD, 'Policy Recommendations for the ICPD Beyond 2014: Sexual and Reproductive Health & Rights for All' http://icpdtaskforce.org/resources/policy-recommendations-for-the-ICPD-beyond-2014.pdf.

© Springer Nature Switzerland AG 2019
P. L. Lau, *Comparative Legal Frameworks for Pre-Implantation Embryonic Genetic Interventions*, https://doi.org/10.1007/978-3-030-22308-3_3

primarily conducted by renowned medical anthropologist, Marcia Inhorn and Pasquale Patrizio.[3] These reasons affect the ultimate goal of the ICPD's agenda: to switch the discourse from fertility and population control, to a deeper appreciation of reproduction (which would include access to reproductive services and technologies).[4]

In fabricating the question of this accessibility raised by Inhorn and Patrizio, I highlight two key factors that need continuous dialogue. First, that the socio-economic patterns of many communities continue to ascribe circumstances often associated with fertility, to women, upon whom the burden of reproduction has been placed. The need to seek out methods that may guarantee the successful conception and birth of a child is therefore acute. The role of ARTs in meeting this need therefore began to change the landscape of reproduction and parenthood, expanding and making acceptable the notions of non-coital reproduction, and changing the dynamics in health and risk assessment.[5] Secondly, men's role, and a need to make a connective effort for their inclusion, in all aspects of reproduction, consistent with the tenets of the ICPD, must be revisited.[6] Although it has been recognized that male infertility is also a contributory factor to the world's infertility statistics (totalling about twenty three percent (23%) of the available global statistics),[7] the stigmatization effects of a failure to successfully birth a child often lies with the woman.[8] To challenge this perception means to challenge an interpretive social framework that sustains a deep-rooted patriarchate: the male's responsibility to provide for his family and ensuring their general health.[9]

By recognizing that these are factors that need to be revisited on an incremental basis, I theorize that the adverse pressure of fertility, conception and pregnancy, are formidable drivers in causing and/or perpetuating couples to pursue, (what may be misconceived in some respect), the best possible methods of gestating and giving birth to a potential child that would carry the best of their combined genetic pool. ARTs, beginning as an altruistic measure to overcome infertility problems, appeared to be the solution. This undertaking is, however, simply the beginning of a larger issue at hand. The provision of newer reproductive technology services (alongside ARTs) now allows a glimpse into these 'best' possibilities, and the ability to then select embryos.

Section 3.2 begins to probe into the main issues of this chapter. Outlining technologies like IVF, used in conjunction with either Pre-Implantation Genetic Screening (PGS) or Pre-Implantation Genetic Diagnosis (PGD), this section introduces the reader to genetic screening technologies in ARTs that have the aim of securing healthy pre-implantation embryos in IVF fertility treatment. Inevitably, the

---

[3] Inhorn and Patrizio (2015), p. 411.

[4] ICPD (n 2).

[5] Brezina and Zhao (2012).

[6] Dudgeon and Inhorn (2004), p. 1379.

[7] Inhorn (2007), p. 1.

[8] Goslinga-Roy (2000), p. 113.

[9] Inhorn (2013), p. 50.

outcome necessitates a process of then selecting the best possible embryo. This section elucidates on the rising societal awareness of genetic screening or testing procedures, which, in them, are relatively harmless in nature simply for the assuaged knowledge relating to possible genetic conditions that may be manifested in future offspring.[10] By drawing distinctions between the clinical application of PGS and PGD, and thereafter focusing on PGD as a complementary mechanism (with the contemporary possibilities offered by gene editing technologies such as CRISPR/Cas9,[11]) the exercise seeks to shed light on the entanglement of embryo selection with possible future genetic interventions. In fact, this phenomenon has already been recently identified by scholars, and has been termed "reprogenetics".[12]

The salient reasons for the reprogenetics discourse are motivated by the identification of specific predicaments that may occur, or are already on the cusp of developing or occurring, because of the advancement of these phenomenal technologies. From the legal perspective, the potentiality of these problems is anticipatory in nature. However, in the mission to champion instrumental values such as justice, personal autonomy and equality (amongst others) in reproduction, particularly in many parts of the world where global infertility problems remain on the rise as an on-going concern,[13] I posit that the avoidance about the propensity of humankind's magnitude for scientific greatness, cannot, and should not, be cast aside as mere aspersions.

Therefore, Sect. 3.2 seeks to underscore by way of comparative analysis, the current state of laws, regulation or other forms of governance of embryo selection vis-à-vis PGD in the selected jurisdictions. This analysis will reveal that the relevant concerns in reproduction generally occupy a diverse and varied space in each jurisdiction's national systems. The level of prioritization to reproduction and its instrumental, associated technologies will also corroborate the hypothesis of my work. It must be highlighted however, that the analysis of ARTs laws or regulation generally, and PGD specifically, is valuable to define the intended outcomes of the next Sect. 3.3, focusing on instances of non-therapeutic treatment.[14]

The resulting product of the comparative analysis in Sect. 3.2 advances into Sect. 3.3, identifying the indicators that make up the differentiation and commonalities of the predominant political, socio-legal or legal, religious and cultural, and ethical and philosophical debates relating to embryo selection in PGD. Some of these plaguing concerns relate to the moral status of the embryos and the "playing God" and pro-choice proclivities; other issues involve the practical, legal conception of possible regulatory dimensions, and how to shape a framework of regulation, grounded in fundamental freedom and liberties for a variety of substantive rights in

---

[10] Norwitz and Levy (2013), p. 48.

[11] Hsu et al. (2014), p. 1262.

[12] Knowles and Kaebnick (2007). Reprogenetics refers to the "creation, use, manipulation or storage of gametes and embryos," as defined in a report by the Hastings Center.

[13] Inhorn (2009), p. 172.

[14] As stipulated in Chap. 1, the overarching concern of genetic interventions and PGD referred to in this book is targeted at the non-therapeutic use of medical and scientific technologies.

reproductive choices; and bioethical concerns primarily dominated by the germ-line modification prohibition, and genetic enhancement. This enables a reflected view of how these differences and commonalities have shaped the legal/regulatory framework of PGD in the jurisdictions, and may be used as a comparison tool to later begin the conversation on regulating pre-implantation genetic interventions in Chaps. 4 and 5.

The concluding remarks in Sect. 3.4 finally considers whether PGD has been complicit in equipping, or will be able to equip, the advancement of reprogenetics that impacts embryo selection. Based on the arguments presented in the foregoing, I hypothesize that we may be entering into a new age of reprogenetics, one that necessitates an ambitious formulation of a robust legal framework in this discipline and a radical revision of how we view the importance of ethics[15] in this discourse.

3.1 Infertility and the Journey to Embryo Selection

The global problem of infertility (affecting at least 186 million people worldwide)[16] is often unaddressed, and only given a paltry space on the shelf of on-going social and economic pontificating, particularly in the national agendas of developing and/ or under-developed states.[17] Inhorn and Patrizio[18] partially address the reasons behind the global infertility problem, in a most critical and comprehensive manner, by providing an extensive demographic demonstration of infertility.[19] Their findings indicate that infertility in largely prevalent in low-resource countries. They particularly note that although there is greater awareness about the problems of infertility, a high occurrence of infertility is prevalent in low-resource environments such as South Asia, the Middle East and North Africa, sub-Saharan Africa, and Central and Eastern Europe.[20]

In this section, I expand on three of these reasons, which are key in the determination of how ARTs came to be a viable solution for the problem (within the context of this research). First, one of the reasons for primary infertility could be attribut-

---

[15] Singer (2016).

[16] Inhorn and Patrizio (2015).

[17] World Health Organization, 'WHO | Infertility Is a Global Public Health Issue' (*WHO*) http://www.who.int/reproductivehealth/topics/infertility/perspective/en/.

[18] Inhorn and Patrizio (2015).

[19] These reasons include both primary and secondary infertility. The World Health Organization (WHO) defines infertility as "a disease of the reproductive system defined by the failure to achieve a clinical pregnancy after twelve months or more of regular unprotected sexual intercourse."(WHO-ICMART glossary). Please see also http://www.who.int/reproductivehealth/topics/infertility/definitions/en accessed on 23/3/2018. However, there are often other 'definitions' ascribed to the meaning of infertility. Inhorn and Patrizio quote a study supported by the WHO and the Bill and Melinda Gates Foundation that defines primary infertility as "inability to have any live birth", and secondary infertility as "inability to have an additional live birth".

[20] Inhorn and Patrizio (2015), p. 412.

able to a variety of demographic studies conducted on infertility. These studies measure childlessness, primary and secondary infertility through different units of analysis, which, Inhorn and Patrizio theorize may be used interchangeably and with no detailed precision. The outcomes of these demographic studies contributes to "demographically based definitions of infertility",[21] using differential methods, demography, and time frames in the study of global infertility. There also appears to be a lack of complete information in relation to male infertility, which Inhorn and Patrizio state contributes to at least "half of all cases of childnessness."[22] As such, it is possible that infertility rates on a global level may be subject to some kind of fluctuating consideration.

Secondly, another reason for infertility, mostly secondary infertility, is reproductive tract infections (RTIs).[23] RTIs are caused by "organisms normally present in the reproductive tract, or introduced from the outside during sexual contact or medical procedures."[24] In countries where there is a high rate of unsafe abortions and poor maternity care, as well as sexually transmitted diseases, the prevalence of RTIs can contribute to secondary infertility and poor outcomes in pregnancy if it is not adequately and properly treated.[25] From their study, Inhorn and Patrizio find that countries in sub-Saharan Africa, South Asia, East Asia and the Pacific, Central and Eastern Europe, and Central Asia are "high-prevalence regions."[26] Part of the initiative of the ICPD has also been to identify and recognize how to 'treat' these problems, vis-à-vis "safe abortion and treatment for the complications of unsafe abortions, including post abortion care; prevention and treatment of sexually transmitted infections and HIV and AIDS; and prevention, timely detection and treatment of cancers of the female reproductive system."[27]

Thirdly, Inhorn and Patrizio associate the lack in infertility treatment and prevention (which also contributes to, or exacerbate the infertility problem), to an unjustified form of population control.[28] Particularly in areas such as the sub-Saharan African region, they believe that infertility has been used an excuse or a 'solution' to overpopulation. Citing work by Rutstein and Shah,[29] Mascarenhas,[30] and Allahbadia,[31] on "demographic dividend",[32] they put forward that infertility is a low

---

[21] Ibid 413.

[22] Ibid 414.

[23] Ibid.

[24] Weltgesundheitsorganisation (2005), p. 11.

[25] Ibid 16.

[26] Inhorn and Patrizio (2015), p. 414.

[27] ICPD (n 2) 16.

[28] Inhorn and Patrizio (2015), p. 414.

[29] Ibid 426.

[30] Ibid 425.

[31] Ibid 424.

[32] Gribble and Bremner (26 November 2012). A demographic dividend has been defined as "accelerated economic growth that may result from a decline in a country's mortality and fertility and the subsequent change in the age structure of the population."

priority issue because of the low resource settings of the countries, inadequate medical facilities and infrastructure, and other life-threatening health issues such as HIV and AIDS.[33] With the expectation that the population in sub-Saharan African countries is expected to rise in the forthcoming decade, the prevalence of infertility could be heralded as an appropriate solution to deal with the "demographic dividend."

The demographic realities of infertility indicated in Inhorn's and Patrizio's study[34] are crucial point-markers to lead discussions on managing a global reproductive health plan, and its coupling with initiatives for improvement on a large-scale level. The identified demographic realities[35] contribute to a global phenomenon in which the need, and subsequently, demand, for ARTs become important interventionist tools in family planning.[36]

Prior to the demographic study conducted by Inhorn and Patrizio, the ICPD in Cairo in 1994 was largely instrumental in contributing to the paradigm shift in understanding population enlivenment. In particular, the Cairo conference shifted its focus to developmental ideas in population, with an emphasis on the necessity to elucidate on human rights considerations and reproductive choices in the arena of reproductive health.[37] Part of the goals developed from the conference pivoted on states' obligations to enhance the field of reproductive health by promoting awareness, knowledge and effective policy-making of reproduction. This would be championed through the magnification of reproductive intervention tools or technologies, as opposed to population and fertility control.[38] In conjunction with these developmental goals, especially in the African regions, which has been identified to have an "infertility belt",[39] the dialogue to address global infertility should be centred on methods such as preventative measures to combat infertility.[40] In turn, the economic

---

[33] Inhorn and Patrizio (2015), p. 415.

[34] Ibid 412–415.

[35] Ibid 413. Inhorn and Patrizio listed six particular demographic realities that contribute to the global infertility issues. In a nutshell, they are as follows (quoted verbatim from Inhorn and Patrizio's study):-

a) Millions of people around the globe suffer from infertility
b) Women in low-resource settings continue to suffer from high rates of secondary infertility
c) Africa continues to suffer from inordinately high rates of infertility
d) High rates of infertility co-exist with high rates of fertility in Africa- a demographic paradox known as "barrenness and plenty"
e) Lack of infertility prevention and treatment services is often justified as a form of population control, particularly in high-fertility settings such as sub-Saharan Africa
f) Those parts of the world with the highest rates of infertility are least likely to offer reliable diagnosis and treatment, including IVF services

[36] Ibid 418.

[37] Etuk (2009), p. 85.

[38] Ibid.

[39] Ibid 86.

[40] Ibid 88. Etuk, a gynaecologist, is of the opinion, as is the opinion of numerous medical professionals, that a large number of problems relating to infertility can be preventable by diagnosis and

and social circumstances often associated with infertility, especially in the case of women, who often bear the stigmatization of infertility, are vested with the capacity to transform disenchantment into positions or a state of affairs deserving of warranted empowerment, understanding and conducive, pragmatic approaches to deal with the infertility problems.

In fabricating the question of accessibility to ARTs *vis-à-vis* the goals and agenda of the ICPD, two important factors must be reflected upon to enable the beginning of a meaningful dialogue in how to achieve access to ARTs. These factors are the crucible in beginning a shift in how we begin to view non-coital reproduction,[41] and challenging reproduction as a commodification exercise of women's bodies and the biological endowment of the body in Foucault's theory of "bio-power".[42]

The first factor relates to the gendered dimension of women's role in reproduction. The stigmatization of barrenness and childlessness inevitably, and unfortunately, falls upon the woman. Gillian Goslinga-Roy presents a beautiful envisioning of the female body and its territorialities,[43] which lends supports to the notion of the burden placed on the woman in reproduction. Traversing the discourse of the female body in biomedicine and feminist theories, Goslinga-Roy finds, quoting Rosalind Petchesky, that it is not uncommon for the female body to be viewed as "imprisoned in the conventional (bourgeois and Lockean) notion of property that involves exclusivity, isolation, objectification, and self interest."[44] On the basis of this reasoning, she argues that some feminist movements have seen fit to view the owner of the body as a "privatized and individualized female self", having a "private capacity to reproduce."[45] Assumptions of this nature, states Goslinga-Roy, have the danger of narrowing the feminist discussions in reproduction to viewing women either as victims, or the saviour, of their own freedom and control over their bodies.[46]

I theorize, like Goslinga-Roy, however, that reproduction is simply not an issue of gender alone. Although it is undeniable that the gender element factors strongly in issues of reproduction, the premise of my movement beyond the gender dimensions also returns to my earlier argument on Foucault's "bio-power".[47] My claim in this section is that if we truly intend to move towards an impactful manner of achieving access to ARTs as a solution to infertility, we must first come to appreciate that human bodies are "maps of power and identity".[48] Foucault's account of the politics of the human body, I claim, provides the manner in which we choose to perceive the imaginings of our place in our interactive personal and social community of beings.

---

early treatment of infections. He propagates an approach on a tiered level, divided into primary, secondary and tertiary prevention mechanisms that may be undertaken by states.

[41] ICPD (n 2).

[42] Foucault (1976).

[43] Goslinga-Roy (2000).

[44] Ibid 121.

[45] Ibid 421.

[46] Ibid 122.

[47] Foucault (1963).

[48] Haraway (2000).

The fact that reproduction, sex and sexuality have been traditionally engendered, as 'female' in nature is the by-product of the social community's underpinnings hastened along by patriarchal state hegemony. In reality, reproduction, or procreation, goes beyond the scope of the sexual and coital. The non-coital methods of reproduction, where life can now begin in a petri dish, and challenges to the traditional feature of family units, serve to illustrate this shift. If we accept that family units are evolving, then it must also be logical to accept that reproduction or procreation is also subject to a similar dynamic shift.

In this regard, the body of woman should no longer be synonymous with the "biologized notion of motherhood."[49] The failure to conceive a child, or experience a live birth, must now be reconceived beyond the biological. Instead, one must be able to expand on the notions of boundaries beyond the biological body of woman through discourse of the "political" body; looking to palpable structures of diversified and dynamism of changing societal networks that can support reproduction. The practice of learning to eliminate blame and fault through revising our understanding of the complexities of the reproductive process would be the first step in throwing off the shackles of gender dimensions in reproduction.

The second factor relating to the ICPD agenda and goals of access to ARTs (and which would also support the first factor mentioned above), would be a more concerted effort to include men in the critical facets of reproduction. By also recognizing that male infertility is a contributory aspect to global infertility rates, and revisiting the man's place in reproduction by expanding their roles and responsibilities, this would enable "the reformulation of expectations, culture, work and reproductive for the large scientific and technical workforce."[50] The main challenge to this is the imbuement of a high degree of patriarchy in the realm of reproduction. The conduct of various medical anthropological researches proffers fascinating accounts of the relevance of men's roles in reproduction.[51] Dudgeon and Inhorn[52] recognize a crucial paradigm shift that made a more connective effort for the inclusion of men in this area, particularly post-ICPD in Cairo.[53] A trenchant analysis, both on a macro and micro level about the way men affect women's reproductive health,[54] yields a more interminable understanding that reproduction, including infertility, should no longer be regarded from the patriarchate viewpoint that it is primarily, or even, solely, the woman's 'problem' or 'concern'. Some fundamental contributing factors, according to Dudgeon and Inhorn, include the growing feminist movement in developing and developed countries that have begun to emphasize and advocate for sexual and reproductive rights,[55] the "denunciation of population

---

[49] Goslinga-Roy (2000), p. 136.

[50] Haraway (2000), p. 44.

[51] Dudgeon and Inhorn (2004).

[52] Ibid.

[53] Ibid 1379.

[54] Ibid 1380.

[55] Ibid 1379., citing Correa and Reichmann (1994) and Petchesky (2000).

control as a motivation for contraceptive research and distribution",[56] the need to reflect on the rise of HIV/AIDS transmission through heterosexuality,[57] and "the failure of family planning and maternal-child health programs to address complex reproductive health issues such as sexuality."[58] These aspects culminate in an important repositioning in social, economic and even political circumstances in various countries. In the viewpoint of Dudgeon and Inhorn, this could be a significant genesis for "a new reproductive health paradigm."[59]

The inclusion of the role of males in reproduction poses to be consequential equipment in the advent of ARTs, especially with the recognition that many problems of infertility are also attributable to male infertility, ranging in around 20% of recorded infertility cases.[60] The increasing emphasis on male infertility also gently propels the 'problem' in reproduction, allowing it to evolve, in some countries, to an acknowledgement of male "reproductive impairments."[61] Inhorn, in her study on male infertility in the Middle East, has provided an account of varicocelectomy[62] surgery undertaken by men who suffer from infertility. Interestingly, the study reveals that one of the reasons, which propel them to undergo such a risky surgery, is the "desire[s] to share the burden of reproductive suffering with beloved wives."[63] In Lebanon, where the study was carried, Inhorn demonstrated, amongst other reasons, men's strong commitment to their marriages, "by sharing somatically in their wives' treatment quests."[64] This significantly turns on the understanding of masculinity within the Middle Eastern dimension; where the masculinity of men are traditionally tied to them being the 'patriarchs' of family units *vis-à-vis* their virility and fertility. A 'failure' in this regard results in them being viewed as weak; Inhorn posits that varicocelectomy as fertility-enhancing surgery became widespread among Middle Eastern men as a way to live up to their ideals of masculinity and "public images as powerful, virile patriarchs."[65] As such, the view of reproduction in these societies, despite being patriarchal in nature, is in unity with concepts of shared responsibilities in all aspects of married and family life. More interestingly, (and I return once more to the Foucauldian idea of "bio-power"), Inhorn also sum-

---

[56] Dudgeon and Inhorn (2004), citing Bandarage (1997) and Dixon-Mueller (1993a).

[57] Ibid 1379., citing Cates and Stone (1992), Dixon-Mueller (1993b), Mbizvo (1996) and Parker et al. (2000).

[58] Ibid., citing Cliquet and Thienpont (1995).

[59] Ibid 1380.

[60] Inhorn (2007), p. 2.

[61] Inhorn (2013).

[62] Binsaleh and Lo (2007), p. 277. Varicocelectomy surgery is a male genital surgery that is the most commonly performed surgical procedure to treat male infertility. One of the causes of male infertility is thought to be abnormalities of the plexus of veins, draining blood from the testicles and causing an abnormal enlargement in the scrotum. Varicocelectomy surgery corrects this occurrence.

[63] Inhorn (2007), p. 1.

[64] Ibid 3.

[65] Ibid 4.

marizes that varicocelectomy surgery is evidence of the male reproductive body being controlled, politicized and disciplined to meet "Middle Eastern societal demands of virility, fertility, and patriarchal continuity."[66]

However, the full given of 'responsibility' in reproduction, whether attributable to both male and female infertility, still remains a point of discussion in many countries; and the reach of the ICPD in Cairo lingers on the edict of "population reduction through family planning."[67] Inhorn theorizes that despite this recognition, the deep-rooted sense of patriarchy in reproduction protracts sustenance because of the interpretive framework that focuses on men's "responsibility" in family planning, healthcare for themselves and their partners/wives, and their role to "protect and ensure the reproductive rights and well-being of others."[68] It is obvious, therefore, that an unabridged overthrow of patriarchal interpretations in reproduction, will be an insurmountable task; but the tenets of the ICPD, at least, can be regarded as a driving force forward in this assimilative endeavour.

Nevertheless, the recognition of these problems has begun to pave the way forward with an identifiable measure of dealing with these concerns: that is, the establishment and creation of available opportunities for infertile couples or individuals to partake in reproduction intervention tools or technologies. The shifting conditions on a global level have therefore armed ARTs, such as IVF to make its impact successfully felt, especially in regions such as the Middle East, and parts of Asia such as Malaysia, Thailand, Singapore, and India. The degree of prioritization of infertility problems in these countries are necessarily varying in nature; pivoting upon factors such as social openness, religion, family structures, and political and policy agendas. These attitudes, however, are also instrumental in determining the general scope of receptiveness to other forms of medical and scientific technologies.

From the birth of the first test-tube baby, Louise Brown, in 1978 via IVF,[69] the development of ARTs has enjoyed monumental growth in many countries around the world. One of the more common and widely used methods of ARTs is IVF,[70] but the use of third-party assisted reproduction such as donor eggs, sperm or embryos,[71] and surrogates or gestational carriers[72] are also gaining increased momentum.[73] The

---

[66] Ibid 5.

[67] Inhorn (2013).

[68] Ibid 50.

[69] Kamel (2013), p. 156.

[70] American Society for Reproductive Medicine (2015), p. 4.

[71] Ibid 14.

[72] Ibid 15.

[73] For example, the practice of "reproductive tourism" *vis-à-vis* commercial surrogacy in Thailand has evolved from the more traditionally 'altruistic' form of surrogacy in the past, to a commercialized 'business' venture targeted at foreign visitors to Thailand. Commercial surrogacy for foreign nationals has since 2016 been banned in Thailand. India, like Thailand, has also been a 'reproductive tourist's' destination for surrogacy, and in a landmark Supreme Court decision in 2016, has now put into place a more formalized and legal framework for the regulation of surrogacy in the country. These are but two instances in Asia that are a reinforcement that the fertility marketplace continues to experience a boom on a global level.

rising importance of ARTs that has enabled the expansion of family units where previously unavailable, however, simultaneously comes with an incontrovertible rhetoric relating to women's bodies in reproduction. In a similar manner as pregnancy and family planning have become matters for state regulation, so, too has the access to and use of various methods of ARTs. In these incumbent decisions that accompany matters of reproduction, especially where the use of ARTs are concerned, it is not surprising that most states see fit that these technologies be subject to some form of regulation. The strength of appropriate ART legislation initially began to focus on matters relating to assessments of safety, standards of care by medical practitioners and caregivers, and ethical considerations, and in some instances, religious concerns.[74] As ARTs continue to improve, and create a dramatic perspective move in how communities approach non-coital reproduction, the discourse shifts into how ARTs may also potentially alter how we view reproduction and the desire to birth healthy children.

From the most basic considerations relating to access in low resource countries,[75] to radical shifts into the potentiality of non-sex, or non-genders,[76] the reproductive discourse of ARTs has encompassed the socio-philosophical-ethical parley as well as the more practical but problematic question of appropriate regulatory intervention.[77] Regulatory measures are crucial as the key to a portal of well-managed resources and facilities of ARTs within a country, but it is only one aspect of the advancement of ARTs that has impacted contemporary societies on a bigger level. In their anthropological work examining the effects of ARTs on the domains relating to "kinship, marriage, and the family, gender, religion and biomedicine",[78] Inhorn and Carmeli surmise that ARTs actually "contribute to the construction of global power relations and new notions of local modernity."[79] In Chap. 1 of this book, I had also posed that the "politicization" of the body[80] enabled regulatory measures and control to be effected over citizens of any given nation through the positivistic aspects of the law. This additional importation of governing the "body" into discourse that relates to forms of scientific or medical technologies is one of the ways in which governmental control may be exerted. It is irrefutable that ARTs in

---

[74] These religious concerns are focused on the Islamic view of assisted reproduction in cases of fertility. As will be demonstrated later, the Islamic view of assisted reproduction is firmly entrenched in how the family unit is created, with a firm grounding on the sanctity of the relationship between husband and wife. In this sense, third party assisted reproduction never has a place in the Islamic way of family life.

[75] Inhorn (2009).

[76] Hughes (2010), p. 15.

[77] Riggan (31 December 2009). This study, although covering "industrially advanced countries", benchmarked against their respective wealth accumulation in the International Monetary Fund, nevertheless provides a semi-comprehensive and informational list of countries that have enacted laws relating to the ARTs industry.

[78] Inhorn and Birenbaum-Carmeli (2008), p. 177.

[79] Ibid 180. The authors quote the work of Bharadwaj (2006), Kanaaneh (2002), Paxson (2006), Raspberry (2009), Roberts (2008) and Tremayne (2006) in making this statement.

[80] Foucault (1963).

themselves are "gendered technologies"[81] (female) because the affectation of infertility imposes its mandates on women more acutely.

With the manifestation of ARTs and infertility issues being targeted for 'correction', the enlargement of further questions also come forward: especially where the 'newer' reproductive technologies such as PGD[82] and PGS[83] and their potentiality to be employed in convergence with other highly developed scientific technologies (such as CRISPR/Cas-9[84]) are concerned.[85] This is particularly where the issues relating to embryo selection become acute. Sidney Callahan also points to loopholes that exist in some jurisdictions, for example, in the United States, which contribute to conflicts about the morality of sex and reproduction.[86] In furtherance of these claims, I therefore argue that a deeper understanding about the complexities of the predominant legal and ethical debates surrounding issues of embryo selection in the "newer" reproductive technologies, allow us a glimpse into how the eventual framework of legislation, or other regulatory measures, have been shaped in the different jurisdictions undertaken in this book's study.

3.2   Genetic Screening Technologies: Pre-implantation Genetic Screening (PGS) and Pre-implantation Genetic Diagnosis (PGD)

Bearing crucial 'power' to intervene in the procreation and reproduction of human beings, ARTs have now permeated levels of sophistication and complexities of a nature previously not contemplated. When it became apparent that genetic screening possibilities could dramatically affect how we choose to view the finiteness of pre-conceptual ideas, the landscape of debates pertaining to ARTs began to shift once again. More commonly known simply as "genetic screening" or "genetic testing", genetic screening or testing began as part of a health assessment to complement the process of ARTs (specifically, IVF). It became clear that the successes or failures relating to IVF could also be attributable to the biological and genetic make-up of the individual parents.[87] In cases of frequent or recurrent loss of pregnancy, or where specific hereditary diseases present in one or both of the potential

---

[81] Inhorn and Birenbaum-Carmeli (2008), p. 180.

[82] Botkin (1998), p. 17.

[83] Brezina and Zhao (2012), p. 3.

[84] Hsu et al. (2014).

[85] Turriziani (2014), p. 595.

[86] Callahan (2009), p. 79.

[87] There are indications that the quality of a woman's eggs are dependent on genetics, but the possible responses in IVF allow for an optimum environment to harvest the eggs; these include external factors such as diet, lifestyle, and the like. The same may be true of the quality of a man's sperm. In any event, the eventual quality and viability of the resulting embryo is a combination of these factors.

parents may be known, genetic screening with IVF is very helpful in ensuring the future implantation of a healthy embryo, free from genetic or other hereditary disorders that may hamper its quality of life in the future.

This section of this chapter focuses on specific legislation or regulation that relates to genetic screening technologies in the respective jurisdictions examined. I believe that the predominant legal and ethical basis for debates that suffuse the contextual space in each of the jurisdictions leads us to a deeper understanding on how the laws and ethics interplay with one another to form a distinct purpose in the development of a regulatory framework. In the same way that ARTs and IVF have changed the landscape of reproduction and parenthood, expanding the notion of non-coital reproduction and the changing dynamics in risk and health assessment, PGS and more so, PGD, now travel a similar path that infiltrates new dimensions of legal and ethical concerns because of its embryo selection effects.[88] Indeed, the range of these concerns appear to impact greatly on the existence and operation of human rights generally.

3.2.1 Distinctions Between PGS and PGD

Within the sphere of ARTs, PGS and PGD have both made an entry as formidable and instrumental complementary technologies in IVF treatment. Both these technologies involve the screening of embryos that have been produced from IVF treatments and are carried out prior to implantation of the embryo into the woman's womb; but the outcomes of the screening tests can be markedly different. Although both technologies target the screening for chromosomal abnormalities in pre-implantation embryos, the main difference lies in the type of abnormalities that can be screened for.

In PGS (also known as Comprehensive Chromosome Screening), the screening is carried out to identify chromosomal aneuploidy,[89] an anomaly that occurs in the numerical structure of chromosomes.[90] Chromosomal aneuploidy results in abnormalities such as Down's Syndrome (Trisomy 21), or Turner's Syndrome.[91] Indications that should be demonstrated before considering PGS include recurrent pregnancy loss, advanced maternal age, and repeated implantation failure[92] but the presence of these factors alone does not necessarily mean that PGS is a suitable

---

[88] Winslow and Kodner (2004), p. 186.

[89] Ly et al. (2011), p. 834.

[90] A cell in a typical human person contains pairs of chromosomes, usually 23 pairs (46 chromosomes in total). There are typically two types of numerical chromosomal abnormalities that may occur in a human cell: first, monosomy, where a chromosome is missing from a pair; and trisomy, where an individual has more than a pair of chromosomes. The phenomenon of either missing or having an extra chromosome is called aneuploidy.

[91] Brezina et al. (2013), p. 37.

[92] Ly et al. (2011), p. 835.

option. Initially developed as a method to overcome the limitations of FISH (Fluorescent *in situ* hybridization),[93] another scientific procedure using fluorescent probes to determine the position of genes in the chromosomes, PGS was seen as a promising technology that could remedy the limitations of FISH.

Recent scholarly studies and experimentations posit that there are also limitations to what PGS can achieve.[94] For example, scientific studies have also shown that the accuracy of PGS may be compromised by factors such as mosaicism[95] in human embryos.[96] Other studies identify the "futility" of PGS in increasing successes in IVF,[97] and question the statistical accuracy of data relating to PGS success.[98] A recent study by the European Society of Human Reproduction also illustrate that the parameters for indications of PGS are expanding, and in this regard, continue to be controversial in nature.[99] These reasons can affect the latitude of PGS' reach and what it is able to achieve in a clinical setting.

By contrast, in PGD, the target of the screening process for chromosomal abnormality is more clearly identified. This would, however, be dependent on specific genetic abnormalities that have been documented in either one or both of the potential parents, and where these couples do not wish to transmit hereditary genetic conditions to their future offspring. This reason therefore has given rise to why PGD has been widely used in the clinical setting as a means to procuring healthy preimplantation embryos. These specific genetic anomalies commonly include single-gene disorders such as sickle-cell anaemia, cystic fibrosis and Huntington's disease, amongst others.[100] The benefit of PGD in these circumstances therefore allows examination of pre-implantation embryos for the specific composition of the cells that relate to these identifiable genetic conditions.[101]

In both these technologies, however, and despite their increased use in fertility treatments, an assessment of accompanying risks does not often commensurate with the reality and emotional turmoil of knowing that an embryo is affected by a genetic mutation. In PGS and PGD, the commonly shared ethical issue relates to the destruction or discarding of such affected embryos, amongst others.[102] But the key difference in the process aspects of these two technologies is the target and specificity of

---

[93] Gozzetti and Le Beau (2000), p. 320.

[94] Gleicher and Orvieto (2017), p. 21.

[95] Munné et al. (1994), p. 373.

[96] Gleicher et al. (2016), p. 54.

[97] Gleicher et al. (2014), p. 22.

[98] Gleicher and Orvieto (2017), p. 3.

[99] Brezina et al. (2013), p. 38.

[100] Chial (2008), p. 192.

[101] This is contrasted to PGS, which identifies chromosomal aneuploidy, and does not target the identification of specific genetic mutations or disorders.

[102] This is an important component in the later examination of PGD laws in the various jurisdictions, as the legal definition of what constitutes an embryo is often at the heart of whether PGD is legally permitted.

screening, and this is what many scholars caution leads to the heart of reprogenetics.[103]

The discourse relating to PGD is also inundated with its own dominions of clinical, legal and ethical complexities. But these dominions are likely to expand if we think about the applicability of PGD with CRISPR/Cas9, dishing up very important considerations for the future that should be taken into account, simply because the globalization phenomenon will continue to affect even developing, and more crucially, under-developed countries. As such, the *leitmotif* of the PGD use discourse is incalculably the ethical challenges that are being faced. Some of these challenges have been highlighted by Jeffrey Botkin[104] and remain lasting topics of discussion even today. Two other issues that are also likely to arise with PGD: germ line therapy and genetic enhancement, offer possibilities for the future. These particular possibilities offered through the use of PGD will also be explored more substantially in Chap. 5 of this book.

John Robertson, several years later, wrote on the extended uses of PGD for HLA-matching purposes,[105] for the screening of embryos for susceptibility to cancer and late on-set diseases,[106] and for gender selection.[107] Like Botkin, Robertson raised extremely relevant issues, which continue to warrant the importance attributed to them in our present day. For example, the nonpareil of gender selection vis-à-vis PGD, has become the open secret of the "reproductive tourism" agenda in South-East Asian countries such as Thailand,[108] and with countries like Malaysia and India, where the technology operates under the radar of inadequate legal and/or regulatory frameworks, the exposure to a new set of pressing issues, such as the perpetuated victimization of marginalized members of society, or the creation of an 'undignified' demand or a free market of sorts for the outputs of reproduction, impact in ways that are more serious than had been initially envisaged at the beginning of the 1990s era.

These warranted concerns in the uses and extended uses of PGD are merely the tip of the iceberg in bioethical circles. At this present juncture, PGD has become an established technology that has proven its usefulness in an identifiable set of circumstances, although the ever-changing landscape that is technological

---

[103] Reprogenetics refers to the "creation, use, manipulation or storage of gametes or embryos," as defined in a report by the Hastings Center. But the potentiality of reprogenetics is drawn specifically to potential human enhancement at the pre-implantation embryo stage using germ-line choice technology such as CRISPR.

[104] Botkin (1998).

[105] Robertson (2003), p. 465. HLA is a protein or a genetic marker that is present in most cells of the human body. In HLA type-matching procedures, a potential donor's HLA is examined to determine if it could be matched to the HLA of a patient (suffering from a HLA-associated disease) for the purpose of a bone marrow or code blood transplant. Some known examples of HLA-associated diseases are often linked to autoimmune conditions such as lupus, multiple sclerosis and Grave's Disease, amongst others.

[106] Ibid 467.

[107] Ibid 468–469.

[108] Natipodhi (2014).

advancements in the age of the twenty-first century will continue to dominate the treatise in bioethics. However, these fledgling worries have now escalated, with the realization that unparalleled advances in genomic understanding and technologies, centered on the perils of possibilities of genetic enhancements, which may be carried through germ-line gene therapy; unquestionably altering completely the multifaceted ideas of a human being. PGD, as a technology that is able to identify specific genetic mutations in the human genome, makes it a powerful mechanism in the sphere of future medical treatment if combined with revolutionary technologies like CRISPR.

3.2.2   Current State of the Laws/Regulation for PGD in the Selected Jurisdictions

3.2.2.1   The United States

From the selection of the jurisdictions in this study, the regulatory landscape in the United States (US) is likely to reveal the most tumultuous considerations in terms of PGD. As will also be demonstrated in Sect. 3.3 of this chapter, one of the biggest reasons for this haphazard approach to regulation of PGD in the US could be attributed to the politicization of issues in reproduction.[109] In particular, the divisive ethical discussions that have inflamed the abortion debates since *Roe v Wade*[110] has inevitably spilled over to the considerations of PGD services. The negative effects of congregating PGD in the same vein as the abortion debates is a flawed misconception, because it seeks to point towards the continuously divergent and indeterminable status of an embryo, as opposed to the exercise of personal liberties and fundamental freedoms in accordance with the US Constitution.

As such, it may be surprising, considering the status of the US on the global stage that the sphere of PGD remains unregulated at this point of time, despite the conversation having been raised from as early as the late 1990s period.[111] The feature of the US constitutional model is such that the purview of regulating matters such as PGD would fall under the scope of state jurisdiction.[112] This is similar to the position of Australia; but the political and technological environment in the US is much more largely saturated across the entire country. The presumed 'supremacy' of the legislative process in the US and the reluctance of the Congress to either enlarge or impose restrictions hinges on the constitutional interpretation of specific fundamental

---

[109] Hudson (2006), p. 1638.

[110] 'Roe v. Wade, 410 U.S. 113 (1973)' (*Justia Law*) https://supreme.justia.com/cases/federal/us/410/113/. Accessed 2 May 2018.

[111] Botkin (1998).

[112] Sholley (1951). Article I Section 8 of the US Constitution lays out the powers of Congress in terms of the legislative overview is concerned. However, the jurisdiction of the individual states are more specifically recognised and laid out in Article IV.

rights and liberties contained in the US Constitution. On this basis, the US is an unusual loner in the Western sphere that has yet to regulate the use of PGD, whether at the federal or the state level.[113]

In terms of viewing the regulatory front in the US, ARTs in themselves are very poorly regulated, and if this is the case, it is not surprising therefore that PGD is also not adequately addressed at the federal level. As highlighted by some scholars, the non-binding guidelines or ethical considerations that may be issued by professional associations such as the American Medical Association, the American Society for Reproductive Medicine, the American College of Surgeons, amongst others, are not treated with the deference that may imbue a legislative act or pronouncement made by Congress or the Supreme Court. Bioethicist Arthur Caplan comically refers to this open-ended landscape as the "Wild West" of reproductive medicine.[114] At this present time, PGD in the US is therefore used for therapeutic or medical, as well as non-therapeutic or non-medical purposes because of the legislative lacuna that exists within the national or state regulatory framework.[115]

From the legal perspective, anything in reproductive medicine in the US may therefore be possible so long as it is not expressly outlawed. The same cannot be said of the national moral and ethical debates on PGD, which reveals divergent and conflicting moral judgments on the acceptability of embryo selection and manipulation. Often times, the follow up debates in PGD inevitably turn to issues relating to genetic interventions (such as CRISPR) and sex selection for non-medical reasons; and the magnitude and breadth of scope a legal formulation should account for. Some American scholars, such as Michelle Bayefsky, have also identified a lack of clarity *vis-à-vis* guidelines issued by professional associations.

For example, the Ethics Committee (Ethics Committee) of the American Society for Reproductive Medicine issued its guidelines that did not favour non-medical sex selection in 1999.[116] In 2015, the ethical guidelines were revised regarding sex selection, particularly that the Ethics Committee did not have a consensus on the permissibility of non-medical sex selection, and therefore, "clinics are encouraged to develop and make available their policies on the provision of non-medical sex selection."[117] To lend further argument to the effect of these guidelines, in 2013, the Ethics Committee addressed the issue of using PGD for serious, adult onset conditions as ethically justifiable if "the conditions are serious and when there are no known interventions for the conditions, or the available interventions are either inadequately effective or significantly burdensome."[118] The cumulative effect of these guidelines, although addressing PGD use in some circumstances, however, do not provide a clear answer on how the permissibility of PGD may be determined in

---

[113] Bayefsky and Jennings (2015), p. 11.

[114] Ibid 12.

[115] Bayefsky (2015), p. 7.

[116] Bayefsky (2016), p. 43.

[117] Ethics Committee, American Society for Reproductive Medicine (2015), p. 1418.

[118] Ethics Committee, American Society for Reproductive Medicine (2013), p. 54.

other situations.[119] In the meantime, guidelines issued by other professional bodies, such as the American Congress of Obstetricians and Gynaecologists (specifically the ACOG Committee Opinion No. 360)[120] indicated support for sex selection "for the purpose of preventing serious sex-linked genetic diseases", but not for other personal or family reasons. The American College of Medical Genetics does not specifically address PGD, but has issued a policy statement dealing with non-invasive prenatal screening for fetal aneuploidy.[121] As Bayefsky notes, the indication is that there must be "validated associations between a mutation and the severity of a disorder."[122] While there is certainly no lack of guiding instruments that address PGD, the lack of clarity in each creates an uncertain terrain in terms of determining a rational and concerted consensus in the uses of PGD in the US. Bayefsky further points out reasons why this may be the case in the country, citing reasons such as "embryo politics",[123] and the lack of a government sponsored healthcare system that covers IVF and by this extension, PGD as well.[124] In light of the US having been positioned as a forefront in proffering IVF treatments for a long time, and the wealth of both brilliant and intellectual talent in the field of biomedical research and reproductive technologies,[125] the lack of regulatory oversight for a technology such as PGD does raise some cause for concern.

In this respect, and on an ironic level, somewhat similar to the actuality of PGD use in Malaysia and Thailand, this landscape of uncertainty has manifested itself *vis-à-vis* a spate of court cases in the US, involving legal theories of wrongful birth,[126] wrongful life[127] and even wrongful conception[128] cases. All these cases essentially fall under the umbrella of negligence or medical malpractice, but the connective element between these cases is prenatal testing technologies whether by screening or diagnosis, that was not detected by medical professionals that provided the service.[129] Certainly, there can be no reasonable acceptability to professional or medical negligence; but if the legal dominion regarding prenatal testing technologies

---

[119] Bayefsky (2016), p. 43.

[120] American Congress of Obstetricians and Gynecologists, Committee on Ethics (2007), pp. 475–478.

[121] Gregg (2016), p. 10.

[122] Bayefsky (2016), p. 43.

[123] Ibid 44. The issue relating to "embryo politics", I contend, is one of the elements that relates to the politicization of reproduction within the US constitutional environment and how it may fit within the framework of interpreting fundamental rights and liberties. In Sect. 2.3 of this chapter, I delve into this issue more extensively.

[124] Ibid. In addition to this, the recent contestation to the Affordable Care Act in the United States has further indicated further erosion in insurance coverage in the country. On the basis that a basic healthcare system appears to be declining in the country, the rung of priority that puts reproductive health and access to fertility treatments on the national level is very low.

[125] Thompson (2016), p. 128.

[126] Crockin (2005), p. 693.

[127] Ibid 694.

[128] Ibid.

[129] Ibid 693.

such as PGD is not well developed and therefore, would be subject to market forces, affordability levels of potential parents, the access to and standards of the provision of such technologies, and the like, then these legal cases are likely to further exacerbate the persistence of problems within the country and the individual states.

3.2.2.2   The United Kingdom

Unlike the US, the state of governance for all manner of fertility and ARTs in the UK is premised under the umbrella of the Human Fertilisation and Embryology Authority (HFEA), an independent governmental body established pursuant to the Human Fertilisation and Embryology Act of 1990 (HFEA Act 1990).[130] The HFEA oversees all regulatory aspects of fertility, ARTs, licences and such in the UK, and from the period of its inception, has appeared to manage the upkeep of regulatory approaches, consistent with the development of fertility technologies over the past two decades. The HFEA Act 1990 was more comprehensively revised in 2008, resulting in the new Human Fertilisation and Embryology Act of 2008 (HFEA Act 2008),[131] which also encompasses a variety of supplementary regulations that has demonstrated the UK's more progressive approach to regulation.[132] Some of the more recently significant supplementary legislation to the HFEA Act 2008 include the Human Fertilisation and Embryology (Mitochondrial Donation) Regulations 2015,[133] and the Human Fertilisation and Embryology (Parental Order) Regulations 2018, for which there was a consultation in March 2018.[134] In view of these revised regulations that have reflected the progress of medical and scientific technologies in fertility treatments and its instrumentalities, the embryo selection phenomenon *vis-à-vis* PGD may be considered the 'older' but no less important aspects of the HFEA's provisions by law.

   Although PGD was allowed in the UK pursuant to the powers, functions and issue of licences and approvals by the HFEA following the HFEA Act 1990, the revised HFEA Act 2008 now restricts the circumstances under which PGD may be obtained by potential parents. There has been some debate regarding the role of the HFEA in the UK; granted, the lack of clarity in some aspects of legislation and the discretionary approaches employed by the HFEA has raised questions about

---

[130] Department of Health, 'Human Fertilisation and Embryology Act 1990 - an Illustrative Text' (9 November 2007) http://webarchive.nationalarchives.gov.uk/content/20130107105354/http://www.dh.gov.uk/en/Publicationsandstatistics/Publications/PublicationsLegislation/DH_080205.

[131] Human Fertilisation and Embryology Act 2008 (Chapter 22) 120.

[132] Strategy and Information Directorate Human Fertilisation and Embryology Authority, 'The HFE Act (and Other Legislation) - HFEA' http://hfeaarchive.uksouth.cloudapp.azure.com/www.hfea.gov.uk/134.html.

[133] 'The Human Fertilisation and Embryology (Mitochondrial Donation) Regulations 2015' https://www.legislation.gov.uk/ukdsi/2015/9780111125816/contents.

[134] The consultation for the revised regulations to the HFEA Act 2008 was to reflect changes that may enable a single person to apply for a parental order.

certainty and clarity; resulting also in several high profile court cases in the UK.[135] However, it cannot be denied that the centralization of functions that relate to fertility treatments has allowed for a more streamlined approach.

The current PGD legislation, as it stands in the UK, appears to be much more compendious. Like its predecessor, the HFEA Act 1990,[136] the revised HFEA Act 2008 recounts circumstances under which licences or approvals may be granted by the HFEA to carry out fertility services. Section 11[137] of the HFEA Act 2008 governs the scope of these licences. In particular Section 11(1)(a)[138] relating to these relevant licences for treatment, storage and research refers specifically to authorized activities outlined in paragraph 1 of Schedule 2.[139] These activities that may be authorized by the HFEA (for the provision of fertility services) include treatment services for "procuring, keeping, testing, processing, or distributing embryos" (Para. 1 Section 1(a))[140] and "other practices designed to secure that embryos are in a suitable condition to be placed in a woman." (Para 1 Section 1(d)).[141]

Paragraph 1ZA of the same Schedule 2[142] in the meantime, refers expressly to embryo testing. The paragraph begins in exclusionary language that states embryo testing is not permitted, but there are exceptions to this general rule (Paragraph 1, sub-paragraph 1). These exceptions for the prohibition against embryo testing include the following purposes:

(a) Establish whether the embryo has a gene, chromosome or mitochondrion abnormality that may affect its capacity to result in a live birth,
(b) In a case where there is a particular risk that the embryo may have any gene, chromosome or mitochondrion abnormality, establishing whether it has that abnormality or any other gene, chromosome or mitochondrion abnormality,
(c) In a case where there is a particular risk that any resulting child will have or develop-

    (i)   a gender-related serious physical or mental disability,
    (ii) a gender-related serious illness, or
    (iii) any other gender-related serious medical condition,

establishing the sex of the embryo
(d) In a case where a person ("the sibling") who is the child of the persons whose gametes are used to bring about the creation of the embryo (or of either of those persons) suffers from a serious medical condition which could be treated by umbilical cord blood stem cells, bone marrow or other tissue of any resulting child, establishing whether the tissue of any resulting child would be compatible with that of the sibling, and

---

[135] 'House of Lords - Quintavalle v Human Fertilisation and Embryology Authority' https://publications.parliament.uk/pa/ld200405/ldjudgmt/jd050428/quint-1.htm.

[136] Human Fertilisation and Embryology Act 1990 (Chapter 37) 48.

[137] Human Fertilisation and Embryology Act. p. 8.

[138] 'Illustrative Text: Human Embryology and Fertilisation Authority Act 1990. As Amended: An Illustrative Text' 18.

[139] Ibid 75.

[140] Ibid.

[141] Ibid.

[142] Ibid 76–77.

(e) In a case where uncertainty has arisen to whether the embryo is one of those whose creation was brought about by using the gametes of particular persons, establishing whether it is.[143]

Although the express words "pre-implantation genetic diagnosis" is not used, the purposes for embryo testing under Paragraph 1ZA essentially covers the most common purposes for pre-implantation testing. The ultimate outcome of such testing would obviously be to ensure that only the healthiest embryos that are free from any genetic disorders described in the sub-paragraphs are implanted into the woman. However, these exceptions to the general rule in Paragraph 1ZA are also conditional upon additional circumstances mentioned in Paragraph 1ZA(2).[144] In the latter, the exceptions in sub-paragraph (1)(b),[145] 1(c)[146] and 1(d)[147] are further subject to the following consideration:-

(2) A licence under paragraph 1 cannot authorize the testing of embryos for the purpose mentioned in sub-paragraph (1)(b) unless the Authority is satisfied-

(a) in relation to the abnormality of which there is a particular risk, and
(b) in relation to any other abnormality for which testing is to be authorized under sub-paragraph (1)(b), that there is significant risk that a person with the abnormality will have or develop a serious physical or mental disability, a serious or any other serious medical condition.

(3) For the purposes of sub-paragraph (1)(c), a physical or mental disability, illness or other medical condition is gender-related if the Authority is satisfied that-

(a) it affects only one sex, or
(b) it affects one sex significantly more than the other.

(4) In sub-paragraph (1)(d) the reference to "other tissue" of the resulting child does not include a reference to any whole organ of the child.

The comprehensive nature of the revised HFEA Act of 2008, coupled with the continuing frame of regulations that are, from time to time, issued by the HFEA, demonstrates at the very least that PGD in the UK appears to be somewhat well regulated. Section 11(1)(aa)[148] of the HFEA Act of 2008 also lists out other activities that encompass non-medical fertility services which may be authorized by the HFEA, but this is generally of a much narrower scope that permitted medical fertility treatments.

One of the other most controversial of the use of PGD services, involving the sex selection of embryos, is also encompassed within the scope of the HFEA Act of 2008. Paragraph 1ZB of Schedule 2[149] of the HFEA Act of 2008 expressly prohibits the selection of embryos based on sex, unless the circumstances of the proposed

---

[143] Ibid.

[144] 'Illustrative Text: Human Embryology and Fertilisation Authority Act 1990. As Amended: An Illustrative Text' (n 138) 77.

[145] Ibid 76.

[146] Ibid.

[147] Ibid 77.

[148] Ibid 18.

[149] Ibid 77.

selections falls under the exceptions in Paragraph 1ZA.[150] In essence, this would mean that the sex selection of embryos would likely only be permitted for avoidance of a X-linked or Y-linked genetic, chromosomal or mitochondrial disorder as envisaged in Paragraph 1ZA. On this basis, we can already see that the express prohibition of sex selection of embryos and the force of law that is wielded by the HFEA Act of 2008 coupled with the grant of authority, powers and functions of the HFEA is a strong influencing factor about the manner in which PGD has been used in the UK.

3.2.2.3   Australia

The Australian position on embryo selection *vis-à-vis* PGD appears to be inconsistent because of its constitutional and political governance system. In these instances, the relevant states regulate the applicability of PGD by state and territory laws that may differ from one state to another. This is however not simply by coincidence or by complete independence from the influence of federalism.

On a federal level, national guidelines issued by the Australian National Health and Medical Research Council (NHMRC) insofar as ARTs are concerned is the most comprehensive guidelines that indirectly binds state practices of ARTs (including PGD) to a minimum professional standard. These guidelines, known as the Ethical Guidelines On the Use of Assisted Reproductive Technology in Clinical Practice and Research was recently revised[151]; the revised guidelines (the NHMRC Ethical Guidelines) are still presently being developed. These guidelines, together with others issued by the NHMRC, in concurrence with the Australian Health Ethics Committee (AHEC), are issued pursuant to the National Health and Medical Research Council Act of 1992 and therefore derive its force of application from the act.

The framework in itself is robust through governmental intervention in what appears to be a multi-layered system of governance, which includes the roles of the AHEC and the NHMRC. The fertility industry appears to be similarly represented *vis-à-vis* the Fertility Society of Australia (FSA), which established the Reproductive Technology Accreditation Committee (RTAC). The RTAC is the body responsible for accreditation of ARTs clinics across Australia (similar to the role of the HFEA in the United Kingdom); the conditions of accreditation ensures that fertility clinics comply with all relevant legislation issued by the government regarding the use of ARTs, and this includes the NHMRC Ethical Guidelines. On this basis, the NHMRC guidelines operate as a strong influencing factor on PGD legislation in the different Australian states. The federal framework of legislation also incorporates the Prohibition of Human Cloning for Reproduction Act 2002,[152] and the Research

---

[150] Ibid 76.

[151] National Health and Medical Research Council (2017).

[152] Prohibition of Human Cloning for Reproduction Act 2002 2016 (No 144, 2002).

Involving Human Embryos Act 2002.[153] At present, Victoria, New South Wales, South Australia and Western Australia have all enacted the relevant legislation that regulates ARTs (some of which include PGD). Taking inspiration from the NHMRC Ethical Guidelines, it is apparent that there is a similar thread of consideration in these state legislations regarding the applicability of PGD (Table 3.1).

The preliminary observations that can be made with regards to embryo selection or PGD laws in Australia is twofold: firstly, that it is allowed (in Victoria, South Australia and Western Australia) only for medical purposes, that is, examining for and/or avoiding genetic diseases or abnormalities in pre-implantation embryos. The remaining Australian states do not appear to have specific legislation that relates to embryo selection *vis-à-vis* PGD, although New South Wales has enacted legislation that governs ARTs generally. Secondly, the review of literature in the Australian legal landscape reveals two dominating concerns in PGD technology: sex selection,[154] and the creation of a saviour sibling.[155]

As a general rule, sex selection appears to be strictly frowned upon and prohibited in Australia, unless for exceptional circumstances where pre-implantation embryos may be stricken with inheritable chromosomal disorders linked to either the X or Y-chromosomes.[156] This is expressly mentioned in Section 8.13 of the NHMRC Ethical Guidelines, where "the condition, disease or abnormality affects one sex significantly more than the other."[157] The concern about sex selection, however, is not concentrated in this determination of avoiding genetically inheritable diseases. The concern arises where parents intend to select the sex of the embryos for non-medical reasons, such as family balancing. In the same manner that the use of ARTs had initially been debated upon, non-medical sex selection has been the subject of the Australian reproductive debates in the last several years.

The AHEC has specifically considered the manifold factors relating to considerations of autonomous, parental, non-medical sex selection,[158] and it is not surprising that the outcome of deliberations falls on the side of prohibiting non-medical sex selection.[159] Because of this prohibition, it is not surprising to find that some of the largest numbers of couples who visit Thailand or Malaysia for sex-selective PGD are Australians.[160] Although the NHMRC Ethical Guidelines do not have the same force and impact as a promulgated legislative act, the incumbent instrumentalities of the ARTs business in Australia are highly dependent on the compliance with, and

---

[153] Research Involving Human Embryos Act 2002 2016 (No 145, 2002).

[154] Feikert (30 April 2012).

[155] Cordelia (2004), p. 121.

[156] US National Library of Medicine, Genetics Home Reference, 'What Are the Different Ways in Which a Genetic Condition Can Be Inherited?' (*Genetics Home Reference*) https://ghr.nlm.nih.gov/primer/inheritance/inheritancepatterns.

[157] National Health and Medical Research Council (2017), p. 69.

[158] Ibid 70–71.

[159] Please see Section 8.14 of the NHMRC Ethical Guidelines, which emphatically states that sex selection for non-medical purposes is not supported by the NHMRC and AHEC.

[160] Smith (3 August 2014).

Table 3.1 Legislation pertaining to PGD (embryo selection) in the Australian states

State/ territory	Legislation	Relevant section on applicability of PGD	Prohibitions (specific to gametes and/or embryos)
Victoria	The Assisted Reproductive Treatment Act 2008 [VIC] 2008 (No. 76)	Section 10(2)(a)(iii): "… risk of transmitting a genetic abnormality or genetic disease to a child…"	*Part 3—Offences relating to use and storage of gametes and embryos and other matters* Division 1—Prohibited procedures (Sections 26–30) (including Section 28—ban on sex selection; and Section 30—ban on destructive research on embryos created for treatment purposes) Division 2—Storage (Sections 31 to 34) Division 3—General offences (Sections 35–37)
New South Wales	The Assisted Reproductive Technology Act [NSW] 2007 (No. 69) 43 + Assisted Reproductive Technology Amendment Bill [NSW] 2016 17	[No specific provisions about genetic testing]	*Part 2—ART providers* Division 3—Use of Gametes (Sections 16 to 29)
South Australia	The Assisted Reproductive Treatment Act [SA] 1988 14 + The Assisted Reproductive Treatment Regulations [SA] 2010 4	Section 9(1)(c)(ii): "if there appears to be a risk that a serious genetic defect, serious disease or serious illness would be transmitted to a child conceived naturally."	*The Act—Part 2* Section 9—Conditions of registration *The Regulations—Section 8 further conditions of registration* Section 8(2)(a): "a condition requiring the person to comply with the NHMRC guidelines."
Western Australia	The Human Reproductive Technology Act [WA] 1991	Section 14(2b)(ii): "where the diagnostic procedure is for the genetic testing of the embryo, there is a significant risk of a serious genetic abnormality or disease being present in the embryo."	*Part 1—Preliminary* Division 2—Specific offences (Sections 5A to 7) Section 7—Offences relating to reproductive technology *Part 4A—Prohibited Practices* Division 1—General Division 2—Human cloning Division 3—Other prohibited practices (including Section 53L Offence—heritable alterations to genome)

fulfilment of the NHMRC Ethical Guidelines as the legitimacy of ART services are highly regulated through the encompassing role of the NHMRC.

The other controversial discussion relating to PGD is its use with Human Leukocyte Antigen (HLA) tissue typing,[161] for purposes of 'creating' a saviour sibling. (This is not to be confused with pre-implantation HLA tissue typing.[162]) In addition to ethical considerations that hinge on the instrumentalization or commodification of children,[163] and particularly, what is referred to as the "spare parts" children, the other more pressing question that needs to be determined is the limits or extent to which genetic materials, tissue, blood, samples and the like, may be 'harvested' from the saviour sibling.[164] There are divergent schools of thought on the ethical and legal proprieties of the saviour sibling dilemma in Australia. Robert Boyle and Julian Savulescu particularly highlight why the sustainability of the argument of harms that may potentially befall a future saviour sibling cannot be reasonably founded[165]; as is also the view of Malcolm K. Smith, who propagates for a safe relaxation of legal rules relating to saviour siblings.[166] But a more balanced perspective is offered by Thomas Cordelia, who takes into account a variable consideration of consent derived from the *Gillick* competence test, and the invasion into the potential child's bodies.[167]

Whatever the case may be, the existence of a legal and regulatory framework governing embryo selection *vis-à-vis* PGD and other reproductive technologies in Australia are, at the very least, established. Although the framework in Australia generally is not as progressive in the manner of the HFEA in the UK, the present structure does support safety, access, a minimum of acceptable standards, and also enforcement mechanisms.

3.2.2.4   Malaysia

It is not uncommon for fertility treatment services to offer PGD in Malaysia. A foremost fertility centre in Malaysia, the TMC Fertility Centre, is located in a medical centre in Petaling Jaya, a bustling and vibrant city a short distance away from the city centre of Kuala Lumpur. Touted to be one of the best in the country that offers IVF treatments, TMC Fertility Centre boasts highly skilled and qualified medical professionals, state-of-the-art equipment and an extraordinarily high degree of commitment to service.[168] PGD is offered as part of the centre's fertility treatment

---

[161] Liu (2007), p. 65.

[162] Ibid.

[163] Boyle and Savulescu (2001), p. 1241.

[164] Cordelia (2004), p. 123.

[165] Boyle and Savulescu (2001), pp. 1241–1242.

[166] Smith (2013), p. 154.

[167] Cordelia (2004), p. 142.

[168] TMC Fertility Centre (21 February 2013).

services.[169] The centre's website explains the key details of PGD, including the different types of PGD that are used for different purposes and functions. But TMC Fertility Centre is not the only medical centre to do so; in fact, PGD is routinely offered and bundled as part of the services ancillary to IVF treatment, and is also available in other major hospitals in Malaysia as well as specialized fertility clinics and centres. Fertility treatment is 'big business' in Malaysia, and there are hundreds of fertility centres operating all over the country. The commerce of fertility has permeated the realm of industrialization, business and drives of profit, and is no longer seen as matters more suited to be dealt with in the confines of the family unit.

It is also not surprising to discover that because of these fertility services, Malaysia has positioned itself as a competitive destination for the provision for fertility tourism, or reproductive tourism.[170] Reproductive tourism is the phenomenon where people cross international borders in order to access fertility or other reproductive services and technologies that may not be available in their own countries, or due to other reasons such as costs.[171] Thailand is the other Southeast Asian country where reproductive tourism appears to be openly supported by the government.[172] I pose this reproductive tourism phenomenon occurs in countries like Malaysia and Thailand, because there is a lack of, or laxity in enforcement of, any appropriate legal or other regulatory guidelines that relate to commercial reproductive practices.

In Malaysia, the country is in a transitional state insofar as legislation relating to ARTs is concerned (under which the purview of PGD would potentially be governed or regulated). In 2015, the Malaysian government was in the process of drafting a piece of legislation, tentatively titled the proposed Artificial Reproduction and Tissue Act.[173] This act would ensure that laws relating to artificial reproduction are in place, and would potentially encompass IVF, donation and storage of ova and sperm, as well as address the current inadequacies in the out-dated Malaysian Human Tissues Act of 1974.[174] However, at this juncture, there is no other available information on the scope of the proposed act; a recent search with the official portal of the Parliament of Malaysia[175] does not reveal that the proposed act has been tabled just yet.

In the circumstances, the legal landscape relating to assisted reproduction in Malaysia is therefore still unregulated through Parliamentary legislation; with the exception of some guidelines issued by the Malaysian Medical Council[176] and the

[169] TMC Fertility Centre (23 March 2015).

[170] Whittaker and Speier (2010), p. 363.

[171] Deonandan (2015), p. 111.

[172] Stasi (2015), p. 17.

[173] Yuen (29 November 2015).

[174] Human Tissues Act 1974 (Act 130) 1974 (Act 130) 9.

[175] Parliament of Malaysia, 'Official Portal of The Parliament of Malaysia' http://www.parlimen. gov.my/index.php?lang=en.

[176] Malaysian Medical Council (14 November 2016).

Malaysian Ministry of Health.[177] (The guidelines from the Malaysian Ministry of Health, however, are not relevant for this discussion as it deals specifically with standards of facilities used in assisted reproductive technologies, such as laboratories and equipment, and do not cover the more pertinent aspects of regulatory measures as sought here.) On this basis, the Malaysian Medical Council guidelines at present are the most express pronouncement of PGD services in the country.[178] In these guidelines, which target the medical professional and those in the provision of medical services generally, the position is made clear that there is no agreement as to the moral status of the embryo, although the Islamic position on the embryo attaining "life" is outlined in the relevant section of the guidelines.[179] The relevant section also states that there is no agreement that relates to the discarding of embryos being equivalent to abortions.[180] It is also of interest that these specific wordings are used in the guidelines with regards to PGD: "At present, it is best that PGD be used for only severe and life-threatening genetic diseases. It would be unethical to analyse and select the inherited characteristics of embryos…".[181] The tone of neutrality and attempted impartiality through the language of these words, "it is best that PGD be used for…" and "it would be unethical" essentially suggests that the reader would be free to come to his or her own conclusion in the interpretation of what PGD should be used for. However, in a further section of the guidelines, Section 15 indicates "ethically unacceptable"[182] or prohibited practices under the guidelines. These include: "pre-implantation diagnosis to create "designer babies" (those with specific physical, social or specific gender characteristics and not for the reason of avoiding serious medical illnesses)."[183] From this reading, and bearing in mind that these are guidelines issued by the Malaysian Medical Council, (and may be arguable as to their force of application from the legal perspective), there certainly appears to be a clash between the advisory, or discretionary approach in Section 14, with the "ethical" prohibition in Section 15. The force of the law is excluded in this sense, and the guidelines can be interpreted as a form of internal, professional self-regulating mechanism for the medical profession.

However, as Malaysia is a multicultural nation with a dualist legal system, with civil law applicable to all, or in some aspects, only to non-Muslim citizens,[184] and

---

[177] Ministry of Health Malaysia, Medical Development Division (October 2012).

[178] Malaysian Medical Council (2006), p. 14.

[179] Ibid. In this chapter, I also briefly outline the implications of regarding ARTs from the perspective of Shari'a law in Malaysia.

[180] Ibid 15.

[181] Ibid.

[182] Ibid.

[183] Ibid 17.

[184] Examples where civil law would not apply to Muslims would be where such laws are remit to the purview of Syariah law. Syariah law is the law that governs all Muslims in Malaysia in relation to specific matters, whereas civil law applies to all other citizens of the country. However, there are many aspects of Malaysian civil law that likewise apply to Muslims too where there has been promulgation made at the federal law (for example, various branches of private law, dimensions in public law and constitutional law, laws relating to criminal offences, etc.). Syariah law is likely to

with Shari'a law, applicable to all Muslims. Shari'a law, or Islamic law, is derived from Quranic interpretations, and is applicable to all Muslims in the country. From a variety of interpretation of Islamic scholars, the Quran does not prohibit ARTs in the form of IVF, provided the gametes belong to both the husband and wife themselves. Marriage and family is held in the highest of regard in Islamic tradition. In the Islamic interpretation therefore, the paramount importance lies in "marriage and purity of lineage"[185] and completely prohibits the insertion of another man's sperm (vis-à-vis a fertilized embryo) into the womb of a woman who is not his wife, as that would, in the Quran, result in a child born out of wedlock.[186] Hence, IVF is permissible only to the extent that the couple is legally married; any other forms of ARTs or IVF, including donation of third party gametes and surrogacy, is completely forbidden.[187] In the meantime, the literature regarding PGD in accordance with Islamic laws and Quranic interpretations is almost non-existent in Malaysia. I make the presumption that Islamic traditions regarding ARTs would similarly apply to PGD (bearing in mind that I am not a scholar of Islam or the Quran), but until a *fatwa* is formally pronounced by the Department of Islamic Development Malaysia, this remains to be seen. Under Islamic traditions and law, this may possibly then involve *ijtihad*, or the exercise of critical reasoning and independent judgment in interpreting the *surah* of the Quran or the *hadiths* of Prophet Muhammad, although *ijtihad* is now seen as contravening the manner in which Islam is to be ruled and practiced.[188]

In the meantime, although there is yet no indication of any "designer babies" born of PGD in Malaysia, in the manner envisaged in Section 15,[189] the global discourse on such a phenomenon is still exuberant. Clarity with regards to the position of PGD, and certainly, in the future, the applications of technologies such as CRISPR, would be welcome in the proposed Artificial Reproduction and Tissue Act. It remains to be seen when such an act would be enacted in Malaysia. Until such time, PGD and certainly, possible genetic interventions in the vein of CRISPR would remain unregulated and could be subject to misuse absent the framework of legal pronouncement, supervision and enforcement.

3.2.2.5 Thailand

Prior to 2015, Thailand's position regarding PGD appeared to be similar to Malaysia's. ARTs generally (including IVF, PGD and surrogacy) were largely unregulated save for guidelines issued by the Thailand Medical Council. The

---

have jurisdiction and governance over Muslims in matters relating to family, children, religion and religious practices, marriage, divorce and such issues.

[185] Ahmad et al. (2016), p. 362.
[186] Ibid.
[187] Ahmad et al. (2016).
[188] Rhode (15 June 2012).
[189] Belluck (4 August 2017).

question of the medical council's powers and functions, like in Malaysia as well, appeared to be limited in nature and allowed a form of self-regulation over the practices of the medical profession. Fertility clinics, particularly in Bangkok, the capital city, were eager to offer PGD services. In the case of Thailand, PGD was most popular for purposes of sex selection[190]; it was not uncommon several years ago to find placards, and notices outside fertility clinics advertising that they carried out sex selection in IVF processes.[191] This appears to be the key ethical issue that dominates the PGD discourse in the country.[192]

In addition, Thailand's position as a reproductive tourism hub in Asia,[193] with competitively low prices, highly sophisticated technological equipment, and skilled medical practitioners, also contributed to its burgeoning fertility industry.[194] Before 2016, the applicable guidelines issued by the Thailand Medical Council were likely the most authoritative 'regulation' regarding assisted reproductive technologies. The guidelines, titled Thailand Medical Council Notification No. 21/2544 On Service Standards for Assisted Reproductive Technologies (No. 2) of 2001, supplemented the earlier Thailand Medical Council Notification No. 1/2540 of 1997, by issuing additional statements to Section 4.1 and 4.2 of the earlier guidelines. These additional statements point to the use of PGD that "may be conducted for certain disease determination as deemed necessary and appropriate, and it must not be done for purposes of gender selection."[195] Like the Malaysian Medical Council guidelines on assisted reproduction, it was also clear that these also did not bear any legislative force; the use of discretionary language as part of the guidelines showed an almost *laissez faire* surrender of the reins of regulation to the profession itself, if they so chose to do it. Therefore, it was not surprising that PGD, and specifically, PGD for sex selection, and an exerted and concerted effort in promoting Thailand for reproductive tourism, continued and grew on a monumental scale.[196]

The reproductive tourism landscape in Thailand, however, altered substantially in 2015. At that point of time, the surrogacy case of Baby Gammy made international headlines. A young Thai woman, Pattaramon Chanbua, agreed to become a surrogate mother for an Australian couple, David and Wendy Farnell, through the mediation of a private surrogacy agency in Thailand. The birth of the surrogate babies subsequently sparked an intense international debate on commercial

---

[190] Natipodhi (2014).

[191] I had visited the capital city of Bangkok in 2016 and 2017, and trawled numerous fertility clinics in the city, both in person and on their websites. In most of the clinics, sex selection in PGD was the service that was most commonly highlighted to potential clients.

[192] Natipodhi (2014).

[193] Harryono et al. (2006).

[194] Stasi (2015).

[195] Thai Law Forum, 'Thailand Medical Council Regulations on Surrogacy and IVF | Thailand Law Forum' http://www.thailawforum.com/medical-surrogacy-regulations/ accessed 11 April 2017. Please see Section 1 of Guidelines No. 21/2544 of 2001.

[196] NaRanong and NaRanong (2011), p. 336.

international surrogacy,[197] because one of the babies, Gammy, had been born with Down's Syndrome, and the surrogate parents refused to care for the child, and abandoned him. Recent developments in the case indicate that Gammy has been granted Australian citizenship[198] as a means of ensuring entitlement and security for necessary medical treatment. Due to the shocking nature of the case, Thailand's military-appointed cabinet at the time took swift steps to approve new laws to ban commercial surrogacy in the country.[199] Gammy's case is only one of the possible many commercial reproductive cases that receive coverage in news portals and social media, and impacted tremendously on an international level because it raised such grave ethical concerns and serious policy questions, and revealed a stark legislative gap in a country that boasted a reputation for being Asia's medical tourism hub. The intensely sad truth in Gammy's case in Thailand was that commercial surrogacy had, indeed, been banned in the country, since 1997, but the scandal clearly exposes Thailand's extremely slipshod 'law' enforcement.

As a result thereof, Thailand has now passed the Protection for Children Born Through Assisted Reproductive Technologies Act.[200,201] It was admittedly the intention of the Thai National Legislative Assembly (the military-installed Parliament) that the act was to clarify the position relating to children born of assisted reproductive technologies, the parenthood obligations, and also to ensure that commercial surrogacy for foreigners was completely banned.[202] It was also highlighted that the Social Development and Human Security Minister for Thailand made a statement claiming that the new act was "aimed at protecting surrogate babies and suppressing human trafficking."[203,204] What is relevant in this context is the fact that Thailand had responded to an international scandal by enacting the appropriate regulatory and legislative measures that would curb any further fall-out from the incident. It is also apparent that the new Act, because of its prioritization on provisions relating to surrogacy in Chap. 4,[205] would be largely inadequate in terms of PGD and its eventual regulatory enforcement.

Under Chapter 2, Section 18, of the Act, a medical practitioner "shall perform a necessary research concerning possible genetic diseases of the embryo. This

---

[197] Pardes (14 January 2016).

[198] Farrell (19 January 2015).

[199] Australian Associated Press (20 February 2015).

[200] Protection for Children Born Through Assisted Reproductive Technologies Act 2015 (167/2553).

[201] It is highlighted in this chapter that the dissection of the provisions of the act is based on an unofficial translation of the Thai text for the act. trans. Messrs. Jurs & Consult.

[202] Buchitchon (2016), p. 1610.

[203] Caamano (2016), p. 598.

[204] The Nation (28 November 2014).

[205] Juslaws & Consult, 'Unofficial Translation- Act Providing for the Protection for Children Born Through Assisted Reproductive Technologies' 4. Please see Section 7(8) of the Act accordingly, where the powers of the Committee for the Protection of Children Born Through Assisted Reproductive Technologies include the presentation of "suggestions regarding the notifications issued by the Medical Council concerning the providing of ART services" under the Act.

research shall not include the selection of sex and shall not have similar characteristics."[206] In Chapter 5 of the Act, which deals the mechanisms of control over the use of assisted reproductive technologies, Section 36 stipulates that "it is prohibited to anyone to create an embryo with other purposes than to assist the birth of a child for the legally married barren couple."[207] The Act also makes reference to the applicability of the provisions to all matters relating to assisted reproduction in conjunction with guidelines issued by the Thai Medical Council.[208] Nevertheless, the Thai Medical Council has not yet issued any revised guidelines to reflect the provisions of the new Act, and on this basis, it would appear that the Medical Council's earlier guidelines of 2001 would therefore prevail. It is also of interest to note that the Act itself does not specifically mention the words "pre-implantation genetic diagnosis", but the wordings of Section 18 imputes a mandatory requirement on the practitioner: "*shall* [emphasis own] perform necessary research concerning possible genetic diseases of the embryo."

By this implication, and subject to an alternative interpretation, which is yet to emerge, it may be imputed that embryo selection *vis-à-vis* PGD is not strictly outlawed. The contrasting provisions within the regulatory framework is not surprising; particularly if the intention of this new Act, as highlighted above, was to ensure a more well-regulated commercial surrogacy framework. In the meantime, the realities in practice regarding embryo selection, as well as the manner of enforcement, certainly seem reinforce the supposition that this is not strictly a prohibited act whatsoever.

3.3   The Predominant Debates in the Selected Jurisdictions

Following the analyses of current PGD legislation or guidelines in the selected jurisdictions, the emerging political, socio-legal or legal, religious and cultural, and ethical and philosophical debates reveal some of the challenges of regulating technologies (which will be addressed more specifically in Chap. 4). There are three main points by which I address these: first, from the perspective of the political environment and its inter-sectionalism with the legislative and judiciary; secondly, from the perspective of religious or non-secular undertones that contribute to "embryo politics"; and thirdly, from the perspective of morality and moral judgment. These points, which are emphasized as the lenses by which to examine the predominant debates in the selected jurisdictions, are able to provide us with a better picture on the historical and potential development of PGD laws in the respective jurisdictions, as well as put forward a reasoned regulatory consideration in accordance with the level of importance attributed in each dimension.

---

[206] Ibid.

[207] Ibid 7.

[208] Ibid 2.

3.3.1 Political and Socio-Legal or Legal Debates

From the perspective of political environments and its relationship with the socio-legal or legal landscape, the judiciary and public policy, this chapter puts forward that the US position appears to be one with the most politically-charged undertones in matters relating to reproductive technologies like PGD. As accurately pointed out by Bayefsky, the "political, economic and social conditions particular to the United States are mutually determining and reinforcing."[209] One of these conditions is the proximity between PGD and the abortion debates.[210] The abortion debate in the US still continues despite the resolution by the Supreme Court *vis-à-vis Roe v Wade*,[211] and is one of the most highly charged and deeply partisan political issues in the country. Besides the political platforms on which the question of abortion has been used as a campaigning tool, the other reason for this may be entrenched in the interpretation of the status of the embryo. The fact that PGD involves a selection from a group of viable embryos, with the rest subsequently being discarded, has been likened to the consequences of abortion in terms of the destruction of embryos. Granted, the discarding of, and the destruction of, embryos may be semantically similar in the view of pro-lifers, but the question is also whether the words "discard" or "destruction" necessarily holds the same in the vein of killing, taking a life, or murder as is put forward by the pro-life American rhetoric. More interestingly, as Stacie Taranto also notes, the abortion discourse has had a long history in becoming "the single most important litmus test in American politics,"[212] demonstrating the historic tussel between Democrat and Republicans, and internal political reforms that sought to reverse the New Deal measures.

The unique position of the Supreme Court (the judiciary) is similarly engaged in these discussions, because of constitutional scrutiny that may entail within the sphere of individual freedoms and liberties. In cases like *Buck v Bell*[213] and *Skinner v Oklahoma*,[214] the Supreme Court has shown itself willing and able to interpret and uphold reproductive or procreative liberty. In further cases like *Roe v Wade*,[215] *Griswold v Connecticut*,[216] *Planned Parenthood of Southeastern Pennsylvania v Casey*,[217] the interpretation by the Supreme Court indicates its willingness to extend

---

[209] Bayefsky (2015), p. 7.

[210] Ibid.

[211] 'Roe v. Wade, 410 U.S. 113 (1973)' (n 110).

[212] Taranto (22 January 2018).

[213] 'Buck v. Bell, 274 U.S. 200 (1927)' (*Justia Law*) https://supreme.justia.com/cases/federal/us/274/200/case.html.

[214] 'Skinner v. Oklahoma Ex Rel. Williamson, 316 U.S. 535 (1942)' (*Justia Law*) https://supreme.justia.com/cases/federal/us/316/535/case.html.

[215] 'Roe v. Wade, 410 U.S. 113 (1973)' (n 110).

[216] 'Griswold v. Connecticut, 381 U.S. 479 (1965)' (*Justia Law*) https://supreme.justia.com/cases/federal/us/381/479/case.html.

[217] 'Planned Parenthood of Southeastern Pa. v. Casey, 505 U.S. 833 (1992)' (*Justia Law*) https://supreme.justia.com/cases/federal/us/505/833/.

the realm of privacy laws as a penumbral fundamental right in accordance with the Constitution, hinging on either the due process or equal protection clause in the Fourteenth Amendment of the Constitution.[218] PGD, however, despite falling under the heading of reproductive liberty, entails an additional aspect of 'selection', and the implications of selection, particularly where genetic interventions are concerned, raises a new set of enquiries on how the Supreme Court may then interpret the extent to which reproductive or procreative liberty under the sphere of privacy, may extend to the selection of genetic traits. The concern here, therefore, is based on both ethical (the status of the embryo) and legal (the interpretive powers of the Supreme Court) grounds.

Countries like Australia and the UK, however, have demonstrated its ability to find a balance between competing political and economic priorities and the necessity to regulate, whether at a minimum level, or on a more extensive basis, these kinds of medical and scientific technologies. Observation though it may be, it is possible that the doctrine of separation of powers in Australia and the UK are likely much more emphatic than in the US. Australia's strong ties to federalism and much more robust legal framework on reproductive technologies and issues, including abortion, are highly regulated. The UK has the benefit of agency oversight from the HFEA in all matters relating to reproductive fertility and technologies. In fact, it is the presence of these institutional structures that likely contribute to a clearer demarcation in the manner reproductive matters are treated, and fairly neutral from the clutches of political tinkering.

In Malaysia and Thailand, the political and economic environment is also just as important a contributory factor to its legal enforcement issues. Malaysia and Thailand do have some regulations for PGD and reproductive technologies on a broader scale; the issues in these Asian countries, however, are centred on the laxity of enforcement and the lack of clarity attributed to the meanings in legislative acts or professional guidelines. In addition, the economic position in Malaysia and Thailand vis-à-vis the fertility tourism industry bears a direct relationship on each respective country's incomes and developmental growth. The revenue generated from fertility tourism on an annual basis also similarly impacts the political veracity of these countries, and further contribute to the infrastructure and facility development in Malaysia and Thailand. It is therefore not surprising that the ability of this economic import is immeasurably tied to the political sphere, and spills over into how legislative enforcement may operate. From the political standpoint, Malaysia and Thailand have dealt with internal political struggles for over a decade. With a recent historical government change in Malaysia, in May 2018,[219] and the new ruling government's promise to its citizens vis-à-vis its 100 Days Manifesto,[220] the government's performance is pegged to the improvement of economic welfare, governance of previously government-linked bodies, and reforms for minimum health standards and educational opportunities. The Thai political situation, in late

---

[218] Sholley (1951).

[219] Heydarian (11 May 2018).

[220] The Star, 'Pakatan Harapan 100 Days' https://www.thestar.com.my/ph100/.

November 2018, began to escalate as it faces the potentiality of new elections this year (2019),[221] following promises of new polls numerous times after a military junta had been installed in its leadership position. To date, it is still grappling with suitable prime ministerial candidates, internal party struggles, and the remnants of populist policies of its previous premier, Thaksin Shinawatra. It is not surprising that these pressing matters have taken centre stage, and relegated other matters (such as reproductive issues) into the foreground.

3.3.2 Religious and Cultural Debates

From the perspective of religious or non-secular undertones that contribute to "embryo politics",[222] the demonstrability of Malaysia and Thailand would rank most likely to utilize religious and non-secular grounds of reasoning for regulation, or the lack thereof. As also further indicated in Chap. 5, the legal framework of Malaysia is a unique one because its constitutional democracy exists *vis-à-vis* a dualist legal system. Thailand is also unique in this sense because of its apparent national deference to the teachings of Buddhism, and its viewing of Buddhism as a religion, as opposed to a way of life. The interpretative facets of religion and non-secularism in the daily lives of Malaysians and Thais demonstrate its pressing influence even in the fields of scientific and medical technologies.

However, this is also where the indicative challenges of legal pluralism are apparent. For example, in Malaysia, the co-existence between common law and Shari'a law inevitably creates a form of division (in some aspects of life) between Islam, and everything else non-Islam. The force of Islam in the daily lives of practising Muslims, for example, therefore have the capacity to influence the understanding and operability of laws based on the socio-religious norms inflicted through the teachings of Islam. Russell Sandberg contends that "a major failing of the concept of legal pluralism has been the inability to distinguish legal norms from other forms of social control."[223] This may be true regarding the status of the *fatwa*,[224] and how a *fatwa* is interpreted in an Islamic country through critical thinking and independent judgment, also known as *ijtihad*.[225] The position in Malaysia regarding the interpretation of the *fatwa* appears in conflict with some legal rules and norms, and particularly so in actual practice and the reality of daily living. This is, in fact, provided for expressly vis-à-vis the Administration of Islamic Law Federal Territory Act 1993,[226] and the express statement that "if there is any conflict between the *fatwa* by the *mufti* and the rulings by the judge, it only occurs due to the difference

---

[221] Hunt (15 November 2018).

[222] Bayefsky and Jennings (2015).

[223] Sandberg (2016), p. 137.

[224] Zakhiri et al. (2016).

[225] Rhode (2012).

[226] Zakhiri et al. (2016), p. 7.

of *ijtihad* applied to the various *fatwa*."[227] In fact, pursuant to teachings of Islam in the Q'uran, there appears to be some fundamental religious reasoning behind issues relating to reproduction, such as third party gametes, surrogacy, adoption and the like.

I argue that the existence of legal pluralism in this manner in Malaysia may make it difficult to regulate PGD because of the necessity to respect and balance the pluralistic legal system with the religious cultural beliefs of individuals who are Muslims.[228] This may also conflict with the considerable need to increase Malaysia's visibility, from both the developmental and economic perspective. The recent political regime change in the country[229] shows Malaysia's dexterity in its journey to transition from a developing[230] to a developed nation. However, this transition is also likely to engage a balancing exercise between religious and cultural values, with more open and liberalized democratic processes.

In Thailand, the religious grounding within the purview of the legal system is more homogenous in nature, with 96% of the total population following the teachings of Buddhism, or identifying themselves as being Buddhist. The royal family or monarchy of Thailand is often also seen as the patrons of Buddhism in Thailand, and on this basis, the practice of Buddhism in Thailand is unique in itself, and may differ from the practice of Buddhism in other countries. Unlike Malaysia, however, the evocation of Buddhism in Thailand, despite infusing daily lives with its social norms, meets an apparent conflict with the requirements for economic and monetary growth and investment into the country. This deeply revering religious aspect of Thailand is contrasted against its implicit role as a Southeast Asian hub for reproductive or fertility tourism, and the need to further encourage revenue from its fertility services as part of its economic growth.[231]

The motivations, therefore, for a less stringent regulation of PGD in Thailand, is not therefore focused on a purely religious aspect, and even lesser of an ethical consideration, but on a more practical nation-building exercise necessary to propel the country to a more developed status. Although the teachings of Buddhism play a fundamental role in the daily lives of the Thai people, it appears that the prioritization of the revenue stream in fertility tourism outweighs that of the religious dimension.

---

[227] Ibid.

[228] Muslims make up the majority of the Malaysian population, amounting to approximately 70% of the total population.

[229] Heydarian (2018).

[230] United Nations, Department of Economic and Social Affairs Development Policy and Analysis Division, Department of Economic and Social Affairs, 'Country Classification- Data Sources, Country Classification and Aggregation Methodology' (United Nations Secretariat) http://www.un.org/en/development/desa/policy/wesp/wesp_current/2014wesp_country_classification.pdf.

[231] Natipodhi (2014), p. 8.

3.3.3 Ethical and Philosophical Debates

From the perspective of morality and moral judgments, it would appear that ethics would be an influencing factor in the framing of these moral judgments. With the exception of a concentrated and small saturation of ethicists and scholars given to consider the role of ethics in the sphere of reproductive technologies such as PGD, Malaysia and Thailand particularly do not appear to engage immersively in these ethical debates, although semi-regulatory ethical committees do exist within these countries. The role of ethics, however, in the US, UK and Australia occupies a central place in the legislative inquiry.

From the analyses provided above, the voluminous existing literature in these jurisdictions indicate the primary ethical considerations that shape the regulatory framework; these ethical guidelines are often times issued by professional bodies or associations within the field of medical and scientific technologies. These include the American Society of Reproductive Medicine and the American Society of Medical Genetics in the US (amongst others), the HFEA in the UK, and the NHMRC in Australia. The main ethical concerns that dominate the discussion in these 'Western' countries are manifold, but in this section, I highlight three common ethical concerns that have given rise to the shaping of regulation in these countries. These concerns relate to, firstly, the destruction of human embryos and its equivalency to the destruction of embryos in the abortion field; secondly, the issue of autonomy and choice exercised by parents over future offspring, and how this brings the argument dangerously close to eugenics practice; and thirdly, the 'designer baby' or germ line modification concerns in embryo selection. In Chap. 2, I had provided a very brief landscape of some of these concerns, namely those raised by Botkin[232] and Robertson.[233]

Firstly, in connection with the discarding of embryos in PGD, and its nexus to the destruction of embryos in abortion, I believe that the vast interpretive differences in the status of an embryo and when it achieves 'personhood' has contributed to this. Unfortunately, the debate on the status of the human embryo is likely to never be fully resolved, although legal definitions in various jurisdictions may provide some guidance. Having mentioned this, however, the determinative, individual mind-set of what an embryo is, and when it achieves the status of a functioning human being, differs in the medical, religious, and cultural dimensions. On this basis, Robertson has pointed out why PGD is objectionable in some circles, citing that its effects "replay the debates over abortion"[234] because embryos or foetuses are deemed as 'persons'. Robertson also pointed out the other objection to PGD, the selection element, which leads into the second and third ethical concerns highlighted in this section. He recounts Leon Kass's articulation of this view, that "human reproduction

---

[232] Botkin (1998).

[233] Robertson (2003).

[234] Ibid 466.

is a 'gift'"; the movement towards embryo selection has a consequentialist concern that may result in a eugenics-based world and 'designer babies'.[235]

Secondly, in respect of the issue of autonomy and choice exercised by parents over future offspring, and how this brings the argument dangerously close to eugenics practice, I also echo Botkin's concern about the need to set acceptable parameters or limits of PGD use,[236] whilst respecting the parent-child relationship dynamics and measurability of personhood and autonomy.[237] As highlighted in Chap. 2, I dealt extensively with regards to the elements of parental autonomy and personhood as a reflective effort to channel the uses of PGD in a more egalitarian manner. On a more philosophical level of reasoning, the 'coercive' element of PGD and genetic interventions is a suspicious discomfort. I. Glenn Cohen has provided an illuminating rumination of the seemingly voluntary nature of genetic enhancements. Cohen postulates that there is still a measure of coercion involved, which he insists is "not an oxymoron,"[238] stating that parental decisions in themselves is a form of coercion.

In *Homo Juridicus*,[239] Alan Supiot writes about the fact that human beings, people, in general, are not, in themselves truly rational beings that derive their sense of the world and reason inherently; instead, they gain this rationality and the emotiveness of simply *living*, by their interaction and access to meanings of rationality shared with other human beings.[240] There is much worth in this particular train of thought, which gives fruit to the notion that Cohen raises. 'Coercion' in this aspect, tied to the formation of the rationality of thought processes may never be a truly autonomous concept in the pivots of human evolution. The virtue of being human, with its diverse pronunciations on the concepts that unassumingly form the tenets of personality, interaction, integration and belonging to society generally, are undoubtedly the influence of a larger scheme of things. Therefore, the shift to individual decision-making, and by this invocation, to parental decision-making in the case of prenatal life, become 'coerced' by the views and understanding of what worldly views dominate in the minds of parents.

Thirdly, and finally, the 'designer baby' and germ line modification concerns raises equal worries which may in the future trigger a form of "genetics epidemic."[241] The revolution of DNA science that is faced in our corporeal existence "will make it possible for humans to actively manage our evolutionary process for the first time in our species' history…"[242] Rightly so, the Foreign Affairs journal identifies that responses to this prospect are both extremely exciting to some, and disturbingly frightening for others; a conceptual dialogue is sorely needed on a global level to

---

[235] Ibid.
[236] Botkin (1998), p. 21.
[237] Ibid 23–24.
[238] Cohen (2014), p. 652.
[239] Supiot (2007).
[240] Cohen (2014), p. 653.
[241] Metzl (10 October 2014).
[242] Ibid.

repel possible destabilizing conflicts between these views.[243] Leon Kass has pointed out that essentially, the workings of the human mind in some individuals, kitted out with the convolutions of pre-conceived beliefs and societal conditioning, experience what is commonly referred to as the wisdom of repugnance, or more simply, the "yuck" factor.[244] This "yuck factor", often dismissed as being devoid of scientific rationality, and viewed as a deep-rooted sense of disgust that often cannot be explained with reasonable validity through expansive homily, nevertheless, may lend some capacity to the veritable tolerance of "unnaturalness" through the lens of a proper claims framework.[245] Kass laments that bioethicists have misconceived the reach of emerging technologies and operate under the naïve delusion that "compassion, regulation and a respect for autonomy"[246] will alleviate humankind's fears of these inchoate machineries.

Sheehan provides a relatively concise manner of construction which differentiates "interference with nature" (the human germ line) and "playing God",[247] both of which are concepts that lead to the "unnaturalness" of genetic intervention technologies, as distinct as they may be as individual claims of objections. He presents the "playing God" claim as one that is "more focused on the character of the agents involved, whether it is a particular clinician or a legislature enacting policy."[248] In short, he puts forward the argument that the pivotal role of these agents takes on the management of decisions, which go "beyond their proper authority".[249] The assumption, therefore, appears to be rooted in a non-secularized view that the 'appropriate' or 'right' agent to take on these decisions, would therefore be God, or a divine being which expands its omniscient powers in the creation of human beings. The "interference with nature" claim, however, can be differentiated *vis-à-vis* the "yuck factor" mentioned above. It is easy to put forward a statement, that a particular thing, for example, sub-dermal implantation (a form of body modification procedure that implants carved or moulded silicone embedded under the skin), is radically an "interference" with what Nature has given to a person, simply because in his or her own most "natural" form through the journey of human birth, human beings do not manifest the dramatic, raised effect of the skin that sub-dermal implantations palpably demonstrate. The reason *why* this is an interference, as evinced by the "yuck factor" argument, is one that usually cannot present a judicious logic, save that which is experienced through the emotive heart and unfailing "gut feeling".

---

[243] Ibid.

[244] Kass (1997), p. 17.

[245] Sheehan (2009), p. 177.

[246] Kass (1997), p. 22.

[247] Sheehan (2009), p. 179.

[248] Ibid.

[249] Ibid.

3.4 Comparisons of the Debates: Shaping the Framework of Regulation

If we bring together the wealth of analyses from the preceding sections, attempting to address the potent conceivable combination of PGD and genetic interventions may become a less complicated task. The knowledge and understanding from research, development, debates and its contribution to policy and law-making is a valuable exercise in shaping the future framework of regulaticn. The predominant debates raised do indeed demonstrate a veritable movement into understanding how regulation may be shaped.

In her work on the ethical and legal analysis of pre-implantation embryo testing in Europe, for example, Judit Sándor provides a critical analysis on the landscape of selective reproduction. In framing the kind of questions that legislators and policy makers need to address, she states that the "law is one of the most influential contributors to the work of delineating boundaries in the field of biotechnology."[250] The permeation of ethical dimensions into the realm of the legal landscape in Europe, spanning issues such as the juridical status of the human embryo,[251] a selection of traits in embryos,[252] medical versus non medical indications of use,[253] and the selection of an embryo to match an existing sibling, that is, the savior sibling dilemma,[254] have necessitated some legislative responses to the phenomenon of pre-implantation embryo testing.

In one of his pieces of work, I. Glenn Cohen also provides how we reconstruct the taxonomy relating to genetic enhancements, but more importantly, he address the in-depth specificity of the role law or legal regulation is able to play, to address the volume of differing deportments[255] once clear distinctions are made, because "enhancement is not a unitary phenomenon."[256] Against this backdrop of these distinctions, (for example, between biological and non-biological enhancements,[257] enhancement after birth or before birth,[258] which can further be categorized into enhancement by selection versus enhancement by manipulating embryos or fetuses which have already been implanted,[259] and delineations between enhancements versus treatments,[260] amongst several other categories highlighted by Cohen), Cohen postulates on the kinds of role that the law is able to take, for instance, by straddling

---

[250] Sándor (2015).

[251] Ibid 355.

[252] Ibid 356.

[253] Ibid 360.

[254] Ibid 361.

[255] Cohen (2014).

[256] Ibid 652.

[257] Ibid 646.

[258] Cohen (2014), p. 648.

[259] Ibid.

[260] Ibid 650.

the relevant considerations of mandates, subsidies or taxes,[261] on the assumption that the law shall permit enhancements of a genetic nature; or perhaps, an outright legal prohibition,[262] which would, in turn, raise questions of the efficacy of a legal system at not only detecting prohibited enhancements,[263] but also to adequately police, monitor and enforce an assumed penalty vide sanctions for the breach of such prohibitions.

Debates further assist to move along other issues that warrant further reflection: of the assumptions that technological innovations are safe, of the "luck egalitarian" concept,[264] and the problem of distribution of equality and fairness in the context of availability and accessibility. These matters of contention are important factors in the way forward because it identifies the problems that may occur if PGD with genetic enhancement technologies, whether voluntary or otherwise, become the norm of expectation in society. The appropriate response then, to these problems, may necessitate a formulation of applicable regulations, whether at the State level or *vis-à-vis* the relevant medical council guidelines or codes of practice; what is apparent though, is that the surmounting distribution of potential problems, like all other social ills, must be addressed, first, through the State's non-derogable obligations to protect human and individual rights, and secondly, through a control mechanism to ensure that technologies of this nature are not subject to abuse by individuals themselves.[265] This, by no means, indicate the exhaustive element of issues to consider: I highlight these points as converging interests that have found resonance with numerous scholars, all of whom have provided incredibly valuable insight into what makes the technological advances desirable, or not.

Safety, always a paramount factor, must be the first salient consideration; because, as with any medical procedures, any form of technology that cuts through the practice of medicine and the provision of treatment or therapy, can be fraught with risks to the human body. Chatterjee provides a simple view that speaks volumes, which begins by framing the essential purpose of medicine.[266] He states, "While safety concerns are undoubtedly real, they are unlikely to have much of a practical impact. The incentives to develop new treatments with minimal side effects are in place. This is not to say that unexpected effects might not be encountered. But, in general, newer medications will continue to be safer, and the safer the medication, the less relevant this concern."[267] In reflecting the history of medicine and treatment through the course of time, there is a measure of truth in Chatterjee's statement. The pertinent question is not whether medicine or technologies that intervene in the human body are safe, but whether they are *safe enough*. The same was

---

[261] Ibid 652.

[262] Ibid 653–654.

[263] Ibid 654.

[264] Ibid 658.

[265] The pertinent aspects of legal regulation, or any other form of regulatory form, will be discussed in Chap. 4.

[266] Chatterjee (2004), p. 698.

[267] Ibid 970.

true of medical interventions such as vaccinations, reproductive technologies such as IVF, surrogacy, and treatments for different forms of cancer, for instance. Coupled with the element of informed consent, a cardinal rule for all medical practitioners, and the freedom of choice given to individuals whether they wished to be treated or not, the idea of safety as an inhibiting factor in genetic technologies, enhancement or therapy, may find itself dwindling in support. The presence of risk in any element and aspect of human life, living or simply, existing, is unavoidable; what is within the confines of control, however, is the decisions made by persons themselves. We must not discount the possibility of human rationality in these circumstances, and the desire for survival is often unsurpassed by the desire to engage in the recklessness of behaviour that will put human life in jeopardy (in this limited context).

The next point of controversy is one that attempts to balance the given of nature, against the distributive aspects of justice and fairness in society. Scholars like Sandel temper the balance by beginning discussions about the "genetic lottery",[268] and Cohen provides a marker at the starting point of the "luck egalitarian" concept,[269] both of which are compelling arguments in originating and bringing into existence the cogitation of equality and distributive justice, examined both at the level of societal hierarchy as well as the provision of availability and accessibility of these genetic technologies. The deliberation of human nature is an inescapable matter; the cardinal recognition is that human beings are at the mercy of a genetic lottery. Absent genetic editing technologies, what is needed to be confronted is the nature of this genetic lottery: there is a wealth of religious perspectives, such as in Sunni Islam, which believe that what nature or *Allah* has given, must be unaltered. The *hadiths* of Muhammad in the majority Sunni Islam world gave credence to these proclamations, and although the original wordings of the Surah through the pontification of translations and numerous interpretative exercises may have undergone differing points of views, it has been accepted that the Surah[270] by various religious interpretations, also prohibits other forms of permanent physical modification.

However, the stance taken here is that religious views of this nature must be capable of being adaptable with changing times and the evolutionary process of human advancement, through means and methods that improve the quality of life. The moral status of nature, and what it means to be human, and how human beings choose to exist in the world, is no longer as simple as it used to be. The natural world, as we used to know it, is now a breeding ground for 'artificiality', ways of living that were never present before, and this does not necessarily constitute aspects of 'good' or 'bad', but simply a change in the way of life. With this realization, Sandel raises thought-provoking questions that mostly pinpoint to our sense of ethics in understanding our own human nature and existence.[271] Sandel further points to the non-secular origins of our concept of human nature, identifying that the sense

---

[268] Sandel (2004), p. 51.

[269] Cohen (2014), p. 658.

[270] Surah An-Nisaa 4: 117-120 of the Quran, which has been interpreted to religiously prohibit any attempt to change the creation of God (Allah).

[271] Sandel (2004), p. 58.

of human giftedness, is one bestowed by an omniscient and omnipotent God; and from this religious standpoint, any form of interference by human beings into this giftedness, is to play to the confusion and misunderstanding of the roles we embody in our existence.[272]

It is, therefore, an understanding of this crucial beginning, which infinitely connects us to the genetic lottery beyond our corporeal control. Because of this random distribution of talent, a disturbing need for mastery and control of the human function, become the masters that rule the day, and the availability and access to genetic enhancement technologies undermine what many, including Sandel believe to be the sum total of human effort,[273] progressing to a point which even criticizes athletes for "playing naked".[274]

It is also possible that questions of taxation and subsidization of enhancement technological advances by the state will come into fruition. In the same manner that the Patient Protection and Affordable Care Act received tremendous criticism in the United States (which later gained the moniker 'Obamacare' initially coined by the Republican contenders to the legislation) because of the individual mandates upon Americans to obtain a compulsory, minimum level of health insurance,[275] (notwithstanding any form of subsidies levied by the United States government on obtaining the health insurance) the enlargement of taxation obligations and subsidy benefits for enhancement technologies within the scope of the "separate spheres", is likely to provoke similar sentiments of dissatisfaction. More specifically, a large section of taxpayers, whose contributions to the national economy of the State undoubtedly strengthen the coffers of the national treasury, are likely to react in a dissident manner, in the event subsidies for enhancement technologies, allocated to a section of society based on the "separation spheres", are not similarly available to them because of the calculation or measurement scope of the level of societal 'deficit' determination.

These are but the several factors that aim to contribute to and shape the framework of regulating not only PGD, but also biomedical technologies. These are conversations that should be continued alongside the existence of institutional and infrastructural facilities that assist in framing the regulatory dimension in some jurisdictions.

---

[272] Ibid 61.

[273] Sandel (2004).

[274] Ibid 61. "Playing naked" is a term used to describe situations, for example, where athletes, especially professional ones, take to the playing field in a particular sport, free of amphetamines, steroids, or other stimulants that are capable of enhancing their performance. The concerns, therefore, about "playing naked" become an obstinate remedy to excel, to fully explore one's talents; and perhaps even lead to the second type of coercion mentioned by Chatterjee.

[275] Jensen (2012), p. 31.

3.5 Reprogenetics As an Influence in Regulation Shaping

The predominant debates behind the legislative or regulatory influences in the selected jurisdictions provide us with useful context in viewing how regulation for PGD is shaped. This signals to us at least one telling common factor between all the jurisdictions: the manner in which reproduction and birthing is viewed, and the level of importance that it bears on societal and familial structures. In the introduction to this chapter, I hint at reprogenetics as an influence in the field of ARTs, and the bid to select the best possible embryos for implantation. The development of ARTs and the manner in which future offspring may come to birth *vis-à-vis* IVF treatments necessitates a more detailed conversation on the effects of reprogenetics and how it influences the shaping of regulation or legislation. In the same manner that the law has responded to the evolution of ARTs and the instrumental components that characterize the services and practices of ARTs, the law, too, must evolve with the tides of reprogenetics that have come to dramatize the landscape of reproduction and reproductive technologies.[276]

The age of reprogenetics that has now come to characterize our reproductive landscape additionally means that regulators must also appropriately frame reasons for governing these technologies. The choices that are conferred on potential parents, and particularly where these choices border on the less 'socially acceptable' reasons for embryo selection, means that the existence of a tussle between parental autonomy and governmental intervention is inevitable. In Chap. 2, I propose that a re-interpretation of parental autonomy is necessary, particularly within the sphere of liberal eugenics.[277] On this basis, a similar vein of interpretation of parental autonomy, in my view, is also necessary within the context of embryo selection. There is already a substantial library of literature on reproductive decision-making and embryo selection, and some key selections of scholarly work has indicated a leaning towards allowing parents to make decisions relating to embryo selection.[278] My claim in this context lies upon a balancing act of purposes in embryo selection, differentiating between the 'want' and the 'need', the therapeutic versus the non-therapeutic within the sphere of reprogenetics. There is no longer a justifiable reason for us to plausibly deny the existence of reprogenetics and its leverage on fertility practices and its legislative forms.

The reprogenetics phenomena can be easily categorized as part of the liberal eugenics movement[279] because its veritable outcome entangles the concept of selection; in these instances, embryo selection. The premise that I make here is not to object to the benefits that may be offered by therapeutic genetic interventions, considered from the viewpoint of the Rawlsian idea of the general goods of society.[280]

---

[276] Silver (2000), p. 375.

[277] Lau (2018).

[278] Gavaghan (2007), p. 168.

[279] Agar (1998), p. 137.

[280] Jaede (2017), p. 18.

My objection would lie against a free-for-all, wholesale liberty of choices, for whatever reasons that may be desired, therapeutic or non-therapeutic, that is completely unfettered, unchallenged, and unquestioned. In this regard, the supposition that is made here is that there are boundaries of permissibility that must be negotiated on a national level at the very least, to keep the embryo selection phenomenon in check. The existence of current legislation (in some of the selected jurisdictions) that governs practices and services relating to PGD and PGS, for example, is an illustration of this very fact: the multi-faceted dimensions in reproduction and its changing social paradigms will always necessitate continued debates.[281]

Within the historical annals of ARTs, IVF was once considered a non-therapeutic, non-essential treatment in reproduction, until the infertility phenomenon became recognized as a medical condition that warranted medical intervention in a hope for couples to procreate and found a family.[282] This is not to say, however, that the various appurtenances in IVFs have also found concrete clarity and certainty; the shifts in medical and scientific developments for interventions into the human body via a host of reproductive technologies continue to impact upon the governance of bodies in the political and legislative sphere.[283] Indeed, Lori Andrews and Nanette Elster further identify some of the problematic concerns that are still inconsistent in the discourse on reproductive technologies governance in the United States.[284]

Taking into account the predominant debates that have surfaced, it is prudent to impart grounds of reasonableness to the embryo selection phenomenon, and to uphold values of non-commodification and the protection from vulnerability and safety.[285] The legislative extensions from existing rules on PGD, I claim, is the most appropriate form and function by which embryo selection may be ethically governed. In these circumstances, it would be possible for existing PGD legislation to provide more clarity to dedicated provisions for embryo selection, taking on the recognition that the discourse must move beyond the viewing of women's bodies as foetal carriers and the "theoretical site in reproductive genetics".[286]

It is precisely the fact of the change in social and cultural factors that may, in some jurisdictions, afford women primarily, and parents specifically, the progression of choice over their own bodies and reproduction, that the positioning of the law cannot be that of a passive bystander, but an active participant in the process to enhance and support procreative liberties whilst managing to invest in the important associations between ethics and legal legitimacy.

Where clarity cannot be provided *vis-à-vis* legislative pronouncement because of a gap in the law, and recourse is had to the interpretive functions of the court systems, the possible perils that may be faced is a disconnect between the judicial understanding and reasoning of the aims of reproductive technologies and by that

---

[281] Andrews, and Elster (2000), p. 45.

[282] Ibid 40.

[283] Foucault (1977).

[284] Andrews and Elster (2000), p. 48.

[285] Ibid 45.

[286] Ettorre (2000), p. 403.

extension, embryo selection; and the practical realities of the burdens placed on women's bodies, the psychological elements of making choices, and the technical and clinical application of these technologies.

References

Agar N (1998) Liberal eugenics. Public Aff Q 12:137

Ahmad N, Lilienthal G, Hussain M (2016) Law of assisted reproductive surrogacy in Malaysia: a critical overview. Commonw Law Bull 42:355

American Congress of Obstetricians and Gynecologists, Committee on Ethics (2007) ACOG Committee Opinion No. 360: sex selection. Obstet Gynecol 109:475–478

American Society for Reproductive Medicine (2015) Assisted reproductive technology- a guide for patients. http://www.fertilityanswers.com/wp-content/uploads/2016/04/assisted-reproductive-technologies-booklet.pdf

American Society for Reproductive Medicine, Ethics Committee (2013) Use of preimplantation genetic diagnosis for serious adult onset conditions: a committee opinion. Fertil Steril 100:54

American Society for Reproductive Medicine, Ethics Committee (2015) Use of reproductive technology for sex selection for nonmedical reasons. Fertil Steril 103:1418

Andrews LB, Elster N (2000) Regulating reproductive technologies. J Legal Med 21:35

Assisted Reproductive Technology Act [NSW] 2007 (No 69) 43

Assisted Reproductive Technology Amendment Bill [NSW] 2016 17

Assisted Reproductive Treatment Act [SA] 1988 14

Assisted Reproductive Treatment Act [VIC] 2008 (No 76)

Assisted Reproductive Treatment Regulations [SA] 2010 4

Australian Associated Press (20 February 2015) Thailand bans commercial surrogacy | World News, The Guardian. https://www.theguardian.com/world/2015/feb/20/thailand-bans-commercial-surrogacy

Bayefsky M (2015) The regulatory gap for preimplantation genetic diagnosis. Hastings Center Rep 45:7

Bayefsky M (2016) Comparative preimplantation genetic diagnosis policy in Europe and the USA and its implications for reproductive tourism. Reprod Biomed Soc Online 3:41

Bayefsky M, Jennings B (2015) Regulating preimplantation genetic diagnosis in the United States: the limits of unlimited selection. Springer, Berlin

Belluck P (4 August 2017) Gene editing for "Designer Babies"? Highly unlikely, scientists say. The New York Times. https://www.nytimes.com/2017/08/04/science/gene-editing-embryos-designer-babies.html

Binsaleh S, Lo KC (2007) Varicocelectomy: microsurgical inguinal varicocelectomy is the treatment of choice. Can Urol Assoc J 1:277

Botkin JR (1998) Ethical issues and practical problems in preimplantation genetic diagnosis. J Law Med Ethics 26:17

Boyle RJ, Savulescu J (2001) Ethics of using preimplantation genetic diagnosis to select a stem cell donor for an existing person. Br Med J 323:1240

Brezina PR, Zhao Y (2012) The ethical, legal, and social issues impacted by modern assisted reproductive technologies. Obstet Gynecol Int 2012:1–7. https://www.hindawi.com/journals/ogi/2012/686253/

Brezina PR, Ke RW, Kutteh WH (2013) Preimplantation genetic screening: a practical guide. Clin Med Insights Reprod Health 7:37

Buchitchon S (2016) The Protection of Children Born from Assisted Reproductive Technology Act 2015: scientific advances, ethics and concerns over the use of human embryo. Adv Sci Lett 22:1610

Buck v. Bell, 274 U.S. 200 (1927) (*Justia Law*) https://supreme.justia.com/cases/federal/us/274/200/case.html

Caamano JM (2016) International, commercial, gestational surrogacy through the eyes of children born to surrogates in Thailand: a cry for legal attention. Boston Univ Law Rev 96:37

Callahan S (2009) The ethical challenges of the new reproductive technologies. In: Morrison EE (ed) Health care ethics: critical issues for the 21st century, 2nd edn. Jones and Bartlett Publishers, Burlington

Chatterjee A (2004) Cosmetic neurology the controversy over enhancing movement, mentation, and mood. Neurology 63:968

Chial H (2008) Rare genetic disorders: learning about genetic disease through gene mapping, SNPs, and microarray data. Nat Educ 1:192

Cohen IG (2014) What (If Anything) is wrong with human enhancement? What (If Anything) is right with it? Tulsa Law Rev 49:645

Cordelia T (2004) Pre-implantation testing and the protection of the savior sibling. Deakin Law Rev 5:121

Crockin SL (2005) Reproduction, genetics and the law. Reprod BioMed Online 10:692

Deonandan R (2015) Recent trends in reproductive tourism and international surrogacy: ethical considerations and challenges for policy. Risk Manag Healthcare Policy 8:111

Department of Health, UK (9 November 2007) Human Fertilisation and Embryology Act 1990 - an Illustrative Text. http://webarchive.nationalarchives.gov.uk/content/20130107105354/http://www.dh.gov.uk/en/Publicationsandstatistics/Publications/PublicationsLegislation/DH_080205

Dudgeon MR, Inhorn MC (2004) Men's influences on women's reproductive health: medical anthropological perspectives. Soc Sci Med 59:1379

Ettorre E (2000) Reproductive genetics, gender and the body: "Please Doctor, May I Have a Normal Baby?". Sociology 34:403

Etuk SJ (2009) Reproductive health: global infertility trend. Niger J Physiol Sci 24:85

Farrell P (20 January 2015) Baby Gammy, Born into Thai Surrogacy Scandal, Granted Australian Citizenship | Australia News, The Guardian. https://www.theguardian.com/australia-news/2015/jan/20/baby-gammy-born-into-thai-surrogacy-scandal-granted-australian-citizenship

Feikert C (30 April 2012) Sex selection & abortion: Australia. Library of Congress. https://www.loc.gov/law/help/sex-selection/australia.php

Foucault M (1963) Naissance de La Clinique Une Archéologie Du Regard Médical. Presses Universitaires de France, Paris

Foucault M (1976) The history of sexuality Volume I: an introduction. Pantheon Books, New York

Foucault M (1977) Discipline and punish: the birth of the prison. Vintage Books, Random House, New York

Gavaghan C (2007) Defending the genetic supermarket: the law and ethics of selecting the next generation. Routledge-Cavendish, Abingdon

Gleicher N, Orvieto R (2017) Is the hypothesis of preimplantation genetic screening (PGS) still supportable? A review. J Ovarian Res 10:21

Gleicher N, Kushnir VA, Barad DH (2014) Preimplantation Genetic Screening (PGS) still in search of a clinical application: a systematic review. Reprod Biol Endocrinol 12:22

Gleicher N et al (2016) Accuracy of Preimplantation Genetic Screening (PGS) is compromised by degree of mosaicism of human embryos. Reprod Biol Endocrinol 14:54

Goslinga-Roy GM (2000) Body boundaries, fiction of the female self: an ethnographic perspective on power, feminism, and the reproductive technologies. Fem Stud 26:113

Gozzetti A, Le Beau MM (2000) Fluorescence in situ hybridization: uses and limitations. Semin Hematol 37:320

Gregg AR (2016) Noninvasive prenatal screening for fetal aneuploidy, 2016 update: a position statement of the American College of Medical Genetics and Genomics. Genet Med 18:10

Gribble J, Bremner J (26 November 2012) The challenge of attaining the demographic dividend. Population Reference Bureau (PRB). https://www.prb.org/demographic-dividend/

Griswold v. Connecticut, 381 U.S. 479 (1965) (*Justia Law*) https://supreme.justia.com/cases/federal/us/381/479/case.html

Haraway DJ (2000) A Cyborg Manifesto: science, technology, and socialist-feminism in the late twentieth century. In: Badmington N (ed) Posthumanism. Macmillan Education, London

Harryono M et al (2006) Thailand medical tourism cluster. Harvard Business School Microeconomics of Competitiveness

Heydarian RJ (11 May 2018) A peaceful revolution in Malaysia. Al-Jazeera. https://www.aljazeera.com/indepth/opinion/peaceful-revolution-malaysia-180511140532987.html

Hsu PD, Lander ES, Zhang F (2014) Development and applications of CRISPR-Cas9 for genome engineering. Cell 157:1262

Hudson KL (2006) Preimplantation genetic diagnosis: public policy and public attitudes. Fertil Steril 85:1638

Hughes JJ (2010) Humans should be free of all biological limitations including sex. Am J Bioethics 10:15

Human Fertilisation and Embryology (Mitochondrial Donation) Regulations 2015

Human Fertilisation and Embryology Act 1990 (Chapter 37) 48

Human Fertilisation and Embryology Act 2008 (Chapter 22) 120

Human Fertilisation and Embryology Authority S and ID, 'The HFE Act (and Other Legislation) - HFEA' http://hfeaarchive.uksouth.cloudapp.azure.com/www.hfea.gov.uk/134.html. Accessed 15 Jan 2018

Human Reproductive Technology Act [WA] 1991

Human Tissues Act 1974 (Act 130) 1974 (Act 130) 9

Hunt L (15 November 2018) Thailand's politics heat up ahead of elusive election. The Diplomat. https://thediplomat.com/2018/11/thailands-politics-heat-up-ahead-of-elusive-election/

ICPD, 'Policy Recommendations for the ICPD Beyond 2014: Sexual and Reproductive Health & Rights for All' http://icpdtaskforce.org/resources/policy-recommendations-for-the-ICPD-beyond-2014.pdf

Illustrative Text: Human Embryology and Fertilisation Authority Act 1990. As Amended: An Illustrative Text

Inhorn MC (2007) Masculinity, reproduction, and male infertility surgery in the Middle East. J Middle East Women's Stud 3:1

Inhorn MC (2009) Right to assisted reproductive technology: overcoming infertility in low-resource countries. Int J Gynecol Obstet 106:172

Inhorn MC (2013) Why Me? Male infertility and responsibility in the Middle East. Men Masculinities 16:49

Inhorn MC, Birenbaum-Carmeli D (2008) Assisted reproductive technologies and culture change. Ann Rev Anthropol 3:177

Inhorn MC, Patrizio P (2015) Infertility around the globe: new thinking on gender, reproductive technologies and global movements in the 21st century. Hum Reprod Update 21:411

Jaede M (2017) The concept of the common good. Working Paper, University of Edinburgh, p 18

Jensen EM (2012) The individual mandate, taxation, and the constitution. J Tax Invest 30:31

Juslaws & Consult, 'Unofficial Translation- Act Providing for the Protection for Children Born Through Assisted Reproductive Technologies'

Kamel RMA (2013) Assisted reproductive technology after the birth of Louise Brown. Gynecol Obstet 3:156. https://www.omicsonline.org/assisted-reproductive-technology-after-the-birth-of-louise-brown-2161-0932.1000156.php?aid=16043

Kass L (1997) The wisdom of repugnance. The New Republic 17

Knowles LP, Kaebnick GE (eds) (2007) Reprogenetics: law, policy and ethical issues. John Hopkins University Press, Baltimore

Lau PL (2018) The Genius & The Imbecile: disentangling the "Legal" framework of autonomy in modern liberal eugenics, from non-therapeutic gene enhancement use in gene editing technologies. In: Current debates in international relations and law, vol 4. IJOPEC, London

Liu CK (2007) "Saviour Siblings"? The distinction between PGD with HLA tissue typing and pre-implantation HLA tissue typing: winner of the Max Charlesworth Prize Essay 2006. J Bioeth Inq 4:65

Ly KD, Agarwal A, Nagy ZP (2011) Preimplantation genetic screening: does it help or hinder IVF treatment and what is the role of the embryo? J Assist Reprod Genet 28:833

Malaysian Medical Council (14 November 2006) Guideline of the Malaysian Medical Council MMC Guideline 003/2006 assisted reproduction

Metzl JF (10 October 2014) The genetics epidemic. Foreign Affairs. https://www.foreignaffairs.com/articles/united-states/2014-10-10/genetics-epidemic. Accessed 14 Jan 2017

Ministry of Health Malaysia, Medical Development Division (October 2012) Standards for assisted reproductive technology facility - embryology laboratory and operation theatre

Munné S et al (1994) Chromosome Mosaicism in human embryos. Biol Reprod 51:373

NaRanong A, NaRanong V (2011) The effects of medical tourism: Thailand's experience. Bull World Health Org 89:336

National Health and Medical Research Council (2017) Ethical guidelines on the use of assisted reproductive technology in clinical practice and research. Australian Government NHMRC, Canberra

Natipodhi P (2014) Practice of sex selection in Asian Region. Working Paper Series Asian Law Institute

Norwitz ER, Levy B (2013) Noninvasive prenatal testing: the future is now. Rev Obstet Gynecol 6:48

Pardes A (14 January 2016) How commercial surrogacy became a massive international business. Vice. https://www.vice.com/en_us/article/how-commercial-surrogacy-became-a-massive-international-business

Parliament of Malaysia, 'Official Portal of The Parliament of Malaysia' http://www.parlimen.gov.my/index.php?lang=en

Planned Parenthood of Southeastern Pa. v. Casey, 505 U.S. 833 (1992) (Justia Law) https://supreme.justia.com/cases/federal/us/505/833/

Prohibition of Human Cloning for Reproduction Act 2002 2016 (No 144, 2002)

Protection for Children Born Through Assisted Reproductive Technologies Act 2015 (167/2553)

Research Involving Human Embryos Act 2002 2016 (No 145, 2002)

Rhode H (15 June 2012) Can Muslims reopen the gates of Ijtihad? Gatestone Institute. http://www.gatestoneinstitute.org/3114/muslims-ijtihad

Riggan K (31 December 2009) G12 country regulations of assisted reproductive technologies. The Centre for Bioethics and Human Dignity. https://cbhd.org/content/g12-country-regulations-assisted-reproductive-technologies

Robertson JA (2003) Extending preimplantation genetic diagnosis: the ethical debate: ethical issues in new uses of preimplantation genetic diagnosis. Hum Reprod 18:465

Roe v. Wade, 410 U.S. 113 (1973) (Justia Law) https://supreme.justia.com/cases/federal/us/410/113/

Sandberg R (2016) The failure of legal pluralism. Eccles Law J 18:137

Sandel M (2004) The case against perfection. Atlantic Monthly 293:51

Sándor J (2015) The ethical and legal analysis of embryo preimplantation testing policies in Europe. In: Sills ES (ed) Screening the single euploid embryo. Springer International Publishing, New York

Sheehan M (2009) Making sense of the immorality of Unnaturalness. Camb Q Healthc Ethics 18:177

Sholley JB (1951) Constitution of the United States of America, Cases on Constitutional Law. Bobbs-Merrill

Silver LM (2000) Reprogenetics: third millennium speculation: the consequences for humanity when reproductive biology and genetics are combined. EMBO Rep 1:375

Singer P (2016) The human genome and the genetic supermarket. In: Ethics in the real world: 82 brief essays on things that matter. Princeton University Press, Princeton

Skinner v. Oklahoma Ex Rel. Williamson, 316 U.S. 535 (1942) (*Justia Law*) https://supreme.justia.com/cases/federal/us/316/535/case.html

Smith MK (2013) The Human Fertilisation and Embryology Act 2008: restrictions on the creation of "Saviour Siblings" and the relevance of the harm principle. New Genet Soc 32:154

Smith J (3 August 2014) Mother Country: the harrowing truth behind Thai "fertility Tourism". The Independent. http://www.independent.co.uk/voices/comment/mother-country-the-harrowing-truth-behind-thai-fertility-tourism-9644517.html

Stasi A (2015) Maternal surrogacy and reproductive tourism in Thailand: a call for legal enforcement. Ubon Ratchathani Law J 8:17–36

Strategy and Information Directorate Human Fertilisation and Embryology Authority, 'The HFE Act (and Other Legislation) - HFEA' http://hfeaarchive.uksouth.cloudapp.azure.com/www.hfea.gov.uk/134.html

Supiot A (2007) Homo Juridicus: on the anthropological function of the law. Verso, Brooklyn

Taranto S (22 January 2018) How abortion became the single most important litmus test in American Politics. The Washington Post. https://www.washingtonpost.com/news/made-by-history/wp/2018/01/22/how-abortion-became-the-single-most-important-litmus-test-in-american-politics/?noredirect=on&utm_term=.b4015648272b

Thai Law Forum, 'Thailand Medical Council Regulations on Surrogacy and IVF | Thailand Law Forum' http://www.thailawforum.com/medical-surrogacy-regulations/

The Nation (28 November 2014) Thailand Portal 'Commercial Surrogacy Bill Passes First Reading with 177 to 2 Votes' http://www.nationmultimedia.com/national/Commercial-surrogacy-bill-passes-first-reading-wit-30248734.html

The Star, 'Pakatan Harapan 100 Days' https://www.thestar.com.my/ph100/

Thompson C (2016) IVF global histories, USA: between rock and a marketplace. Reprod Biomed Soc Online 2:128

TMC Fertility Centre (21 February 2013) 'Home' (TMC Fertility Centre) http://www.tmcfertility.com/

TMC Fertility Centre (23 March 2015) 'Pre-Implantation Genetic Diagnosis (PGD)' (TMC Fertility Centre) http://www.tmcfertility.com/treatment-options/pre-implantation-genetic-diagnosis-pgd/

Turriziani JV (2014) Designer babies: the need for regulation on the quest for perfection. Law School Student Scholarship 595. http://scholarship.shu.edu/cgi/viewcontent.cgi?article=1595&context=student_scholarship

United Nations, Development Policy and Analysis Division, Department of Economic and Social Affairs, 'Country Classification- Data Sources, Country Classification and Aggregation Methodology' (United Nations Secretariat) http://www.un.org/en/development/desa/policy/wesp/wesp_current/2014wesp_country_classification.pdf

US National Library of Medicine, Genetics Home Reference, 'What Are the Different Ways in Which a Genetic Condition Can Be Inherited?' (Genetics Home Reference) https://ghr.nlm.nih.gov/primer/inheritance/inheritancepatterns

Weltgesundheitsorganisation (ed) (2005) Sexually transmitted and other reproductive tract infections: a guide to essential practice

Whittaker A, Speier A (2010) "Cycling Overseas": care, commodification, and stratification in cross-border reproductive travel. Med Anthropol 29:363

Winslow ER, Kodner IJ (2004) Ethics and genetic testing. Semin Colon Rectal Surg 15:186

World Health Organization, 'WHO | Infertility Is a Global Public Health Issue' (WHO) http://www.who.int/reproductivehealth/topics/infertility/perspective/en/

Yuen M (29 November 2015) Act to ensure country has regulations on artificial reproduction – Nation. The Star. https://www.thestar.com.my/news/nation/2015/11/29/birth-of-a-new-law-soon-act-to-ensure-country-has-regulations-on-artificial-reproduction/

Zakhiri M et al (2016) Legal position of Fatwa: observations from selected jurisdictions. Seminar on Law and Society (SOLAS 2016), School of Law, Universiti Utara Malaysia

Chapter 4
The Regulatory Framework in Biomedical Technologies

Abstract This deals with regulatory frameworks for biomedical technologies generally, and pre-implantation genetic interventions more specifically. In doing so, the chapter highlights some of the challenges in attempting to effectively regulate biomedical technologies that move at the pace of lightning speed. In addition, the chapter proposes the use of a combination of regulatory approaches, in complementarity with existing legal frameworks, to consider building a more flexible and reflexive form of governance for biomedical technologies. Further clarity may also be had in examining the development of the regulatory frameworks in the abortion debates and earlier pre-natal testing technologies. These are discourses that bear a close nexus to pre-implantation genetic interventions and may impart the values and modalities in these debates to complement the role of the law or legal framework in a regulatory environment.

This chapter reemphasizes the technological advances in reproductive technologies and Pre-Implantation Genetic Diagnosis (PGD) highlighted in Chap. 3, and advances the necessity for some kind of legislative intervention by the state (in this context, referring to either federalized systems of government whether at the state or federal level, as well as centralized systems of governments); or, at the very least, some form of quasi-legal, regulatory or governance mechanisms for these particularized PGD technologies. There is no doubt why these technologies should be regulated, and why we should be concerned with the legality or the "lawfulness" of such regulatory mechanisms. Section 4.1 herein intends to shed enlightenment on reasons for regulation. In this connection, the far reaches of technological advances, especially insofar as medical technologies is concerned, draws comparisons to, and pushes the boundaries of what we used to believe was humanly feasible. This section also explores the foundations of regulatory theory, and whether it may operate effectively in regulating biomedical technologies like PGD. These warrant some basic discussions on self-governance regulatory mechanisms and the symbolic legislation theory; and what will be apparent is that some of the selected jurisdictions in this book, do indeed, employ a mixed variable of regulatory approaches.

The transitional aspects of the chapter then move on to the more concerning issue on "laws" and its reflexive role in regulating new technologies in Sect. 4.2. I posit that the 'regulability' of new and emerging technologies requires a reflection on

© Springer Nature Switzerland AG 2019   123
P. L. Lau, *Comparative Legal Frameworks for Pre-Implantation Embryonic Genetic Interventions*, https://doi.org/10.1007/978-3-030-22308-3_4

whether the law alone is an appropriate tool by which to regulate these technologies. The discourse may appear to be of an ontological nature at this point, but the specificities of how new technologies have come into existence, and subsequently, used, necessitate the consideration of how technologies may be shaping (at the very beginning), a self-regulatory framework; or whether this may perhaps be occurring on a reverse basis. In the larger scheme of things, what must also be considered is whether we should regulate these new technologies on a parallel basis with fresh and new scientific advances and discoveries; or whether we should look at the main focus beyond transnational practices that emphasize specific concerns in biotechnologies and determine applicable regulatory necessities at that point of convergence. Following this examination, a reflection on the challenges of regulating biomedical technologies is also emphasized. These challenges are numerous in nature, but this chapter intends to focus on primary issues that concern the scope of such regulation, the legitimacy, effectiveness and enforceability of such regulation, and how this is able to operate in constitutions that are either cosmopolitan, or legally pluralistic, in nature. I attempt to identify and fill in the regulatory gaps, and shed light on these legal challenges that need to be addressed.

Upon a determination on the relationship between the regulation/governance of new technologies, and the main issues posed in the domain of pre-implantation genetic interventions, Sect. 4.3 broadens the scope of the range of regulatory models that are available for consideration in this constantly evolving landscape of scientific technologies. In this regard, the recognition put forward is that the regulation of technologies must be duly accompanied by accountability, which can be obtained through an acceptance of the universality of this discourse and an effective enforceability mechanism that affirms the claim that technologies have particularly shaped the sphere of biotechnology. In postulating a well-rounded framework of regulation, Sect. 4.3 looks to other examples in the biomedical field, such as the abortion debates, and the early prenatal testing technologies. These almost 'neighbouring' technologies to genetic interventions may serve as useful analogies to formulate a greater understanding of regulating genetic interventions. It is possible to glean from these 'neighbouring' discussions the following matters: one, that the normative desideratas of regulatory design through the consideration of values are important in that framework. Two, it is useful to identify a more flexible system of governance that takes into account a wider range of stakeholders and interested parties. These measures therefore may lend support to how the transition of a previously unregulated sphere of technology may now be adjudged. I will demonstrate that these areas are more closely related to the domain of this research than may have been apparent, because of the shared underlying system of values that they each seek to embody and protect within the framework. These models of governance can be useful indicators to the future regulation of biomedical technologies.

Finally, in Sect. 4.4, I demonstrate that the underlying purpose for regulatory governance is equivalent to state control, but this should not be viewed from a negative aspect. Instead, the positivist aspect of laws through state intervention via legislation is necessary, not simply on its own, but worked in conjunction with other forms of regulation. In particular, I combine the foundational components of regulatory theories with a proposition for structural models for biomedical regulation to be operable on a more practical basis. In beginning to recognize the need for constitutional scrutiny as is the case in any form of legislative process and inquiry, this concluding section to the chapter provides a pre-cursor into the following Chaps. 5 and 6 that delve more deeply, respectively, into the role played by international biomedical laws, and the dynamics of constitutional rights in the selected jurisdictions.

4.1 Regulatory Theory in the Field of Biomedical Technologies

4.1.1 Why Should Biomedical Technologies Be Regulated?

Almost 50 years ago, when the wonders of the Internet and the simultaneous ease and disfavor of the modern technological age had yet to penetrate the beginning of reconstruction of the New Deal,[1] and the adapting lives of humankind, Sir Arthur C. Clarke[2] penned a series of witty essays[3] that attempted to glimpse into a future that was not yet present. Clarke's uncanny revelatory glimpses into technological rapture, a phenomenal future of humankind, could have been easily mistaken, in the late 1960s period, as a phantasmagoria of fiction. Yet, the survival of Clarke's well-known formulated adage called 'Clarke's Three Laws'[4] remains a philosophical citation in popular culture today. Clarke's relevant third 'law' states, "any sufficiently advanced technology is indistinguishable from magic."[5] This statement should not be mistaken as the ramblings of a futurist, and instead, sheds light on how ignorance, misinformation and a resistance to change, would summarily dismiss possible advances of technologies, as conjectural 'magic'.

It is along this vein that I attempt to now elucidate briefly how the conceptualization of 'magic' found its place in the historical annals of technological advancement. On a much more impassioned level, Clarke's differentia of 'magic' in his third law can possibly be seen as the increasingly transgressed trajectory of Natural Laws.[6] A logical apprehension of the erosion of natural laws is a well-founded basis that a hedonistic digitalized environment would have the effect of displacing our fundamental and basic Rule of Law, in favour of the Rule of Technology.[7] Hence,

---

[1] History 'The 1960s - Facts & Summary' (*history.com*) http://www.history.com/topics/1960s.

[2] Anthony J. Pennings, 'Arthur C. Clarke's Three Laws of Innovation' *Writings on Digital Strategies, ICT Economies, and Global Communications* (1 July 2012), http://apennings.com/political-economies-in-sf/arthur-c-clarkes-three-laws-of-innovation/.

[3] The Guardian, 'Profiles of the Future by Arthur C Clarke – Review' *The Guardian* (4 March 2011) https://www.theguardian.com/science/2011/mar/04/profiles-future-arthur-clarke-review.

[4] Clarke (1985).

[5] ibid.

[6] Finnis (2015), p. 199.

[7] Brownsword and Yeung (2008), p. 5.

the metaphorical providence of 'magic' in Clarke's third law, drifts well beyond a superficial phraseology, compelling us to extricate an understanding of the breadth and depth of the things we cannot understand, or do not yet have the capacity to understand.

Clarke has meaningfully provided how the lack of a full and complete under-standing of the workings of technologies, would lead a person to surmise on its impossibilities. Indeed, the attribution to Clarke's third law has permeated popular culture to such a great extent, that science fiction enthusiasts have described how his third law helps to create the underpinnings of science fiction as we view it today: "it strikes at the heart of what technology is: a way for humans to do thing previously believed not just implausible, but impossible."[8]

This is not to postulate that a made-up kind of platitude that transmogrifies itself, as 'law', should be perilously treated as the gospel truth. Instead, the presentation of the wonders of new and emerging technologies is the rudimentary basis for embark-ing on a discourse about its "regulability".[9] Lawrence Lessig's formidable work, *The Code*, on a "newly salient regulator"[10] that occupies the space of governance in connection with cyberspace, for example, is an excellent source of reference. Although the frontiers of cyberspace may very well be different than that of new and emerging technologies envisaged in this book, the commonality lies in the fact that these technologies are the new global rhetoric that necessitates powerful consider-ation and debates. This is because, as Brownsword correctly identifies, "the slide towards the technologically enhanced state creates a new risk of total control."[11] It is also this very same reason that perhaps drove towards a crucial inter-disciplinary conference on 'Regulating Technologies' held in London in 2007.[12] Despite a col-lective voice in the technology discourse domain about our inevitable march towards a dystopian, Orwellian-type future,[13] I posit that we should, instead, be calculably cognizant of the affectations of these technologies in the present and now. There is no doubt that at the time of writing this, we have entered a renewed Golden Age,[14] with advances in medicine, biotechnology, neuroscience, nanotechnology, genetics and a myriad of scientific fields lauding gloriously over the Grecian ages.

---

[8] Esther Inglis-Arkell, 'Technology Isn't Magic: Why Clarke's Third Law Always Bugged Me' *io9* (28 April 2013) http://io9.gizmodo.com/technology-isnt-magic-why-clarkes-third-law-always-bug-479194151.

[9] Lessig (1999), p. 43.

[10] ibid 6.

[11] Brownsword and Yeung (2008), p. 5.

[12] ibid 3.

[13] Jonathan Freedland, '1984 by George Orwell, Book of a Lifetime: An Absorbing, Deeply Affecting Political Thriller' *The Independent* (2 July 2015) http://www.independent.co.uk/arts-entertainment/books/reviews/1984-by-george-orwell-book-of-a-lifetime-an-absorbing-deeply-affecting-political-thriller-10360789.html.

[14] Johan Norberg, 'Why Can't We See That We're Living in a Golden Age?' *The Spectator* (20 August 2016) https://www.spectator.co.uk/2016/08/why-cant-we-see-that-were-living-in-a-golden-age/.

At the same time, our gravitational ease into this technological age is also one that breeds ambivalence; that, as the world connects humankind through a greater reach, it also has the capacity to cause divisiveness and inequalities. This chapter does not aim to wander into the benefits and disadvantages of living in a technological Utopian fantasy in the manner that we currently do; but instead, puts forward how we may still enjoy the fruits of technological labour within a regulatory repository. For this to manifest itself, Brownsword cites Julia Black's definitional significance of the term 'regulation'[15] as follows: "the sustained and focused attempt to alter the behaviour of others according to standards or goals with the intention of producing a broadly identified outcome or outcomes, which may involve mechanisms of standards-setting, information-gathering, and behaviour modification."[16] Therefore, for avoidance of doubt, a kind of regulatory framework mentioned in this chapter refers to a range of modalities that may comprise forms of 'regulation', including the law (legislative pronouncements), 'soft law' in the form of guidelines in the specific context of medical and health technologies, international covenants, agreements and treaties, and even differing governance models employed on the basis of shared values or principles in achieving particular objectives; amongst others. Hence, 'regulation' in this space is promulgated on a broad level as the means by which to effectively and safely govern existing, new and emerging technologies, not necessarily solely through government or state interventions, but in concert with other appropriate regulatory or professional bodies.

The most cerebral answer to the pre-emptive question on why biomedical technologies should be regulated (and also older technologies that may have outlived their effectiveness, or have evolved to a present point in need of re-examination), is simply because of the intensely saturated debates about the risks, safety and future of these technologies within the working circles of scientific research, development and endeavours,[17] and its impact on the lives of human beings. More importantly, the organic facets of what makes our evolved modern societies democratic (to the greatest extent possible, at least), within the operability of a constitutional legal system, necessitates the exercise of protective rights, remedies and limitations (if applicable). Lessig sums this gargantuan task most beautifully, by stating that the framework of a constitution[18] must have a capacity "that structures and constrains social and legal power, to the end of protecting fundamental values— principles and ideals that reach beyond the compromises of ordinary politics."[19] For similar reasons, Brownsword espoused the numerous questions that indubitably plague the minds of lawyers, regulatory theorists and other similar individuals alike.[20] Heralding

---

[15] Brownsword and Goodwin (2012), pp. 24–25.

[16] ibid 25. See also Black (2005), p. 11.

[17] Mandel (2009), pp. 75, 76.

[18] Lessig (1999), p. 5. It is interesting at this juncture to note that Lessig did not mean "constitution" to refer to a legal text in the manner of the United States Constitution. Instead, Lessig refers to an alternative interpretation of "constitution" that is equated with a "way of life".

[19] ibid.

[20] Brownsword (2011), p. 207.

a significant shift in the way we view a regulatory environment, Brownsword succinctly points out that there is a cardinal need to "frame" our inquiries[21] about the complex nature of regulating technologies. There is no doubt that the issue of "framing" the appropriate inquiry, (as a basis for determining the underlying questions that drive our normative concerns), should take center-stage in the serpentine setting of the regulatory configuration.[22] Contemporary scholars of political thought and opinion believe that "while issue frames may affect the content of one's beliefs, they also affect the importance individuals attach to particular beliefs."[23] In the landscape of our highly politicized environment where globalization has reached out its artful fingers to all corners of the earth, "objective political communication"[24] has as much to do with a community of moral rights, as does an architectural structure ensued of legal and quasi-legal concepts.

Within the context of reproductive technologies, such as PGD, the need for a paradigm shift becomes increasingly pressing, as infertility problems continue to escalate on a global scale,[25] and access to these reproductive technologies, particularly in low-resource countries, need to be expanded on a more equal footing, and fed into national discourse and policies in recognizing a right to reproductive health, freedom and autonomy. Propped up against this already complicated background addressing infertility and the reproductive sphere, genomic modification technologies, such as the CRISPR-Cas/9 system and Transcription Activator-Like Effector Nuclease (TALEN),[26] Zinc Finger Nucleases (ZFNs)[27] and germinal choice technology,[28] (all of which have the munificent ability to modify or edit the genomes of living organisms in the pursuit of disease eradication, and prolonging health and wellness, for example), compounds the adversity of reproduction. Pointing to the underlying reasoning for regulating technologies, particularly these reproductive and genomic modification technologies, may also require an esoteric dissection of the expedition down the rabbit hole; taking into account the dynamics of challenges, values, politics and policy, and how these feature into the essentially legal, or quasi-legal framework in which emanates the root concerns of power and knowledge.[29]

The fundamental journey of this chapter, from this point onwards, is to provide an earnest account of the aspects of regulatory design for existing, new and emerging biomedical technologies. This chapter's expedition is to explore regulatory or governance models for biomedical technologies, and even technologies that may no longer be new, but whose use has developed beyond the initial vestiges of its operability. In this process, it will factor, *inter alia,* inevitable regulatory challenges,

---

[21] ibid 208.

[22] Nelson and Oxley (1999), p. 1040.

[23] ibid 1041.

[24] Nelson and Oxley (1999).

[25] Inhorn (2003), p. 1837.

[26] Nemudryi et al. (2014), p. 22.

[27] Carroll (2011), p. 773.

[28] Stock (2005), p. 27.

[29] Foucault (1977).

other fundamental questions in the sphere of a community of moral rights, and an ideal set of desiderata for an efficient employability of a secure, normative framework. Posited against the backdrop of the Collingridge Dilemma,[30] and a dyadic tug-of-war between the consideration of normative social norms and the trickle-down effect of top-down "legal exclusivity",[31] the chapter intends to provide suggested guidelines of regulation or governance, based on the convergence of values and constitutional principles central to regulatory design.

4.1.2 The Foundations of Regulatory Theory

In this section, it would be prudent to briefly trace the foundations of regulatory theory, asking questions such as why regulation would be necessary in the first place. The framework of regulatory theory itself, however, is much broader and would be impossible to capture in a brief expostulation, as it encompasses disciplines such as law, politics, economics and public policy, amongst others. Nevertheless, the motivations for a quick view at the foundational aspects of regulatory theory are "necessary to frame the right questions"[32] (in relation to biomedical technologies) and "to inform the way in which those questions are answered."[33]

It is not surprising that most regulatory theory literature takes their starting point in the field of economics.[34] Canonical work on regulatory theory, and regulation, such as those by Ian Ayres and John Braithwaite,[35] and Robert Baldwin, Martin Cave, and Martin Lodge,[36] deal with multi-disciplinary approaches in regulation that cross-cuts law, politics, economics, business and commerce, and public policy.

---

[30] Liebert and Schmidt (2010), p. 55. See also: Brownsword and Yeung (2008). I found it compelling to give mention to the issue of the Collingridge Dilemma, which was raised by Browsword in *Rights, Regulation and the Technological Revolution*. The basis of the Collingridge Dilemma is essentially a trade-off process: between regulating technologies while they are still new, and their future consequences may be projected; or waiting to see how the technologies will develop over a given period of time, but then losing the ability to control the manner in which it may be regulated.

[31] Brownsword (2011), p. 208.

[32] Level Crossings, Law Commission Consultation Paper No. 194, Scottish Law Commission and Law Commission Discussion Paper No. 143, 'Regulatory Theory' (2010) https://www.scotlaw-com.gov.uk/files/5312/8024/5698/regulatory_theory.pdf.

[33] ibid.

[34] This is particularly accurate because of the global economic slump and boom in the late 1960s period, which led to the approach of "regulation theory", a derivative of Marxist economic theory as a means of dealing with capitalism, a shift in the industrial regime, and development in society. However, regulatory theory is not the same as "regulation" theory, although it may be said that the underlying purposes of these theories address the linkage between legal, social, cultural and political contexts.

[35] Ayres and Braithwaite (1992).

[36] Baldwin and Cave (1999).

Baldwin et al cover risk-based regulation, principles-based regulation, multi-level regulation, and regulatory impact assessment.[37]

In the work of Peter Drahos,[38] he emphasizes that in addition to the formal requirements of the legal system, regulation should move through a broad view that includes "non-legal forms of norm-making, along with the idea that private sovereignty over such norm-making mattered to regulatory outcomes."[39] Thus, moving "beyond the narrow or juridical view of regulation… leads to a theory of regulation with much more empirical content."[40] However, he also emphasizes that despite this broader view of regulation, the role and position played by the state remains in the picture, and they then become "part of a network of regulation in which the tasks of regulation are redistributed in various ways among actors within the network."[41]

To give an example of how this would operate in a hypothetical reality: Drahos takes us through a familiar reimagining of *Leviathan*,[42] which has undoubtedly become a cornerstone in the formation of contemporary understanding of political and legal history and legal jurisprudence. In essence, the philosophy put forward by Hobbes in *Leviathan* is that commands issued by the sovereign is the law; and through the course of time, these commands issued by the sovereign became rules, and rules were synonymous with law, as did regulation, a means by which a state would carry out a certain thing *vis-à-vis* the sword of the law.[43] In contemporary legal societies, Hobbes' philosophy cannot fully work because scholarly research in regulatory theory has revealed findings that regulation goes beyond the scope of the law and what lies beyond it: these findings relate to meta-regulation, self-regulation mechanisms, and the decentralization of the state.[44]

The changing topography of a regulatory framework as part of, and as an alternative to the traditional models of law, has had significant influence on the role of governance and regulation within societies. Drahos illustrates this by reference to Foucault's work regarding power relations,[45] implicating that "power circulates rather than being a thing possessed by a sovereign."[46] The conception of regulation has changed in such manner that experienced scholars in the research of regulatory theory have put forward a variety of regulations that have been described to "challenge rulish presumptions."[47]

---

[37] ibid.

[38] Drahos (2017).

[39] ibid 3–4.

[40] ibid 4.

[41] ibid.

[42] Thomas Hobbes, *Leviathan or the Matter, Forme, & Power of a Common-Wealth Ecclesiasticall and Civill.* (Andrew Crooke at the Green Dragon in St Paul's Church-yard 1651) https://socialsciences.mcmaster.ca/econ/ugcm/3ll3/hobbes/Leviathan.pdf.

[43] Drahos (2017), p. 12.

[44] ibid 14.

[45] Foucault and Gordon (1980), p. 98.

[46] Drahos (2017), p. 17.

[47] ibid 16.

If we believe this to be true, then it is likely too that the compliance of processes in a democratic society would be a consequential outcome of regulation. As a concept that is relational to regulation in general, compliance "bridges the world of the regulated and the world of the regulator."[48] Valerie Braithwaite questions why regulation would necessarily evoke images of the state or government wielding vast amounts of power and control, as opposed to "valuable community oversight,"[49] the latter of which becomes engaged through our social world. Instead, she pus forward that recognizing a regulatory purpose (from the psychological dimension) instead would help in the determination of the outcomes of regulation; she quotes Christine Parker and John Braithwaite that regulatory purpose is "to steer the flow of events."[50] This flow of events as a regulatory purpose has the capability to influence the "future legitimacy and trustworthiness of the regulatory system more generally,"[51] especially within the environment of a "regulatory community".[52]

An interesting perspective culminating from this "flow of events" is its connection to regulatory processes; whether through the idea of regulation reacting in ways that mimic the rules of laws, or the pyramidal structure that elevates conflicts with regulations through a process of escalation, or even through what is known as the emotion theory[53] and its premise that "psychological processes may not be responsive to processes of intervention based on rational or procedural fairness."[54] The study of processes, whether processes of interventions, or influences from other processes "that enable us to understand those events as part of a pattern"[55] can be crucial in the regulatory framework, because it "creates a link between regulatory theory and what appears to be true—namely, that we live in a world were processes from the most microscopic to the most macroscopic are everywhere."[56] By this token, a study of these processes as part of enhancing a regulatory framework would enable better responses of intervention *vis-à-vis* suitable and appropriate laws.

---

[48] Braithwaite (2017b), p. 28.

[49] ibid 26.

[50] ibid 27.

[51] ibid 29.

[52] ibid. Braithwaite further refers to the work of Meidinger (1987), pp. 355–386. A 'regulatory community' is one where there are groups of regulatory actors comprised through "different subcultural groups with their own values, normal, beliefs and processes." The idea of a 'regulatory community' is that it is powerful enough to either support by extension, or even undermine, regulatory authorities.

[53] Hume (1896).

[54] Drahos (2017), p. 17.

[55] ibid 16.

[56] ibid 18.

4.1.3   Regulatory Approaches in Biomedical Technologies

At this stage, we should therefore begin by asking what kinds of regulatory theory should be applicable to biomedical technologies. For example, in risk-based regulation, ascertaining risks and its impact would be the focal consideration; in terms of biomedical technologies, this would therefore translate to questioning the kinds of risks that human beings would be exposed to, the level of the risks, the impact of the risk upon the assumption that a mishap takes place, and control through safety, quality provision standards and information by medical or health professionals, amongst others. If a risk-based approach is taken, then it is also likely that it would be combined with an approach that considers suitable responses in the event regulations are not complied with, for example, the imposition of sanctions (punishment and reprieve), although this has been described by some scholars are being too heavy handed in nature.[57] The Organization for Economic Co-operation and Development (OECD) also recognized the importance of biomedicine and health innovation, and issued a report in 2010 regarding the impact of biomedicine on the changing nature of health innovation.[58] I view this as an important component in the determination of regulatory approaches because the global considerations in the development of biomedicine significantly impact governance and regulatory systems,[59] commercialization ventures that result from these biomedical research and development,[60] intellectual property concerns,[61] and financing, research grants, infrastructure,[62] amongst others. This is not to say, however, that there is no form of regulation over biomedical technologies at all; within the purview of many jurisdictions, there are forms of biomedical regulations but these are very poorly saturated and very much Western-centric. From the international perspective, there are a limited number of international instruments that deal specifically with the human genome and the interventions that may be performed on the human genome.[63] These include UNESCO's Universal Declaration on the Human Genome and Human Rights, (UDHG)[64] and the Universal Declaration on Bioethics and Human Rights (UDBHR).[65] In the European sphere, the Council of Europe's Convention on the Protection of Human Rights and Dignity of the Human Being with regard to the

---

[57] Ayres and Braithwaite (1992).

[58] OECD, 'Biomedicine and Health Innovation: Synthesis Report' (2010) http://www.oecd.org/health/biotech/46925602.pdf.

[59] ibid 14.

[60] ibid 15.

[61] ibid 17.

[62] ibid 18.

[63] These international instruments will be more specifically expanded on in Chap. 5.

[64] UNESCO, 'Universal Declaration on the Human Genome and Human Rights (11 November 1997)' http://portal.unesco.org/en/ev.php-URL_ID=13177&URL_DO=DO_TOPIC&URL_SECTION=201.html.

[65] UNESCO, 'The Universal Declaration on Bioethics and Human Rights (2006)' http://unesdoc.unesco.org/images/0014/001461/146180E.pdf.

Application of Biology and Medicine (the Oviedo Convention)[66] and the European Medicines Agency (an agency of the European Union) report on genome editing technologies used in medicinal products[67] as well as the European Commission Joint Research Center Science for Policy report that encompasses an overview of EU national legislation on genomics,[68] are useful for ascertaining the spheres of regulation over biomedical technologies. The World Medical Association's non-binding Declaration of Helsinki: Ethical Principles for Medical Research involving Human Subjects (Declaration of Helsinki)[69] also imparts an ethical component to the consideration of human rights within these fields, and targets the primary concerns related to physicians and medical research involving human subjects. However, the reach of these international instruments has been less than effective, where human rights considerations are concerned. This has therefore prompted questions regarding the 'constitutionalization' of international law and the plausibility of that attempt.[70]

More specific to the advances and developments in biomedical technologies, a very much more fluid and responsive approach is necessary. The main premise that I make at the end of this chapter is that regulatory design of a framework for biomedical technologies should ideally converge the law, negotiated, shared values[71] and constitutional principles. This can be achieved by modernizing the conceptual framework of biomedical regulation through the inclusion of other forms of regulation (besides the law).

Part of the work of Shawn Harmon is focused on modernizing biomedical regulation,[72] and in addressing some of the disparities that exist between medical and health innovation, and the problems of regulation, he contends that "responsible

---

[66] Conseil de l'Europe, *Convention for the Protection of Human Rights and Dignity of the Human Being with Regard to the Application of Biology and Medicine: Convention on Human Rights and Biomedicine* (Editions du Conseil de l'Europe 1997) http://193.205.211.30/lawtech/images/lawtech/law/convenzioneoviedo.pdf.

[67] European Medicines Agency, 'Report of the EMA Expert Meeting on Genome Editing Technologies Used in Medicinal Product Development' (European Medicines Agency 2018) EMA/47066/2018.

[68] European Commission JRC Science for Policy, 'JRCF7- Knowledge Health and Consumer Safety, Overview of EU National Legislation on Genomics' (European Commission 2018) EUR29404EN http://publications.jrc.ec.europa.eu/repository/bitstream/JRC113479/policy_report_-_review_of_eu_national_legislation_on_genomics_-_with_identifiers.pdf.

[69] World Medical Association, 'WMA Declaration of Helsinki - Ethical Principles for Medical Research Involving Human Subjects' (19 October 2013) https://www.wma.net/policies-post/wma-declaration-of-helsinki-ethical-principles-for-medical-research-involving-human-subjects/.

[70] Kleinlein (2012), p. 703.

[71] This does not mean that I do not accept the universality of some common values and legal norms that are recognized in international law. However, the rhetoric of the "Asian values" principles which may, in the past, sometimes been unfairly categorized as the negative aspects of cultural relativism, can provide some insight into the contemporary motivations that influence regulation-shaping in some Asian countries.

[72] Harmon (2016), p. 680.

research innovation" (RRI)[73] used with "legal fore-sighting" (LF)[74] assist in the promulgation of better regulation.[75] In RRI, Harmon advocates the importance for research undertakings to "identify, and reflect broad values, which serve as signals to stakeholders and publics about what will inform actions and decisions."[76] This is consistent with the premise that I put forth in this chapter. He cites some of the fundamental values that have been highlighted in international biomedical conventions, a plethora of values that take into account respect for the "society and the role of science".[77] In terms of individual persons, this encompasses values like human dignity, autonomy, justice, and safety[78]; and on the part of researchers, and medical or scientific professionals providing services, transparency, accountability, and engagement.[79]

Advocating for a combination of RRI with the LF approach,[80] Harmon believed that the supporting research framework in existing, new and emerging fields of biomedicine, can inform the practice of LF by "generat[ing] diverse evidence supportive of more effective and durable governance practices and instruments."[81] He acknowledges that fields of study such as those in the biomedical spheres requires "'lengthened foresight',[82] not only because of the consequences they can have for future generations, but also because of the non-traditional forms of oppression permitted by them."[83]

I agree that this manner of regulatory approach put forward by Harmon is a workable one that may adequately inform and deal with the fluidity of biomedical technological advances. However, in addition to the recognition of values systems *vis-à-vis* the RRI, and with LF, I also believe that constitutional principles and values are equally significant in a regulatory system, and a suitable regulatory approach, or a combination of approaches may be necessary to effectively address the peculiarities that may surround the circumstances of biomedical technologies.

Bearing this in mind, and tracking on the establishment of Harmon's approaches to regulation, I posit that an initial and ideal, well-considered and rational framework

---

[73] ibid 681.

[74] ibid 683.

[75] Although Harmon's article appears to focus on precision or personalized medicine, and biobanking/data sharing, the import of the components he has suggested have the aim of ensuring the biomedical regulation are timely, critical, safe and effective.

[76] Harmon (2016), p. 684.

[77] ibid 685.

[78] ibid.

[79] ibid.

[80] ibid 683. Harmon refers to LF as "a future oriented process aimed at identifying and exploring possible and desirable legal or quasi-legal interventions directed at better achieving valued social and technological ends."

[81] ibid 682., March 10, 2016, and Shawn H.E. Harmon 'Evidence, Engagement and Transparency in Decision-Making', presented at Canadian Centre for Vaccinology, June 10, 2016.

[82] Laurie et al. (2012), p. 1.

[83] Harmon (2016), p. 682.

for regulatory approaches in biomedicine should be open to combining several approaches that may be incorporated into law or legislation: RRI and LF, the responsive regulation approach,[84] and 'smart' regulation approach.[85] In addition to this, I also argue for considering the symbolic legislation theory,[86] as a manner of symbolism, that is, to "demonstrate that .....rights were taken seriously, at least on paper."[87] I believe that it would be worthy to consider the symbolic legislation theory because two of the selected jurisdictions here, Malaysia, and Thailand, appear to employ such a regulatory theory in the culmination of laws relating to PGD.

In the theoretical contribution of responsive regulation approach first put forward by Ayres and Braithwaite,[88] its fundamental applicability lies in the establishment of a method "for creating regulatory policy solutions that "transcend" both public interest oriented calls for effective regulation of businesses and business oriented calls for the dismantling of state based regulation."[89] The approach puts forward that "regulatory policy should take neither a solely deterrent nor a solely cooperative approach."[90] The manner in which responsive regulation works is influential in how regulatory enforcement may work *vis-à-vis* the "impact on compliance".[91] The strategic placement of some form of hierarchy within the framework of this approach embodies methods in which the various reasons for compliance (or non-compliance) of rules may target the categories of compliance, whether for economic, social and normative reasons.[92]

The most well-known aspect of responsive regulation is its pyramid structure.[93] This pyramid structure represents regulatory enforcement strategies of rules: and following the expansion of this initial work by Ayres and Braithwaite, the pyramid structure has been expanded to find connections in other areas such as "restorative justice"[94] and "social movement advocacy" (Fig. 4.1).[95]

In essence, the responsive regulation approach "suggests that governance should be responsive to the regulatory environment and to the conduct of the regulated in

---

[84] Ayres and Braithwaite (1992). This responsive approach also includes the exercise of a 'soft law' approach and self-governance regulatory mechanisms. In Section 4.1.3 of this chapter, p. 185, I provide further illustration on this regulatory approach and how it may benefit the fluid nature of biomedical technologies and yet achieve a necessary balance from a legal perspective.

[85] Scottish Law Commission and Law Commission (2010). The smart regulation approach is a multi-level, multi-lateral regulatory approach that employs a broad range of tools in regulation, and reaches multiple layers of stakeholders that are relevant to the regulatory dynamics of the system.

[86] Van Klink (2016).

[87] Poort et al. (2016), p. 1.

[88] Ayres and Braithwaite (1992).

[89] Parker (2013), p. 2, 2.

[90] Nielsen and Parker (2009), p. 376.

[91] ibid 379.

[92] ibid 378.

[93] Ayres and Braithwaite (1992), p. 39. Please see: Figure 2.3, p. 39.

[94] Parker (2013), p. 4. Please see also: Braithwaite (2002).

[95] ibid. Please see also: Braithwaite and Drahos (2000).

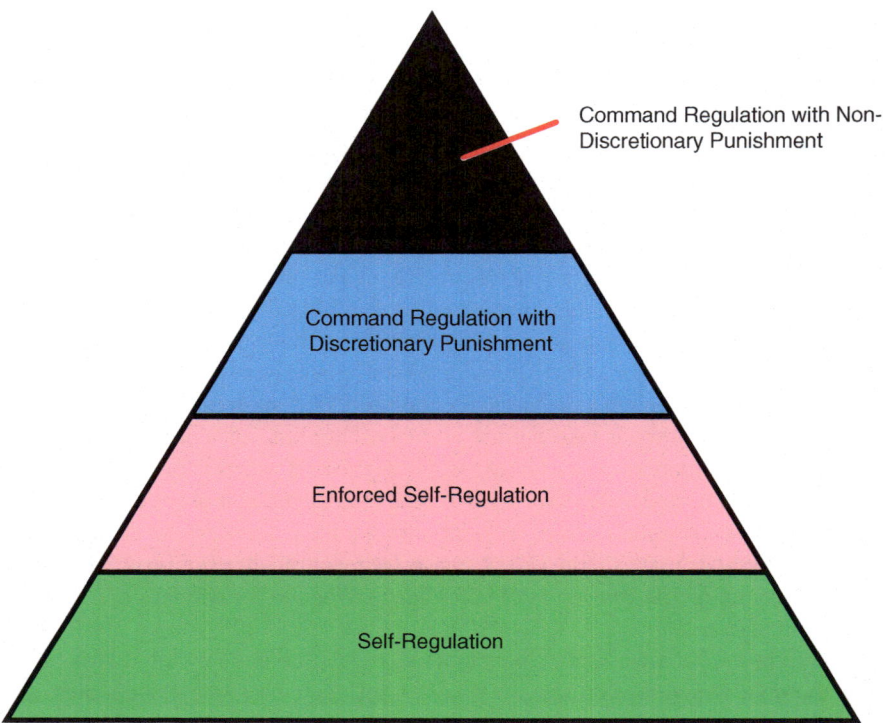

Fig. 4.1 Ayres' and Braithwaite's pyramid structure in the responsive regulation approach, for enforcement strategies

deciding whether a more or less interventionist response is needed."[96] The base of the pyramid foundations is formulated as a blend of persuasion with self-regulation[97] as part of regulatory enforcement strategies. Ayres and Braithwaite contended that a method of escalation upward the pyramid, towards actions at the top, where states enforce the compliance of rules, is likely to encourage the "industry and regulatory agents to make regulation work at lower levels of interventionism."[98] The base of the pyramid therefore encourages a significant portion of the regulatory control to be handled with the least amount of state intervention. In this sense, it is also possible to attribute the regulatory responses to not only states and governments, but also to the actors in civil society.

Although the responsive regulation approach grew as a way to deal with the frustrations in business regulations, its legacy is equally important if we look at how it may be applied in the field of biomedical technologies. Since the base of responsive regulation is hinged on persuasion and self-regulation, this is consistent with the way in which some form of governance has evolved in the biomedical

---

[96] Braithwaite (2017a), p. 117.

[97] Ayres and Braithwaite (1992), p. 40.

[98] ibid 39.

technological arena. Case studies of Western European countries such as Germany, France and the United Kingdom (UK) have demonstrated how self-regulation in human biotechnologies contributes to the symbiotic relationship between state and regulatory actors.[99] These studies indicate that France's model comprises the traditional "command and control" method of governance; but Germany's development of a "hybrid model" sees a combination of the state's direct intervention with a delegated aspect to medical professionals.[100] As illustrated in Chap. 3 regarding the role of the HFEA in the UK, regulation of biomedical technologies that concern fertility have been implemented and handed to an independent regulatory authority. Medical professionals in the community, as well as industry players in the field of biomedical technologies, can be seen as the 'first responders' in the development of regulatory conduct, behaviour and framework because they directly confront the outcomes of biomedical development technologies. In this respect, the "incorporation of voluntary self-regulation"[101] *vis-à-vis* ethical guidelines and codes of conduct that pertain to biomedical technologies can be seen as the first response batch to any concerns that may arise.

"Smart regulation"[102] is also one of the ways in which regulatory approaches may be combined in a field that develops as rapidly as biomedical technologies. Although initially developed by Neil Gunningham, Peter Grabosky and Darren Sinclair in the field on environmental policies, the elements of smart regulation employs flexible and pluralistic approaches that can contribute positively to the way in which an overall regulatory system performs. Loosely linked to the manner in which the responsive regulation approach works, and further built upon the pyramid model by Ayres and Braithwaite, smart regulation, however, looks to how using a variety of policy instruments and enlisting the involvement of a broader range of regulatory actors[103] as a way to "compensate for the weakness of standalone"[104] policies. The other interesting component put forward by smart regulation is that the enforcement mechanism is not simply confined to the state or government regulators, but also to "surrogate regulators".[105] Gunningham and Sinclair explain that these "surrogate regulators" could be contained within each component of the responsive regulation pyramid model; thereby creating an additional escalation pyramid within each component of the pyramid model.

In advocating for the employment of multiple regulatory tools and instruments in smart regulation, Gunningham and Sinclair draw the attention to a detailed explanation of instrument combinations which they contend may also apply to not only environmental policies, but to "other areas of social regulation."[106] These

[99] Engeli and Rothmayr (2016), p. 248.

[100] ibid 249.

[101] ibid.

[102] Gunningham et al. (1998).

[103] Gunningham and Sinclair (2017).

[104] ibid 139.

[105] ibid 135.

[106] ibid 140.

combinations, which may be complementary to each other[107] include the "command-and-control regulation, economic instruments, self-regulation, voluntarism,"[108] and "information strategies".[109] Although smart regulation has not been immune from academic criticism,[110] the engagement with it is worth an exploration in the field of biomedical technologies because of its ability to adapt in a regulatory system. Indeed, fielding the fuel of academic criticism about smart regulation,[111] Gunningham and Sinclair refer to how the shortcomings of smart regulation can be combatted by "merging the smart regulation theory with the policy arrangement approach and the policy learning concept."[112] Although it is theoretically much more optimistic than the reality of implementing this in practice, the manner in which smart regulation may be preserved through a multilateral network with policy arrangements and learning is worth a visit in the biomedical technological sphere. The changing landscape and diversity of progress in biomedical technologies necessitates constructive ways of implementing a creative, but workable and effective regulatory system.

Finally, within the scope of symbolic legislation theory, although described as an "essentially political concept",[113] it appears that in developing constitutions around the world, the theory itself is capable of acting as a 'buffer' in circumstances where there is not yet the readiness to 'commit' to a fully functioning form of legislation. The symbolic legislation theory, in essence, is often associated with negativity because, as its name suggests, the legislation prescribed is "ineffective and [sic] serve other political and social goals than the goals officially proclaimed."[114] In simple terms, such legislation is basically prepared to accept that "the law is not meant to be complied with, but its main purpose is to give expression to values in the political sphere."[115] It is not suggested that the symbolic legislation theory is an effective manner in the sphere of biomedical technologies. What is advanced, is that the understanding of this theory is helpful in defining the contours of the legislative landscape in developing constitutions such as Malaysia and Thailand.

Regarding PGD laws in Malaysia and Thailand, the symbolic legislation theory plays strongly in these jurisdictions; and the governance of various forms of reproductive technologies are meted out *vis-à-vis* guidelines instead of legislation. In

---

[107] ibid 141.

[108] ibid 140.

[109] ibid 141.

[110] ibid 145.

[111] Please see: Baldwin and Black (2008), pp. 59–94; Bocher, M and Toller, AE 2003. Conditions for the emergence of alternative environmental policy instruments, paper presented at the Second European Consortium of Political Research Conference, Marburg, Germany, 18–21 September. Some of the criticism leveled against smart regulation is that it fails to take into account and address "institutional issues, compliance type specific responses, performance sensitivity and adaptability of regulatory regimes."

[112] Gunningham and Sinclair (2017), p. 145. Gunningham and Sinclair refer to van Gossum et al. (2010), pp. 245–261.

[113] van Klink (2016).

[114] ibid.

[115] ibid 22.

specific studies of biomedical laws, scholars have espoused on how the symbolic legislation theory has made a revival because the variance of 'legislative' activity that govern biomedical technologies and activities are, at best, soft law in nature.[116] This is also observed in relation to the international human rights instruments that govern the discourse on biomedical technologies, which provisions are often subject to a variety of interpretations of "the symbolic connectedness between persons and their bodies."[117] Despite the pluralistic nature of these interpretations, scholars believed that it is still possible to develop other ways of governing biomedical laws by looking to the "symbolic aspirations and effects of bio-ethical legislation".[118] If we look at Malaysia and Thailand, the symbolism of governance in biomedical technologies, particularly PGD, has been characterized by political aspects and a fear of normative reprisal on the world stage, and the need to demonstrate that there is, at least, some laws in place to govern such important issues that touch upon interventions into the human body. Further, the additional underlying symbolism in Malaysia and Thailand could also be attributed to potential economic development in the manner of medical or fertility tourism[119] that often generate modest revenues for the respective countries. In this sense, an understanding of the symbolic legislation theory contributes to a deeper analysis of possible future regulatory approaches that may be more meaningful and effective to these countries.

4.2   The Challenges of Regulating Biomedical Technologies

Justice Michael Kirby states that "there are no real experts in the subject of regulating technologies."[120] In the closing chapter called *New Frontier: Regulating Technology by Law and 'Code'*,[121] Justice Kirby underscores a host of prevailing paradoxes on the subject of regulating technologies,[122] which is analogous to the concept of the Collingridge Dilemma faced by regulators, described by Brownsword in the following manner: "[regulators] find themselves in a position such that either they do not know enough about the (immature) technology to make an appropriate intervention or they know what regulatory intervention is appropriate but they are no longer able to turn back the (now mature) technology."[123] Justice Richard Posner theorizes on the possible occurrence in future of the diminishing role of the judge, because "the continued rapid advance in science is going to make life difficult for judges…..[because]…..breakneck   technological   change……will   thrust   many

---

[116] Poort et al. (2016), p. 3.

[117] ibid.

[118] ibid 4.

[119] Harryono et al. (2006). See also: Mohd Mutalip (2012), p. 1.

[120] Brownsword and Yeung (2008), p. 373.

[121] ibid 367.

[122] ibid 373–381.

[123] Brownsword (2011), p. 209.

difficult technical and scientific issues on judges, for which very few of them are prepared because of the excessive rhetorical emphasis of legal education."[124] Former Judge of the European Court of Human Rights (ECtHR), András Sajó emphasized that judicial reasoning in technologies should be concerned about the "interaction between technological change and the relevant social and market reactions to the implications of this change."[125] Emphasis on the framing exercise, and the "dilemma of how to balance old norms in new contexts"[126] is a necessary component in "the judicial process of translating something new into the language of past legal models."[127]

There will always be a tenuous relationship between law and technology; and perhaps what is needed at this juncture is an unconventional manner of viewing how the sovereignty of legal dominion may be expanded to accommodate the vicissitudes of new, emerging and existing biomedical technologies. One of the biggest challenges of regulating biomedical technologies is undoubtedly the scope of such regulation. A prioritization of a hierarchy of needs and responses to biomedical technologies must be formulated with careful and reasoned study. The scope must ideally take into account the intended beneficiaries of the technologies and the providers of such, other notable concerns such as risks, safety, access, costs, commercialization dimensions, the human rights implications and a necessity to ensure either a maintenance or reformulation of social orders, and any other moral or ethical concerns that may accompany the use of technologies. On the assumption that these hurdles can be overcome, the legal and policy dimensions that hold a constitutional order within the workings of a legal system also needs to be addressed. These perspectives are also challenging in themselves, but are vital to the health of a legal system.

4.2.1 Legitimacy

Brownsword's work[128] primarily catechizes how it would be possible to establish legitimacy in environments characterized with a plurality of moral norms and opinions. The legitimate nature of any form of legal pronouncements, promulgated through procedural due process attributed to the democracy of a healthy, functioning system of governance, may often be in conflict with other equally crucial values in such systems of governance.

The challenge of legitimacy, in itself, can be divided into three main segments: firstly, relating to three ethical perspectives[129] which Brownsword refers to as "the

---

[124] Posner (2006), p. 1049.

[125] Sajó and Ryan (2016), p. 3.

[126] ibid 4.

[127] ibid 8.

[128] Brownsword (2008), pp. 31–131.

[129] ibid 31.

bioethical triangle"[130]; secondly, relating to the issue of consent as the primordial apogee of autonomy[131]; and thirdly, relating to the inherently pluralistic nature of different communities across international regions.[132] In essence, Brownsword's main concerns in the ethical sphere is steadfastly built upon the existence of "shared baseline values",[133] where it would be slightly less complicated to accommodate a plurality of moral and social norms, by reference to existing interpretative or application mechanisms put forward by the ECtHR, for instance.[134]

Legitimacy, at the very least, rests on the crucial values of, and a meeting of the minds, in 'consensus'; at the very least, insofar as a basic framework of opinion and values is concerned.[135] This latter reference is what Brownsword refers to ideally as "a community of rights."[136] A community of rights in Brownsword's quixotic sense is a "community that engages in a reflective and ongoing way on the best interpretation of its commitments."[137] Conundrums are likely to be much more articulated at an international level in the absence of a general consensus that gives an aspect of legitimacy to any considered model of regulation. Brownsword drew the difficulties presented by these ethical constituencies from his understudy of the provisions in the UNESCO Universal Declaration on Bioethics and Human Rights (UDBHR)[138] and notes that "each constituency demands that regulators do the right thing; but each has its own gold standard for doing the right thing."[139]

4.2.1.1 The Bioethical Triangle

The first challenge of legitimacy, the 'bioethical triangle'[140] considers three distinct, competing ethical positions, all of which are equally essential determinants in the course of scientific revolution and the incremental, developmental progress of regulation following on its heels. However, the pluralized, reference points of these different ethical perspectives do not make it easy to locate a convergence of opinion, specifically in relation to international bioethical consensus.[141] The lack of any form of consensus, as we are aware, makes any task of drafting provisional regulation, and its subsequent enforceability at the international level, a very provocative one.

---

[130] ibid 35.

[131] ibid 70.

[132] ibid 100.

[133] Brownsword (2008), p. 209.

[134] ibid.

[135] ibid. See also (n 44).

[136] Brownsword (2008), p. 24.

[137] ibid 273.

[138] 'The Universal Declaration on Bioethics and Human Rights' (2006).

[139] Brownsword (2008), p. 35.

[140] ibid.

[141] ibid 31.

The first ethical position Brownsword identifies in the bioethical triangle is the utilitarian perspective promulgated by Jeremy Bentham,[142] which also presents itself in Article 4 of the UDBHR.[143] Bentham, in his work titled *A Fragment of Government*, considered that "it is the greatest happiness of the greatest number that is the measure of rights and wrong."[144] Placed within the context of the regulatory environment of new and emerging technologies, the utilitarian view calls for means that are "most likely to maximize utility or minimize disutility."[145] Hence, the propagation of utility would include factors such as happiness, convenience, and economy, that would have the effect of boosting the pleasures on an individualistic manner whilst satisfying any accrued preferences of the said individual.[146] Disutility is discounted, and viewed as a negative counter in the larger scheme of things, and included pain, suffering, distress, and the like.[147]

The second ethical perspective is represented by the present, existing human rights corpus. Brownsword points to Article 3(2) of the UDBHR,[148] essentially theorizing that human rights considerations are not primarily concerned with the denouement of a regulatory sphere, whether positive or negative in nature[149]; instead, the *leitmotif* of the human rights discourse proffers a healthy respect for the human rights of individual persons, and the general theme that these fundamental human rights cannot be sacrificed at the altar of what may be considered the greater good of a flourishing society.[150] At this juncture, having been presented with the utilitarian view and the human rights perspective, it is easy to assume that the intrinsic tension in the bioethical triangle is created by the loggerhead clashes between these two ethical views. This also raises interesting questions about the purported clash in the provisions of the UDBHR (Article 4 and Article 3(2) as earlier mentioned), and whether there may be interpretative elements from other sources or authorities which may be able to provide us with some clarity as to how the function of these two seemingly-conflicting principles should be dealt with. In addition, Brownsword opines that these clashes may be moot in the context of the emergence of the

---

[142] Burns (2005), p. 46.

[143] 'The Universal Declaration on Bioethics and Human Rights' (2006). Article 4 states: "In applying and advancing scientific knowledge, medical practice and associated technologies, direct and indirect benefit to patients, research participants and other affected individuals should be maximized and any possible harm to such individuals should be minimized."

[144] Bentham et al. (1988).

[145] Brownsword (2008), p. 37.

[146] ibid 36.

[147] ibid 37.

[148] 'The Universal Declaration on Bioethics and Human Rights' (2006). Article 3(2) of the Declaration stipulates that "the interests and welfare of the individual should have priority over the sole interest of science or society."

[149] Brownsword (2008), p. 37.

[150] ibid.

dignitarian perspective,[151] which may have the effect of further complicating the ethical conundrums faced in the bioethics sphere.

The final piece in the bioethical triangle, the dignitarian vista, is characterized vis-à-vis its emphasis on the impact of technologies on dignity of the human body.[152] Often described as a "conversation stopper",[153] the concept of human dignity can be interpreted in ways that are akin to painting with a broad brush, so simultaneously diversified and titillating are the debates surrounding it. As such, any application of technologies that, *au fond*, violates or undermines human dignity is regarded with severity and opprobrium by the dignitarian sector. Some examples that are considered in great disdain include, for instance, human reproductive cloning and human embryonic stem cell research, amongst others.[154] From a more theoretical perspective, the initial "two-way contest"[155] between utilitarianism and human rights corpus has become entangled in "a genuine triangular contest"[156] with dignitarian ideals. Brownsword further demonstrates the interplay of dignity in two (2) distinctive, but clearly opposing roles: firstly, in the context of substantiating individual autonomy (by way of empowerment)[157]; and secondly, by acting as a constraint on autonomy.[158] Although this may sound antithetical in nature, Brownsword further postulates that a disentanglement and closer examination of these roles will allow us to arrive at a deeper, more formulated understanding of how the concept of human dignity operates in both modern human rights and bioethics discussions.[159] However, in context of a regulatory environment, Brownsword quotes Philippe Seguin, that "…legislators must, despite the difficulties, act to ensure that science develops with respect for human dignity and fundamental human rights, and in line with national democratic traditions."[160]

4.2.1.2 Informed Consent

The second challenge of legitimacy, the concept of consent, and much more specifically, informed consent, is another fundamental apogee in the quest to consider and assess the legitimacy of a regulatory environment in new and emerging technologies. United States Supreme Court Justice Benjamin Cardozo in the case of *Schloendorff v The Society of the New York Hospital*[161] stated the "every human

---

[151] ibid 39.
[152] ibid.
[153] ibid.
[154] Francioni (2007), p. 50.
[155] Brownsword (2008), p. 39.
[156] Francioni (2007), p. 51.
[157] Brownsword (2008), p. 41.
[158] ibid.
[159] ibid 42–43.
[160] Brownsword (2008), p. 43.
[161] Schloendorff v The Society of the New York Hospital (1914) 211 NY 125 (The Court of Appeals of New York).

being of adult years and sound mind has a right to determine what shall be done with his own body…". This quote has embodied the seminal discourse on the establishment of the socio-legal doctrine of informed consent in the field of medicine and healthcare. Thus, the doctrine of informed consent is "a voluntary and explicit agreement made by an individual who is sufficiently competent or autonomous, on the basis of adequate information in a comprehensible form and with adequate deliberation to make an intelligent choice about a proposed action."[162] By extension in the medical profession and provision of healthcare services, it presupposes patient autonomy and frank, effective communication by a physician or healthcare providers as the upholder of moral values that would ensure effective decision-making on the part of the patient.

It is vital that a strong sense of individual self-determination and patient autonomy is maintained and highly valued as a consistent theme in these areas. In many circumstances, the principles of autonomy often override the principles of patient beneficence too. This view is also thematic in the United Kingdom, as per indicated by Lord Goff in *Airedale NHS Trust v Bland*.[163] However, the doctrine can only effectively operate if all material facts are disclosed to a patient. In many jurisdictions, informed consent has already been recognized as one of the most important duties a physician is responsible for towards his or her patients.

Specifically why this concept has become an immense cornerstone, "rightly watermarked into the justificatory currency…"[164] of autonomous individuals, (who consider themselves to have shaken off the shackles of patriarchal thinking), is because the representational shift from the historical parochial nature of the provision of healthcare, encourages a more patient-centred approach and instils a greater responsibility for physicians to respect the wishes of their patients. The same can be said of the consideration of this concept in connection with drafting a regulatory framework in the advent of existing, new and emerging technologies; viewed from the bioethical triangle visage[165] mentioned in the previous section above, the position of centrality that the doctrine of informed consent occupies, varies in accordance with the lens view of such ethical positions. As Brownsword correctly points out, 'consent', together with other concepts such as 'harm' and 'precaution', "are not neutral concepts."[166]

The installation of consent within the different spheres of the bioethical triangle would feature most prominently in the human rights perspective. In utilitarian and

---

[162] Aveyard (2002), p. 201.

[163] Airedale NHS Trust v Bland [1993] AC 789, p. 864 per Lord Goff: "…The principle of self determination requires that respect must be given to the wishes of the patient…..the principle of the sanctity of human life must yield to the principle of self-determination…and…the physician's duty to act in the best interests of his patient must likewise be qualified."

[164] Brownsword (2008), p. 71.

[165] 'The Universal Declaration on Bioethics and Human Rights' (2006). See Article 6.1 and Article 6.2 of the Declaration, which provides for the "prior, free and informed consent of the person concerned", whether in connection to any medical intervention or scientific research.

[166] Brownsword (2008), p. 70.

dignitarian scholarship, the nucleolus of the ethical concerns would subsume utility, and dignity, respectively; almost exclusively to the consideration of other concepts that may be relevant in technologies that have ramifications on the human body. It is the human rights aggregation of the bioethical triangle that views informed consent as a clarifying significance "because consent is always parasitic upon rights."[167] For this reason, seeking a heterogeneous consensus for informed consent[168] constitutes one of the obstacles in granting a spectrum of legitimacy in a regulatory anatomy. The Gordian knot of legitimacy is also then further tightened through how we may "judge regulators most harshly if they fail to attend to the consent of rights-holding agents."[169] A further determinant is also necessary insofar as a "community of rights" is concerned[170] because there is a differentiation between "background informational rights and responsibilities"[171] which arises independently of consent, as opposed to a specific right to informed consent.[172]

4.2.1.3 Pluralism

The third and final challenge of legitimacy relates to pluralism in different communities. At the most primitive level, pluralism should not be regarded as a negative component of the make-up of global society. However, within the sphere of a regulatory environment, the prevalent pluralism of communities makes the equivalently fundamental philosophical and ethical values as the motivating factors behind regulation, much more difficult to grasp and arrive at a meeting point. As Brownsword points out,[173] the House of Commons Science and Technology Committee's report on *Human Reproductive Technologies and the Law*, recognizes that "[in] a society that is both multi-faith and largely secular, there is never going to be consensus on… the role of the state in reproductive decision-making."[174] Despite this fact, the report further pushed forward that a difficult in obtaining consensus did not mean that the task of determining the matter at hand should be abandoned. In short, consensus should always be striven for in all circumstances to the best of the abilities of the regulators.

However, some aspects of biomedical research may be prone to more deeply rooted pluralistic tendencies than others, for example, relating to issues of therapeu-

---

[167] ibid 75. Here, Brownsword also refers to Beyleveld and Brownsword (2007).

[168] ibid 102. We will find, that in looking at the challenge of legitimacy in pluralistic environments, where questions about the meaning of 'harm' and the precautionary principle has to be balanced on an equal footing, the positioning of views and perspectives within the bioethical triangle is once again considered.

[169] ibid 98.

[170] ibid 24. See also (n 49) and (n 50).

[171] ibid 92.

[172] ibid 87.

[173] ibid 100.

[174] ibid.

tic human cloning and human embryonic stem cell research.[175] In the light of conversely strenuous circumstances of this nature, where the pluralistic normative conformities of each jurisdiction demonstrates a great variation even within the confines of a common, regional space, Brownsword believes that proceduralism[176] may become the choice of the day in the face of these challenges, where it is possible that procedural due process, instead of substantive due process, may become the guiding factors in the design of the regulatory framework.[177] In some circumstances, the nature of proceduralism in the vein of epistemic democracy, having met the meritorious requirements for "political and epistemic fairness",[178] then imparts upon a decision made upon democratic principles, the bestowment of legitimacy that it requires to operate.

An example of pluralistic interpretation can be seen in the cardinal principles of doing no harm to others, and the exercise of precautionary behavior "in the face of uncertain but potentially serious and irreversible risks".[179] Briefly, the principle of not doing harm to others has been qualified within the scope of the UDBHR under the purview of Article 4.[180] However, the qualification as to the nature of harm, within the wording of the Article, stipulates that "any possible harm to individuals should be minimized."[181] The potential pitfall in the use of the word "minimize" suggests, in my view, that an acceptable margin of error in the calculated risks pertaining to harm would be tolerated. Without the appropriate parameters that operate to determine an 'acceptable' range of error margins, (and particularly where human lives are concerned in these new and emerging technologies), coupled with the divergence of the meaning of 'harm' in the context of the bioethical triangle, where each ethical perspective seeks to value and understand 'harm' to a human person in remarkably different ways,[182] I additionally argue that the consequential uncertainties and volatility of the apparent margin of error, could contribute to a most delicate breakdown, not only from the perspective of clinical practice and ethics, but also in the normative discourse in the community of rights.

---

[175] ibid 101.

[176] Peter (2008), p. 33.

[177] Brownsword (2008), p. 101.

[178] Peter (2008), p. 33.

[179] Brownsword (2008), p. 101.

[180] ten Have and Jean (2009).

[181] Brownsword (2008), p. 102.

[182] ibid 102–105. Brownsword elaborates on the employability of the meaning of 'harm' in the context of the pluralistic perspectives of the bioethical triangle, encompassing the human rights perspective, the utilitarian view and the dignitarian standards. The determination of 'harm' in these varying ethical viewpoints further contributes to the difficulty of pluralism, as an additional layer of determinative considerations.

In Article 17 of the UDBHR,[183] a possible interpretation of the precautionary principle in "access and utilization of biological and genetic resources",[184] through the nonchalant use of the words "due regard" is also a suggestion of a non-emphatic placement of the precautionary principle within the scope of bioethical concerns. As much as the UDBHR should be lauded for its attempts to converge a meeting of the minds in bioethical matters and research, the deliberative wordings of the relevant provisions can be viewed as providing a wide berth of space or margin of appreciation to states, as and when necessary, to make appropriate assessments of "due regard". Similarly, undertaking a critical reading of the operability of the precautionary principle, (at the risk of repetition), would once more depend on the fall of the coin within the space of the bioethical triangle.

4.2.2   Effectiveness and Enforceability

In the field of biomedical technologies, the polarity of debates over the management and uses of the same further contributes to the regulators' justifiable need to legitimize a regulatory overview. It cannot be denied that the legion of wherewithal of new and emerging technologies calls for the research and development of such technologies in safe, secure and carefully monitored ways, pointing towards the imperative inspiration of confidence for the public spectacle.[185] Further challenges in the sphere of the regulatory market, Brownsword explained, does not only encompass the legitimacy aspect above, but also necessitates a consideration of effectiveness of such proposed regulation. In this manner, the stability and foundational basis of the regulatory framework is simultaneously imbued with means and measures that are "morally clean and effective."[186] However, coupled with the inherent pluralism within the community of rights that is almost always naturally prevalent, it is also not surprising that meeting legitimate effectiveness in the regulatory space is subject to its own unique, demanding and equally inhibitive set of complexities. In dealing with these complexities, Brownsword highlights the pilot role of guidelines or modern technologies, *vis-à-vis* a thorough critique of various analytical studies carried out by scholars within the field of regulating modern technologies,[187] which he believes have the capacities to enhance our "regulatory intelligence"[188] to lead to a

---

[183] 'The Universal Declaration on Bioethics and Human Rights' (2006). Article 17 states: "Due regard is to be given to the interconnection between human beings and other forms of life, to the importance of appropriate access and utilization of biological and genetic resource, to respect for traditional knowledge and to the role of human beings in the protection of the environment, the biosphere and biodiversity."

[184] ten Have and Jean (2009).

[185] Brownsword (2008), p. 132.

[186] ibid.

[187] ibid 133.

[188] ibid 134.

formulation of cogent considerations for making interventions in these new and emerging technologies.

In further identifying the challenges of conceiving a regulatory framework, Brownsword further alludes to Stuart Biegel's[189] basic principles,[190] a fairly pliable and useful set of twenty guidelines that point towards the regulation of cyberspace.[191] Although we must take cognizance of the fact that some of these principles are specifically applicable for the use of cyberspace activities and Internet technology,[192] Biegel's regulatory guidelines can be useful in contributing to regulatory intelligence, and establishing the beginning of dialogue for a quasi-prescriptive values outline. Indeed, Brownsword correctly identifies that "regulation is a constant learning process"[193] and that at this point, "a comprehensive and reliable regulatory jurisprudence"[194] is still in the process of being formulated and falls short of a particular combination of regulatory instruments that adequately promotes the effectiveness necessary for the governance of new and emerging modern technologies.

In Brownsword's expansion on Biegel's basic principles, he also correctly deduced that there might be inherent limits of legal effectiveness; and that there are non-exhaustively three major problems in regulatory theory that presents lessons for the future.[195] First, he extrapolates on the tendency for regulators to commit what is known as "regulatory over-reach",[196] where the regulators' enthusiasm "push beyond the terms of the [social] contract"[197] in a politico-legal system. In the context of new and emerging technologies, the plausibility of control in a regulatory space may prove to be difficult because of considerations relating to peculiarities of locality, for instance.[198] Secondly, he expands on Braithwaite's evaluation of the relationship between regulators and the intended subjects of regulation that are often imbued with issues of capturing legitimate aims, and corruption, for example in pharmaceutical businesses.[199] Thirdly, he proposed that regulative intercession sometimes produce unwanted effects due to the nature of self-defeating industrialization prevalent in the economy of modern technologies.[200]

The manifestation of technological effectiveness may also be partially dependent on the way we choose to view the impact of new and emerging technologies within a social structure. An interesting paradigm can be viewed from the perspective of

---

[189] Biegel (2001).

[190] ibid 353–364.

[191] Brownsword (2008), p. 145.

[192] ibid 146.

[193] ibid 138.

[194] ibid.

[195] ibid 139–150.

[196] ibid 139.

[197] ibid.

[198] ibid 140.

[199] ibid 141.

[200] ibid 142.

technological determinism, *vis-à-vis* the redeeming constructivist approach that can lend positive and valuable lessons to the understanding of the basis of technological effectiveness. The narrative presented here is that technological determinism (the way in which we view the relationships between technological endeavours and a variable of human activities and interactions)[201] can profoundly exert a sway on the intended purposes and conduct that make up the fundamental tenets of effectiveness in regulating new and emerging technologies. The traditional confusion about the term 'technological determinism' is attributable to the characterization of technology as a tool that develops and moulds society through a seemingly autonomous trajectory and therefore, becomes the incendiary that propels societal change in unintended ways. The distinct absence of human intervention in this social change is remarkable because of the presupposed imposition of feral brutishness in the way technology lumbers forward unhinged. Bimber provides a fascinating account of technological determinism viewed from the Marxist perspective, and how the interpretation of Marx's theories, in particular, the forces of production, weave an intricate link between economic activity and technology.[202]

However, a refreshed and reinvigorated constructivist approach[203] to technological determinism, taking into account the element of the human agency as a complementary mechanism to the interpretation, can provide us with the basis to ask the right question. In this respect, Allan Dafoe[204] states "the question should not be a dichotomous one of whether technological determinism is right or wrong, but a set of questions of degree, scope, and context: to what extent, in what ways, and under what scope conditions are particular kinds of technology more autonomous and powerful in shaping society?".[205] Dafoe additionally imparts upon this construction the human agency element, by stating that the approach also takes into account pressing questions such as "to what extent, in what ways, and under what scope conditions are particular groups of people able to shape their sociotechnical systems?"[206]

The central idea posited here is that the technological determinism construction of old may no longer be a valid interpretation in this day and age, but that it can, through the constructivist approach, clarify our thinking about the fundamental meaning and aims of new and emerging technologies in modern societies. It may be interesting at this juncture to view the quintessential constructivism of Immanuel Kant, distinguishing between pure and empirical knowledge, stating as follows:

> …. though all our knowledge begins with experience, it by no means follows that all arises out of experience. For, on the contrary, it is quite possible that our empirical knowledge is a compound of that which we receive through impressions, and that which the faculty of cognition supplies from itself (sensuous impressions giving merely the occasion), an

---

[201] Bimber (1990), p. 333.

[202] ibid 335.

[203] Phillips (1995), p. 5.

[204] Dafoe (2015), p. 1047.

[205] ibid 1050.

[206] ibid.

addition which we cannot distinguish from the original element given by sense, till long practice has made us attentive to, and skilful in separating it. It is, therefore, a question which requires close investigation, and not to be answered at first sight, whether there exists a knowledge altogether independent of experience, and even of all sensuous impressions.[207]

Therefore, the constructivist approach to technological determinism, if we so choose, may allow us to find new meaning psychologically, epistemologically, sociologically and historically, in the consideration of the technological burst in society. If we choose to view technologies, and by that implication, new and emerging technologies that aim to improve the functionality and ease of tasks, and not focused on its "specific materiality",[208] taking into account any measurement of empirical and content-specific questions,[209] the contributions to technological intelligence and a way forward in the attempt to design an effective system, can be fruitful.

4.2.3   Legal Pluralism/Cosmopolitanism

The upshot of new and emerging technologies, and also, in many cases, the evolution of older technologies into new and creative ways of use, pose a mismatch and divergent trajectory between the intentions and pursuits of the law, with the meteoric enlargement of these technologies. Indeed, the dangers of this mismatch, particularly if they are large-scale, induce a state of permanent crisis where the regulatory position will never be able to find a path of clarity.[210] From a much more critical viewpoint, it has been correctly pointed out that this therefore violates the Rule of Law.[211] In Lon Fuller's *The Morality of Law*,[212] he sets out eight essential principles of the legality of law, one of which is the principle of congruence.[213] Consistent with the spirit and letter of this principle, this means that the law (or in this instance, the regulatory framework) must be infused with clarity and certainty as to the regulatory position.[214] A mismatch or disconnection of the regulatory environment, from the development of technologies, also possibly violates the principle of constancy, which promulgates stability and knowledge of the applicable expectations from citizens.[215] This is not to say that the regulatory framework or laws cannot be changed at all; instead, a robust mechanism would be necessary to ensure that

---

[207] Kant (2003).

[208] Dafoe (2015), p. 1051.

[209] ibid 1054.

[210] Brownsword (2008), p. 160.

[211] ibid.

[212] Fuller (1969).

[213] Brownsword (2008), p. 161.

[214] ibid 160.

[215] Fuller (1969).

rapidly evolving new and emerging biomedical technologies can be dealt with through calculative supervision, monitoring and opening new lines of debate when necessary to meet the requirements of democratic functions at the societal level.

The democratic functions at the societal level are similarly influenced by the legal environment and cosmopolitan nature of the system. In many jurisdictions such as countries in Africa and certainly, Malaysia and Thailand, the existence of legal pluralism inevitably shapes the manner in which a more sensible tailored regulatory framework must take place. This is consistent with the cosmopolitanism argument that does indeed challenge the framing of a regulatory environment. From the anthropological perspective, cosmopolitanism, at its very heart, aims to "reconcile universal values with a diversity of culturally and historically constructed subject positions."[216] A truly gargantuan ideal, the idea of cosmopolitanism aims "to do justice to the twin ideals of universal concern and respect for legitimate difference."[217] Within the regulatory context, however, the challenges of cosmopolitanism becomes much more acute when juxtaposed against the background of national, regional and international values and plurality.[218] Indeed, Brownsword poses that "on the one hand, the ideal of universal concern demands that there should be no under-interpretation of fundamental value; on the other hand, the ideal of respect for legitimate difference demands that there should be no international over-reaching— even if cultural diversity is not a good in itself, permissible local difference should be treated with respect."[219]

At the core of this discourse, reigns the perhaps misconceived idea of subjugation to a larger sentience beyond the national framework. Alexander Somek's prominent work, *The Cosmopolitan Constitution*,[220] perhaps highlights the main practical difficulty of regulatory cosmopolitanism; to find "effective ways to secure compliance with international or regional regulatory articulations of fundamental values while empowering national regulators who strive to seek compliance with local standards."[221] Somek's third epoch of the division of constitutionalism is what is critical in this dialogue: he identifies this as "The Cosmopolitan Constitution",[222] or as he terms it in the book, "Constitutionalism 3.0". In a nutshell, Somek is of the opinion that this form of 'cosmopolitan' constitutionalism genuinely reaches into roots of political constitutionalism, *vis-à-vis* "the world of a perplexingly diffuse administrative state sans sovereignty juxtaposed with a multilevel system of fundamental rights protection."[223] Does Somek therefore predict the obstreperous conflict between national policy ideals and regional or international normative values? I believe so. This is also precisely the fatiguing juxtaposition that Brownsword

---

[216] Ribeiro (2001), p. 2842.

[217] Brownsword (2008), p. 185.

[218] ibid 186.

[219] ibid.

[220] Somek (2014).

[221] Brownsword (2008), p. 186.

[222] Somek (2014), p. 242.

[223] ibid 22–23.

highlights in his book, through the allegory of patent regimes and the seemingly illusory moral exclusions that restrict patentability across the columnar sphere of Europe and the United Kingdom.[224]

The perhaps irrational fear that is raised by cosmopolitanism is hinged upon the possible fatality of national sovereignty. At the time this is being written, parts of Central Europe are inundated with the twin rhetoric of ideological populism and the rise of illiberal democracies.[225] The existence and longevity of institutions of modern democratic thought, such as universities, institutions of higher learning, nongovernmental organizations, all of which strive for the protection of basic fundamental freedoms and liberties, have come under attack from illiberal ideals founded on narcissism, misunderstandings and an intense misgiving about the role of the European Union.[226]

The precision of this visceral reality, this present phenomenon in Europe is an erudite representation of Somek's main criticism of the "new" cosmopolitan constitution: one that signals "political constitutionalism",[227] as opposed to a legal one. In the space of this "political constitutionalism", Somek credits the rise of "authoritarian capitalism"[228] attributable to "bourgeois Europeanists",[229] and if we indeed, subscribe to this accuracy, then it is also possible that any visions for a proposed regulatory framework for new and emerging technologies, will inversely be viewed as an extension of authoritarian capitalism, through the perpetuation of new movements and developments in science and technology, the enemies of capitalism haters.

As we move into the aspiration (whether born of a misconceived belief or not) that modern societies have become more tolerant, accommodating and similar insofar as the protection and regard given to fundamental human rights is concerned, the inverse may be reflected in terms of our realities. Yuval Harari perceived that we, as humankind, are able to work on collaborative measures (particularly between nations) in a global, monumental scale because of our abilities to conform to a meeting of the minds in the form of "shared fictions."[230] Harari argues that these shared fictions traverse things like religion and money,[231] (he interestingly terms this

---

[224] Brownsword (2008), p. 187.

[225] Uitz (2015), p. 279.

[226] Jan-Werner Müller, 'The Problem with Poland' *The New York Review of Books* (11 February 2016) http://www.nybooks.com/daily/2016/02/11/kaczynski-eu-problem-with-poland/ See also Christian Keszthelyi, 'Government's "Stop Brussels" Campaign Revs Up', *Budapest Business Journal* (2 May 2017) http://bbj.hu/politics/governments-stop-brussels-campaign-revs-up_132259.

[227] Somek (2014), p. 22.

[228] Patterson et al. (2015), p. 667, 676.

[229] Somek (2014), p. 253.

[230] Galen Strawson, 'Sapiens: A Brief History of Humankind by Yuval Noah Harari – Review' *The Guardian* (11 September 2014) https://www.theguardian.com/books/2014/sep/11/sapiens-brief-history-humankind-yuval-noah-harari-review.

[231] Harari (2015), p. 162.

chapter "The Unification of Humankind"), of which, in his book,[232] he provides detailed historical and cultural accounts. Later, in Harari's acccunt of what he terms the "Scientific Revolution",[233] he exposes the colossal consequences of scientific advancements, whilst recognizing that science, in itself, possesses a most unique nature that we may not even begin to fathom; to the point that "science is unable to set its own priorities."[234] of the profundity of humankind's hubris in the scientific quest for immortality.[235]

There is a chance that the quest for human immortality, and perhaps, the span of the Project Gilgamesh[236] as well, is one of the "shared fictions" identified by Harari. And there is certainly a strong reason why these are referred to as "fictions", simply because the realities of globalization, threatening divisiveness on a daily basis, dictate to us otherwise. This kind of "shared fiction", in my opinion, leads one to overestimate one's shared values, creating a cocooned haven in which humankind perceives a kind of singularity, a shared sense of community oneness and unity that detracts from the unforgiving reality and phenomenon of pluralism.

4.3 Analogies of Regulatory Design

In undertaking the elephantine task of regulating existing, new and emerging biomedical technologies, the critical thinking aspects of the theoretical and philosophical foundations for the need of such regulation, also forms part of the crucial narrative in this discourse. In addition to being cognizant of the challenges that regulation will pose in the field of biomedical technologies, it is also useful to then begin about thinking about the suitability of regulatory models with the adaptive capability of meeting the ever-changing needs of technology and its dimensions of use, propriety and other relevant instrumentalities. This section specifically highlights some aspects of the abortion debates and the early prenatal testing technologies as analogies that may be used as models of regulation, as a comparative means of analysis in the discourse. In this analogical exercise, it is also incumbent to identify and control the applicable normative values that we wish to consider in a regulatory framework. In doing so, perhaps it is useful to bear in mind the spirit and intent of addressing the fundamental question of values; this may entail asking the difficult

---

[232] Harari (2015).

[233] ibid 246.

[234] ibid 274.

[235] ibid 266.

[236] Gibbs (2014), p. 168. Project Gilgamesh, deriving its name from the 'Epic of Gilgamesh', the mythical King of Uruk, is an initiative that is focused on the extension of human life and the control of human aging through scientific and medical technologies. Similar to the SENS Research Foundation project on aging, Project Gilgamesh advocates the use of science, cryonics and other methods for radical life extension as its key moral imperatives. See also: http://www.projectgilgamesh.com/what-is-project-gilgamesh/.

questions on the weight that regulators give to values, described as the "key dyadic values",[237] the desiderata that make up the foundational normative framework behind any necessary regulation or laws. Identifying these values, in combination with recognizing the regulatory challenges, may then contribute to the determination of the fundamental principles to be applied in the regulatory space.

Dependent on the balance and weight of interests given to each value, this may not translate to the practice and reality of avoiding Collingridge's Dilemma.[238] The problems that were foreseen by Collingridge in the early 1980s, is certainly the reality of our techno-science period. Indeed, he states "the social consequences of a technology cannot be predicted early in the life of the technology. By the time the undesirable consequences are discovered, however, the technology is often so much part of the whole economics and social fabric that its control is extremely difficult. This is the dilemma of control. When change is easy, the need for it cannot be foreseen; when the need for change is apparent, change has become expensive, difficult and time consuming."[239] In light of this problem, I posit that even the adoption of a specific model of regulatory or governance framework, may not adequately avoid this dilemma raised by Collingridge. However, if we base the initial design of the regulatory framework on normative values and principles of an acceptable standard, instead of specific substance requirements from the very outset, then it may be possible to substantially mitigate the effects of the Collingridge Dilemma.

4.3.1 The Abortion Debates

Throughout the course of the history of women's bodies in the sphere of reproduction, there has been no issue more polarizing and divisive than the abortion debates. In Chap. 5, the human rights framework in each of the selected jurisdictions particularizes more profoundly the development of laws and attitudes to the abortion arguments. On a deeper level, the analysis described in Chap. 5 traces how the abortion debates interlink with some fundamental rights and individual liberties, whether implicitly or explicitly guaranteed under a constitutional system. In this section, however, I trace the polarizing abortion debates on a broader level by questioning its implications on biomedical technologies, and vice versa, and how informing ourselves in these matters will assist in a well-rounded analogical discussion.

Bearing in mind the historical trajectory of the development of laws and attitudes to abortion, the flurry of debates never ceased even after the legalization of abortions in the United States (US), *vis-à-vis* the famous case of *Roe v Wade*.[240] In 1986, the then director of the Hastings Center, Daniel Callahan opined that technology has

---

[237] Trebilcock and Iacobucci (2009), p. 455, 457.

[238] Liebert and Schmidt (2010).

[239] Collingridge (1982).

[240] 'Roe v. Wade, 410 U.S. 113 (1973)' (*Justia Law*) https://supreme.justia.com/cases/federal/us/410/113/.

the capacity to "re-frame" the abortion issues.[241] Although this report is from the 1980s period, Callahan has highlighted the development of abortion to the extent that its relevance continues to permeate contemporary discourse in medical and scientific technologies. In particular, Callahan mentioned several points of note regarding the various implications of legalizing abortions, but I draw on two considerable notes from his discourse, on what I consider to be most relevant for the purposes of this book. First of this is the influence of various interest groups within the spatial domains of a political environment that has allowed the abortion discourse to continue receiving flak[242]; and secondly, the breadth of legal, social and moral implications that "create a new configuration of relevant considerations."[243]

The influence of the pro-life and pro-choice interest groups that began to branch out following *Roe v Wade* has had a tremendous impact in the abortion debates. This was further seemingly strengthened by the views of the then President, Ronald Reagan and his known anti-abortion views. Historical accounts would have us believe that the late President Reagan was vehemently anti-abortion, but other biographical accounts have indicated that he did allow a permissive abortion bill, which was however later withdrawn,[244] 6 years prior to *Roe v Wade*. The lessons that we can take away from this illustration, and as highlighted by Callahan, is that the political forces that shape the landscape of a country's legislative backdrop can be persuasive in nature. The indicative accounts of the political landscape demonstrate how the abortion debates became ingrained into a national social policy agenda that warranted grave concern and legislative control, fuelling the burgeoning pro-life and pro-choice interest groups that continue to wage war for and against abortions.[245] Indeed, John T. Noonan voiced out a scathing concern over the ruling of the Supreme Court in *Roe v Wade*,[246] further lending credence to the polarity of the debates.

As the development of issues in the US regarding abortions continued post-*Roe v Wade*, the contemporary setting over the last 2 years in the country reminds us once more of the difficulties that may challenge the abortion discourse once more with the election of President Donald Trump. President Trump has declared himself

---

[241] Callahan (1986), p. 33.

[242] ibid 33.

[243] ibid 34.

[244] Cannon (1991), p. 812.

[245] Stacie Taranto, 'How Abortion Became the Single Most Important Litmus Test in American Politics' *The Washington Post* (22 January 2018) https://www.washingtonpost.com/news/made-by-history/wp/2018/01/22/how-abortion-became-the-single-most-important-litmus-test-in-american-politics/?noredirect=on&utm_term=.b4015648272b.

[246] Noonan (1977), p. 29.

to be anti-abortion,[247] and there is growing fear that this presidency could throw a spanner into the works of all the progress made in abortion laws in the US.[248]

In Northern Ireland, too, the 'right' to life has been singularly guarded as the sole reason why abortions are illegal. It should be borne in mind that Northern Ireland never agreed nor signed up to the UK Abortion Act of 1967 because of the political divide in the UK. A recent spate of controversies regarding abortion in Northern Ireland has made the headlines of international correspondents; but the historical understanding of Northern Ireland's remarkably astute position on anti-abortion laws because of its deeply religious Catholic and Christian beliefs, is also vital. The lack of support for the Abortion Act of 1967 in the Northern Ireland Assembly also indicates why the act itself does not apply in Northern Ireland.[249] With the exception of the Infant Life (Preservation) Act of 1929, the laws relating to abortion in Northern Ireland have been half-hearted and piecemeal in nature. It was only through controversial cases such as the case of *A, B and C v Ireland*,[250] at the European Court of Human Rights, and the death of Savita Halappanavar[251] in Galway (who died because she was refused an abortion) that prompted a change of law in the Irish republic in 2013.[252] However, the circumstances under which abortions were allowed were extremely limited, and did not allow for abortions in cases of rape, incest or foetal abnormality.

In Australia, the transition for the legalization of abortions across the country has also not been an easy one. In New South Wales and Queensland, abortions are still illegal unless they come within the exceptions stipulated in those states.[253,254] The

---

[247] Lee, Michelle Yee Hee, 'Donald Trump's Claim He Evolved into "pro-Life" Views, like Ronald Reagan' *The Washington Post* (31 March 2016) https://www.washingtonpost.com/news/fact-checker/wp/2016/03/31/donald-trumps-claim-he-evolved-into-pro-life-views-like-ronald-reagan/?noredirect=on&utm_term=.a98e5a4e62f8.

[248] Megan Farokhmanesh, 'How a Trump Administration Threatens Women's Health', *The Verge* (12 December 2016) https://www.theverge.com/2016/12/12/13904032/trump-womens-reproductive-health-affordable-care-planned-parenthood.

[249] Jon Kelly, 'Why Are Northern Ireland's Abortion Laws Different?' *BBC News* (8 April 2016) http://www.bbc.com/news/magazine-35980195.

[250] Case of *A, B and C v Ireland* [2010] Grand Chamber 25579/05.

[251] BBC, 'Abortion "Would Have Saved Wife"' *BBC News* (14 November 2012) http://www.bbc.com/news/uk-northern-ireland-20321741.

[252] Allison O'Connor, 'How the Death of Savita Halappanavar Changed the Abortion Debate', *The Irish Examiner* (28 October 2017) http://www.irishexaminer.com/analysis/how-the-death-of-savita-halappanavar-changed-the-abortion-debate-461787.html.

[253] Tom Gotsis and Laura Ismay, 'Abortion Law: A National Perspective, Briefing Paper No. 2/2017' NSW Parliamentary Research Service https://www.parliament.nsw.gov.au/researchpapers/Documents/Abortion%20Law.pdf.

[254] In other Australian states, abortion is legal up until a period of time, or such other expressly stipulated condition in the relevant State statutes. For example, in the Australian Capital Territory and South Australia, abortion is legal if medically agreed upon by 2 doctors that it would be necessary for the benefit of the woman's physical or mental health. In Victoria, Tasmania, Western Australia and the Northern Territory, abortion is legal up until a specified period of time, after which special considerations may apply.

depth of importance placed on the right to life features prominently in these anti-abortion movements, which makes this a continuous human rights issue that pits "life" against "choice" (privacy).[255] Of particular note in this context is the highly significant Zoe's Law Bill in the Parliament of New South Wales.[256]

The impact of Zoe's Law Bill is highly significant because it would dramatically change the landscape of not only abortions in Australia, but also how 'personhood' was viewed for an in-utero foetus. The effect of Zoe's Law would be to attribute legal status to the child in-utero as a "person"; on this basis, the arguments, on all spectrums of religion, philosophy, science, and the like,[257] is difficult enough as it is to meet any form of reconciliation. The underlying reasoning for Zoe's Law was to amend the Crimes Act of 1900, to include a prohibition against conduct or acts that "causes serious harm to or the destruction of a child in utero."[258]

The background for the introduction of Zoe's Law in 2013 was to counter from a legal standpoint, the death or serious harm caused to an unborn child in utero such as that in the case of Brodie Donegan; in 2009, one Brodie Donegan suffered the loss of her child in utero, posthumously named Zoe, after she had been hit by a drunk driver on Christmas Day.[259] The implications, if Zoe's Law were to be successful, would have the serious effect (whether intended or unintended) of eroding women's rights and access to abortions; although, from the criminal justice point of view, manslaughter or grievous bodily harm inadvertently or negligently caused to an unborn child or foetus, cannot be disputed and must be treated with serious accountability by the law. Besides the contentious nature of defining legal 'personhood' by extending it to a foetus in utero, and re-defining, legally, and by explicit mention in a legislative statute, when life begins, it is posited that this is inevitably a backhanded manoeuvre to curtail women's access to safe and legal abortions, and generally, their reproductive liberties; and I further pose, an imposition of non-secular, religious views upon the general population regarding the conception of life. Amongst some furore, however, in 2013, Zoe's Law Bill was not successfully passed and had been defeated in the Upper House of the New South Wales Parliament.[260]

---

[255] International Women's Development Agency (IWDA) 'Reproductive Rights, Abortion & Zoe's Law: Why Freedom of Choice Is Still Feminism's Biggest Fight' (26 April 2015) https://iwda.org.au/reproductive-rights-abortion-zoes-law-why-freedom-of-choice-is-still-feminisms-biggest-fight/.

[256] Crimes Amendment (Zoe's Law) Bill 2017 https://www.parliament.nsw.gov.au/bills/Pages/bill-details.aspx?pk=2936.

[257] Christina M.H. Powell, 'Being Human: How Should We Define Life and Personhood?' *Enrichment Journal* http://enrichmentjournal.ag.org/201002/201002_134_define_person.cfm.

[258] Crimes Amendment (Zoe's Law) Bill 2017 (n 256).

[259] Jenny Noyes, 'On Zoe's Law, And The Accidental/On Purpose Erosion Of Your Reproductive Rights' *Junkee* (21 November 2013) http://junkee.com/on-zoes-law-and-the-accidentalon-purpose-erosion-of-your-reproductive-rights/21659.

[260] ibid.

However, more recently, the Zoe's Law Bill was reintroduced in 2017, *vis-à-vis* the Crimes Amendment (Zoe's Law) Bill 2017,[261] by Rev. Fred Nile, Australia's Christian Democratic Party Leader and a well-known pro-life activist. The opponents to the bill have described it as "a stalking horse to restrict women's abortion rights,"[262] although Rev. Nile dismisses the claims about Zoe's Law Bill being an anti-abortion bill, and reiterates that safeguards have been put into place for medical exemptions for abortions. In the meantime, the Crimes Amendment (Zoe's Law) Bill 2017 has passed its second reading in the Legislative Council of the Parliament of New South Wales; whether it will be successfully passed in the future remains to be seen.

Returning to the second point of note in Callahan's report,[263] the abortion debates, and particularly in our contemporary settings, becomes more intertwined and expanded in its scope when medical and scientific technologies develop and put forward new ways of treatment, rehabilitation and curative mechanisms in the field of reproduction. As reproductive biomedical technologies became more sophisticated, including within the fields of prenatal genetic testing, the legal, social, ethical and moral implications that initially clouded the abortion debates, grew to encompass embryo selection in reproductive technologies in the same category. The development and progress of neonatal medicine,[264] embryological knowledge[265] and even growing legal and philosophical feminist theories[266] are all developing dimensions that contribute to the changing landscape of the abortion debates. Whatever the circumstances, it cannot be denied that abortion was seen as "an object of political struggle over the terms and future of women's citizenship"[267] and the advancement of rights pertaining to abortion was identified as an effort to "restructure the social and economic order"[268] through constitutional law.

Indeed, such is the propensity of the advancement of biomedical technologies, and certainly now too where the outcomes of gene editing and interventions are no longer simply possibilities, but can someday be translated into reality, that it becomes much more necessary to rein in the speculative fears at the risk of the erosion of the human rights protection that have been put in place. Studying the developmental framework of national policies, attitudes and legislation relating to abortion allows the translation of these findings to be permuted into a future regulatory dimension for biomedical technologies that will address the shortcomings of the past.

---

[261] Crimes Amendment (Zoe's Law) Bill 2017 (n 256).

[262] Kelsey Munro, 'Fred Nile Gives Renewed Push to Zoe's Law to Criminalise Harm to a Fetus', *Sydney Morning Herald* (10 March 2017) http://www.smh.com.au/nsw/fred-nile-gives-renewed-push-to-zoes-law-to-criminalise-harm-to-a-fetus-20170309-guup40.html.

[263] Callahan (1986).

[264] ibid 34.

[265] ibid 35.

[266] ibid 38.

[267] Erdman (2015), p. 39.

[268] ibid.

4.3.2 Prenatal Testing Technologies

Unbeknownst to the future of biomedical technologies, it is possible to pinpoint the reproductive technology revolution from the emergence and progress of neonatal medicine[269] and scientific and scholarly work representations in the field of reproduction. In the late 1970s, the case of Louise Brown[270] thrilled the world, because it indicated the possibilities of overcoming infertility and a new hope in reproductive technologies. As encompassed by the premise in Chap. 3, the concerns that grew in simultaneity with the progress of reproductive technologies were varied, particularly in cases like Pre-Implantation Genetic Screening (PGS) and PGD that led to the embryo selection phenomena. The fear, of course, is that genetic interventions are likely to take the same course of discussion as it already has in contemporary debates.

However, prior to PGS and PGD and the firmament of legislation that governed those activities, the early prenatal testing technologies from the 1970s and 1980s were much more simplified and practically scarce. As part of the incumbent prenatal care that supported pregnancies, the scope of prenatal testing was much more confined to circumstances to ensure the general health and well-being of mothers and their prospective children. These included basic ultrasound procedures, chorionic villus sampling (CVS) and if needed, amniocentesis. The latter prenatal testing technologies were developed as early means to determine genetic abnormalities that particularly targeted the Down's Syndrome genetic mutation. Scholars like Sonia Suter have identified that a general mix of societal norms, professional views from the medical and scientific community, and a growth of knowledge relating to the inheritability of genetic conditions has influenced the manner in which prenatal testing technologies have become *de riguer* in pregnancies and are less likely to raise the concerns that accompany, in the inverse, late on-set genetic testing.[271] Suter claims, however, that the "routinization" of prenatal testing mechanisms contribute to a different set of issues that hinge on informed consent and proper genetic counselling.[272] The question that we should therefore pose at this juncture is whether we may impose the routinization of the earlier prenatal testing technologies, to the future possibilities of genetic interventions in reproductive technologies particularly. The movement from these early prenatal testing technologies into the more sophisticated fields of PGS and PGD are indicative that the identifiable controlling factor in these circumstances is the genetic element in reproduction.

The success of the Human Genome Project,[273] for one, has enabled a multitude of discoveries about the human genome and heritability of human conditions. Suter

---

[269] Callahan (1986), p. 34.

[270] Kamel (2013), p. 156.

[271] Suter (2002), p. 233.

[272] ibid 235.

[273] National Human Genome Research Institute (NHGRI) 'An Overview of the Human Genome Project' https://www.genome.gov/12011238/an-overview-of-the-human-genome-project/.

is correct in the claim that insofar as prenatal testing is concerned, the assumption that is made is that all women who are pregnant will undergo prenatal testing. She says that the question is no longer as simple as whether CVS or amniocentesis is performed, but whether other kinds of genetic screening technologies are instead.[274] Although Suter's work is very much more focused on redrawing the boundaries of routinization in prenatal testing and how the scope of informed consent must be more artfully considered, the findings proves the significance of success of the human genome mapping project. The movement into prenatal *genetic* testing, as opposed to simply prenatal testing, provides us with insight into the contemporary movements into the spatial discussion about genetic interventions at the prenatal level.

Whether prenatal testing involves either invasive or non-invasive treatment methods, the sheer volume of legal and ethical considerations are abound. In 2017, the Nuffield Council on Bioethics issued a report on the ethical issues related to non-invasive prenatal testing.[275] The Chair of the Working Council, Tom Shakespeare indeed recognized that genetic knowledge opens up the realm of personal discoveries and benefits, but like any other advancements made for the purported betterment of humankind, safeguards are necessary to be put into place to prevent "dangers of a genomic future."[276] In brief, the report explores the broad range of ethical approaches that must be taken in the proffering of non-invasive prenatal testing that may cover screening tests for Down's, Edwards' and Patau's syndromes; including specific values relating to choice and autonomy, avoidance of harm, and inclusion and equality.[277] These prenatal testing technologies are intended to be "offered as part of the NHS[278] fetal anomaly screening programme from 2018."[279] Although the report identifies that there is no specific legislation that provides any guidance on prenatal testing of this nature, the fall-back on professional regulators is highly advised and encouraged; some of these may include the parameters of the UK Medical Devices Regulations 2002, the Abortion Act of 1967 and even the Equality Act of 2010.[280] It should be borne in mind however that the report specifically deals with non-invasive prenatal testing, and is therefore different from the scope of PGD[281] and certainly genetic intervention technologies, which may require more invasive methodological treatment. However, the lessons that may be taken from the issuance of the report bear upon the same ethical values and principles that must be

---

[274] Suter (2002), p. 252.

[275] Nuffield Council on Bioethics, *Non-Invasive Prenatal Testing: Ethical Issues* (2017).

[276] ibid.

[277] ibid 2.

[278] The NHS is the UK's National Health Service under the Department of Health and Social Care. It is an executive but non-governmental public body that is primarily tasked with the management, sustainability and development of health care in the UK.

[279] Nuffield Council on Bioethics (2017), p. 2.

[280] ibid.

[281] The scope of PGD services is presently being regulated under the purview of the Human Fertilisation and Embryology Authority in the UK.

applied when considering genetic interventions to be applied on the human body. Considerations may also be had to the Nuffield Council on Bioethics report on gene editing, which ultimately lays out a two urgent ethical issues to consider in terms of gene editing: human reproduction[282] and livestock.[283] I put forward that these reports should not be considered in isolation, and must draw a strategic convergence point to enable a unification of ethical approaches in prenatal testing technologies with gene editing, as is the possibility raised for the future.

Further to the "routinization" of prenatal testing technologies raised by Suter, it is also of equal importance to consider the margins of "medicalization."[284] "Medicalization" refers to the phenomenon where human or life problems become more understood and are therefore treated as "medical" issues There is no acceptable consensus on whether this is good or bad, as sociologist Peter Conrad notes that medicalization should be attributed a neutral value.[285] However, within the context of genetic interventions, we must be careful to consider the differences between a medical or therapeutic treatment or problem, and a non-medical or non-therapeutic one that may have crossed into the boundaries of "medicalization." Michael Morrison has persuasively argued that the medicalization theory has been subject to political machinations in the diagnosis and treatment of individuals, whilst attempting to sideline economic, environmental and other approaches to redress the unequal distribution of health and illness in society.[286] It is easy to make the claim for personal autonomy and individuality, and the free will attributed to decision-making as a guaranteed and fundamental constitutional liberty; but the treading danger of a wholesale acceptance of individual autonomy would also mean that "medicalization" may be given a broad term, erroneously assumed, and be extended into the class settings of a social order.

In fact, part of the work of sociologist, Anne Kerr, that has been focused on PGD in the UK, reveals that PGD is presented to women in the context of routine testing associated with pregnancy. The implication of this is the assumption by most women that doctors do indeed recommend such screening because any responsible mother would also do so. The shift of responsibility to the woman or her husband and/or family is therefore an indirect 'nudge' into this direction, prompting us to question the true nature of autonomy and free choice. Because a failure to undergo such testing results lies upon the woman, the ensuing responsibility for having a disabled child also then falls upon the woman (and/or her family). Translating this similar autonomy into the field of genetic interventions, a hypothetical example would involve prospective parents claiming the necessity of "medicalization" to engage in genetic interventions, as a means of enabling their future offspring to obtain better

---

[282] Nuffield Council on Bioethics, *Genome Editing: An Ethical Review* (2016).

[283] ibid 117.

[284] Maturo (2012), p. 122.

[285] Parens (2013), p. 28.

[286] Morrison (2016), p. 720.

intelligence,[287] or better propensity in sports or other activities,[288] based upon similar assumptions of necessity in the routinization of genetic testing. Another example about how "medicalization" has shifted in such a manner that affects infertility[289] is the work of Ann Bell. Her research findings explore the margins of medicalization in infertility, particularly focused on the groups that may not be within the circle of reproductive care. Spanning participants comprising women of lower socio-economic status and in same-sex relationships, and men who were part of an infertile couple, her findings demonstrate the inequalities of medicalization based on gender, class, and social location.[290]

Instead, Erik Parens suggests that "medicalization" in itself should be considered through a fresh perspective, by reviewing our notions of "medicalization" (not bad) and "over-medicalization" (not good).[291] The contextual nature of genetic interventions certainly requires this reflection, and necessitates drawing the lines to ensure that the world is not "turned into one huge hospital, where everyone is everybody else's humane nurse."[292] This may be a well founded fear, but the medicalization discourse must be addressed in genetic interventions, and iron out any form of complexities, ambiguities and uncertainties that may lead towards a path of frivolous human activities simply to enhance the way in which the experience of life is savoured. Particularly more urgent would be in environments where free choice is not really free choice, but embedded in a context that imposes wider socio-cultural constraints that indirectly shows State endorsement for the view of disabled lives not being worth living. In such instances, this reinforces my view that individual or parental autonomy alone is insufficient to prevent a form of social or market-based eugenic outcomes, and must be coupled with regulatory models of rights-based and constitutional governance that might service to protect the underprivileged.

4.4   Values and Modalities in Regulatory Approaches

Framing the conceptualization of institutional impetus or reform for biomedical regulatory frameworks is not an easy task. In addition to an overwhelming focus on substantive issues, the practicality of the focus, however, also requires that the

---

[287] Sanders (2017), p. 14. The reality however is that at present, there appears to be at least 52 identifiable genes that are linked to human intelligence and these do not work in isolation, but with each other. As such, it is not yet possible at this juncture to simply "edit" the relevant genes that influence human intelligence because an alteration to one gene linking intelligence is likely to influence another linked human intelligence gene. Nevertheless, the prospect of such possibilities in the future should, in any event, be considered.

[288] Sandel (2004), p. 51. Sandel provides illustrative examples of athletes and questions the significance between 'doping' in sports and genetic enhancements to improve athletic prowess.

[289] Bell (2016), p. 39.

[290] ibid 44.

[291] Parens (2013), p. 29.

[292] ibid 35.

substantive nature of these policies be "mediated through the institutions that investigate, enforce and adjudicate..".[293] In a piece of work on competition law by Trebilcock and Iacobucci, they highlight the underlying values that generate the internal workings of institutions, which are not, in each of the criteria, inherently controversial. The provocation, however, lies in the fact that "each value implies an obverse value and indeed interactions with other values, thus rendering the weighting of, or trade-offs among, values a quintessential polycentric and highly contestable exercise."[294]

Trebilcock's and Iacobucci's pronouncement of the key normative dyadic values in a framework poses each value in interaction with another. For example: independence versus accountability, expertise versus detachment, transparency versus confidentiality, administrative efficiency versus due process, and predictability versus flexibility. Trebilcock and Iacobucci further explain the desyncronization between these dyadic values. Indeed, the authors recognize that the representation of these values is also often diffused with the competing concerns that are faced in reality. For example, the value of administrative efficiency is compatible with, but also inverse to the value of due process.[295] Because of the constant tug-of-war that may ensue between the dyadic values, it is recognized that "a complex, subjective, and inevitably highly contentious optimizing calculus is involved."[296] However, in the light of trying to establish pragmatic solutions in the context of biomedical technologies, I pose that it becomes incumbent on the regulatory authority or legislator, as the case may be, to apply existing rules of construction or interpretation *vis-à-vis* the larger constitutional framework. The assessment of the applicability of constitutional values would be markedly different from one jurisdiction to another and would also be dependent on the emphasis placed by each legal system on considerations of fundamental rights and liberties.[297]

For instance, in the context of the US, legislation impacting on fundamental rights in accordance with the protection granted under the United States Constitution would be subject to a determination of constitutionality in accordance with the tiered scrutiny tests, or what is more commonly known as the "levels of scrutiny" first established in *United States v Carolene Products Co.*[298] Although the issue of fundamental rights was not the main question to be determined in the case, Justice Stone, in his judgment of the case, in "the most famous footnote in constitutional law,"[299] Footnote 4 in the judgment, made an express defined point that the courts were to apply stricter standards of review and a higher level of scrutiny to specific

---

[293] Trebilcock and Iacobucci (2009), p. 455.

[294] ibid 457. Foreword, p. vii.

[295] ibid 458.

[296] ibid 459.

[297] Brownsword (2008).

[298] 'United States v. Carolene Products Co. 304 U.S. 144 (1938)' (*Justia Law*) https://supreme.justia.com/cases/federal/us/304/144/case.html.

[299] Exploring Constitutional Conflicts, 'Levels of Scrutiny Under the Equal Protection Clause' http://law2.umkc.edu/faculty/projects/ftrials/conlaw/epcscrutiny.htm.

range of cases, especially where these cases may involve a violation of fundamental rights and liberties.

In the European context, as a comparison of the treatment of fundamental rights at the level of the ECtHR, for example, the proportionality test is used as a means of trying to gauge permissible limits of interference by a particular state into matters relating to fundamental rights or liberties. The proportionality test, originating in Germany and with the exception of the US, is now being used in most countries around the world in constitutional rights adjudication.[300] Justice Aharon Barak defines proportionality as a "legal construction"[301] comprised of four (4) important components: "proper purpose, rational connection, necessary means, and a proper relation between the benefit gained by realizing the proper purpose and the harm caused to the constitutional right."[302] Hence, any law or legislation that seeks to limit the scope of a constitutional right would be subject to the proportionality test "to pass constitutional muster."[303]

Therefore, in determining the weight to be given to a particular dyadic value, whether based on the levels of scrutiny tests prevalent in US constitutional jurisprudence, or the proportionality test advocated in constitutional rights adjudication in other parts of the world, I advance the argument that in the event a regulatory or governance framework on the national or federal level is to be established (for example, in the US), the values in regulatory design can be utilized *vis-à-vis* an operable calculus that employs the varied constitutional doctrines of scrutiny or proportionality, as the case may be, in the same way that issues relating to the constitutionality of legislation affecting fundamental rights is concerned.

Another perspective in fabricating the design of a regulatory framework presents itself in adaptive governance, which has been touted to be an "evolving research framework for analysing the social, institutional, economic and ecological foundations of multilevel governance modes that are successful in building resilience for the vast challenges posed by global change."[304] The fundamental tenet of adaptive governance involves a network of dialogue, discussions, built-in consultative processes and reflections on multiple stakeholder levels, through the work of different special bodies established to pursue this necessary imperative. The main idea behind adaptive governance is aimed not only at creating communicative relationships between legislators, governments and interest groups, but also allows for the extended input of perspectives from specialist experts in a variety of disciplines that cross-cuts international borders. It is acknowledged that protecting the sustainability of, and access to resources necessitates the collaborative efforts of various disciplines ranging from geography, sociology, to political science, policymaking, and law. What is particularly valuable about the exercise in adaptive governance is that

---

[300] Sweet and Mathews (2008), p. 72.

[301] Barak (2012), p. 131.

[302] ibid.

[303] ibid.

[304] Stockholm Resilience Centre, 'Adaptive Governance' (6 December 2010) http://www.stockholmresilience.org/research/research-streames/stewardship/adaptive-governance-.html.

an understanding of "interactions between individuals, organizations and institutions at multiple levels"[305] has the capacity to contribute to how responses to changes and crisis are influenced. This could certainly be very useful in the field of biomedical technologies.

The premise relied upon is that the foundational aspects of adaptive governance can be imported and emulated into the discourse on designing a workable, comprehensive regulatory framework for new and emerging technologies. The justification for this reliance is incumbent upon the recognition of similarities relating to both environmental sustainability and changes, and the nature and reach of new and emerging technologies. For example, the subject matter in both these fields are highly dynamic in nature, imbued with intense fluidity and rapidly changing, and often-demonstrating lofty complexities and uncertainties incapable of keeping up with forces beyond the control of the proponents.[306] The density and rapidness of environmental change and its complexities in the intertwining interactions with multiple stakeholders, particularly in the field of climate change, biosphere conservation and even management and preventive or remedial actions relating to infectious diseases governance, makes it necessary for a regulatory or governance model that is capable of moving in a highly adaptive and flexible manner.[307] The same can be said of new and emerging technologies. The crucial debates regarding the ramifications in pre-implantation genetic interventions is keenly felt in relation to the stakeholders that are most likely to be affected by these technologies (especially in relation to access, costs, insurance coverage, tax reliefs and such), the third parties outside of the legislative, policymaking and scientific research arena. The broad implications on the spectrum of human rights corpus must be masterfully adept at moving along regardless of the extraneous implications of the development of these technologies.

It has been recognized (and correctly so particularly in industries where technologies and circumstances move at a lightning speed), that governance models such as "top-down, state-based orientation, rarely match the relevant scale of [ecological] complexity"[308] and that "centralized governance via top-down directives or command-and-control policies often fails to provide effective solutions for highly contextualized situations."[309]

Following the impact of new and emerging technologies in the rights and rights discourse analysis, the involvement of a diverse but expertized field of stakeholders at multiple levels is crucial. The consequential framework may not be a legally enforceable framework in its stages of infancy as such, but the ellipsis and strength of best practices can begin to serve as focal points of acknowledgement. Scholars have also envisaged that a transition to a more holistic form of governance of this nature may take a considerable amount of time. Much would depend on the kind of

---

[305] ibid.

[306] Chaffin et al. (2014), p. 56.

[307] ibid.

[308] ibid.

[309] ibid.

framework envisaged by the incorporation of adaptive governance practices into existing models of regulation for biomedical technologies, dependent on jurisdictional and industry-driven practices.

An equally important component in the union between laws and regulatory approaches is the structural model of the regulation, through the culmination of different agency models. One potential model is the bifurcated judicial model,[310] which rests on the basis that "specialized investigative and enforcement agencies must bring formal complaints before and seek remedial relief from the courts, subject to normal rights of appeal to appellate courts."[311] The authors then assess the 'performance' of this bifurcated judicial model marked against the normative dyadic values they have identified, citing positive responses relating to accountability, transparency and confidentiality, but lower on the scale of expertise, timeliness and process costs.[312]

The second model, the bifurcated agency model,[313] particularly "seems designed to achieve a reasonable balance amongst the various values."[314] This is because the bifurcated agency model demarcates a separation between specialized, investigative and enforcement agencies, with specialized adjudicative agencies. This, to me, is consistent with the concept of separation of powers in constitutional systems, enabling the dyadic values to be complementary to each other more effectively. For example, similar to the different branches of a government in the separation of powers model, the bifurcated agency model theoretically performs similar functions as the legislature, executive and the judiciary respectively. The model would ideally comprise of a specialized bureau that could undertake review, investigation and enforcement responsibilities to ensure compliance with regulation. Adjudicative functions relating to breach of regulations would be handled by a tribunal with adjudicative functions (this branch could be set up in accordance with legislation, for example, like the HFEA pursuant to the HFEA Act in the UK). Recourse could also be had to the conventional courts system within a jurisdiction, where specific points of law may be addressed after the tribunal system has been exhausted.[315] This model could ensure that there is independence in the exercise of judicial functions through the tribunal system, whilst also ensuring accountability in performing this function, through the courts judicial appeals process, if necessary. Other values such as due process, predictability, and flexibility would also be afforded through the proposed mix of judicial and non-judicial membership of the tribunal, and the recourse to the conventional courts system.

The third and final model is the integrated agency model, in which a single agency bears all responsibilities and functions of investigation, enforcement and

---

[310] Trebilcock and Iacobucci (2009), p. 460.

[311] ibid.

[312] ibid 461.

[313] ibid 462.

[314] ibid.

[315] ibid 463.

adjudication.[316] Accountability and expertise, for example, are high scorers in this model, but the fact that adjudicative functions are also carried out by investigators and enforcement officers inevitably mean that the reality and practicality of adjudicative decisions may be tainted with some form of bias, particularly in a specialized circle such as competition law.[317] These models are not new within the sphere of legal, regulatory or governance systems, but its worth indicates to us the difficulties at a pragmatic level, particularly at stages of implementation and enforcement.

Other forms of models that may be interesting to explore are forms of modalities proposed by Lawrence Lessig. In the summer of 1997, Lessig, at The State of the First Amendment At the Approach of the Millennium Symposium,[318] gave compelling arguments about the effects of regulation and the law's focus on regulatory constraints.[319] At that time (1997), before *Code and Other Laws of Cyberspace*[320] was published, Lessig was of the opinion that "behavior in real space is regulated by [at least] three sorts of constraints."[321] He identifies these to be law, social norms and nature.[322] These are the pre-cursor to his "four regulatory modalities (or modes of regulation),"[323] which "characterizes the activities of regulators back in the real world."[324] In *Code*, he expands on the initial constraints, and refers to this as "the law, social norms, the market, and architecture".[325] Lessig recognizes that each of these constraints is "distinct, yet they are plainly interdependent,"[326] and important in reassessing the limits of constitutional restraints on existing, new and emerging technologies.

Lessig's discourse is qualified with a self-profession that he is a constitutionalist.[327] On this basis, his declaration on liberty is apt: "We build a world where freedom can flourish not by removing from society any self-conscious control, but by setting it in a place where a particular kind of self-conscious control survives. We build liberty as our founders did, by setting society up on a certain constitution."[328] However, Lessig does not simply envisage a written constitution with the weight of legality. Instead, he refers to the constitution in this context as a product of architecture, "constitution as in lighthouse— a that helps anchor fundamental values."[329] Essentially, his argument is not for a "top-down form of control," but "the values

---

[316] ibid 464.

[317] ibid.

[318] Lessig (1997), p. 181.

[319] ibid 181.

[320] Lessig (1999).

[321] Lessig (1997).

[322] ibid.

[323] Brownsword (2008), p. 13.

[324] ibid.

[325] Lessig (1999), p. 87.

[326] ibid 88.

[327] Lessig (2006), p. 4.

[328] ibid.

[329] ibid.

that a [constitutional] space should guarantee." Lessig's model is rich in the inter-weaving complexities between the divergence of laws and technologies, asking "Are we, in the digital age, to be a free society?"[330] In many ways, the outcome of his work mirrors the main premise of this chapter, which requires us to ask, on a very crucial level, as to the kind of values we wish to protect.

4.5   The Role of Law in Regulatory Frameworks

In all these circumstances, however, we should not dismiss the role that law plays, even if the regulatory approaches chosen represent the most effective way for the operation of biomedical technologies to work. The laws, *vis-à-vis* measures of con-trol by the state, or governments, can be mobilized in simultaneity and integrated into present guidelines relating to biomedical technologies. In particular, Barbara Cosens et al identify that "the law can and, in fact, must be made adaptive to facili-tate and even trigger the emergence of adaptive governance and to aid in institution-alizing adaptive governance as it emerges."[331] In biomedical technologies, it is acknowledged that there is an infinite struggle between an effective governance system and "a co-evolutionary race."[332] An action as simple as the provision of information[333] can have a lasting, positive impact on the multiple stakeholders because effective governance requires "good, trustworthy information about stocks, flows, and processes within the resource systems being governed, as well as about the human environment interactions affecting those systems."[334] In complicated sys-tems, the existence of inequalities in bargaining positions and power relations can also affect governance on a larger scale of things, specifically where dispute or conflict resolution is concerned.

This is consistent with Foucault's view on power relations,[335] where he argues that the actuality of power relations in daily societal lives impact tremendously on the co-relation between individuals and institutions. Instead of considering power in this narrow context and viewing it purely from the oppressive point of view, Foucault's philosophy moves towards its relationship with resistance, opining that there are positive effects that can be established between an individual's self-development and his or her relationship to the institutional framework. Insofar as technologies are concerned, the possible ensuing consequences are predominantly between researchers, investors in the technologies, and pharmaceutical (also known as the proliferation of "Big Pharma") or medical technology companies looking to exploit the fruits of such technological advancements, and/or the legislators and

---

[330] Lessig (2001).

[331] Cosens et al. (2017), p. 30.

[332] Dietz et al. (2003), p. 1907.

[333] ibid 1908.

[334] ibid.

[335] Foucault and Gordon (1980).

policy makers in the field. A manner of conflict resolution suggested by Dietz et al.,[336] for example, includes using a "broadly participatory process"[337] as well as the experimentation of various governance models, using ballots and polls, for instance, as a complement to managerial roles in ministerial authorities.[338]

The role that may be played by the law is three-fold: first, the law has the ability to create a disturbance or window of opportunity.[339] This essentially means that crises characterized by, for example, political or economic factors may be treated as an impetus to "trigger the emergence of new approaches to governance"[340] and conversely, force changes in behaviour to the treatment of technologies, for example. Secondly, law also plays a crucial role in eliminating debilitating boundaries and facilitating adaptive processes[341] *vis-à-vis* the action of conducting inquiries into dealing with governmental action in the governance of a particular system. Thirdly, the law can also be instrumental in terms of imparting legitimacy in adaptive governmental processes.[342] The finicky question of administrative processes, bureaucratic red tape and the like, can often be unintended barriers to the resolution of rising problems. The maintenance of an adequate administrative system is necessary, to ensure equity and justice, particularly in terms of access and allocation of resources, and the overall management by the authorities in relation to the implementation and enforcement of regulatory measures.[343]

At the heart of the discourse of 'regulation', from values and principles as a starting frame, and ultimately moving to models/modalities of variable forms of governance, the conclusion that can be made at this juncture is that the applicability of biomedical technologies, and even existing technologies that may have morphed beyond the practicality envisaged from the time they were "new" or "emerging", must continue to be available for the improvement of the general human condition, in particular, relating to all aspects of human health and wellness. Within the context of democratic management of these technologies, which should ideally allow for the advancement, research, development and egalitarian standards of human rights considerations in the technologies, and drawing from the historical foundations of any one particular legal system, the promise of regulation or some form of governance framework is a vital component in ensuring these aims.

Brownsword indeed determined with clarity that it is not easy to be specific about the "characteristics of a regulatory (or regulated) environment"[344]; very much depends on, for example, the significance attributed to a formal legal presence, or

---

[336] Dietz et al. (2003).

[337] ibid 1909.

[338] ibid.

[339] Cosens et al. (2017), p. 32.

[340] ibid.

[341] ibid 33.

[342] ibid 35.

[343] ibid.

[344] Brownsword and Goodwin (2012), p. 27.

whether such systems would benefit from self-regulatory mechanisms.[345] Hence, "when we put together the different strands of (governmental) law, (governmental) regulation, and (non-governmental regulatory) governance, we have the ingredients- ingredients that are still largely normative- that make up the particular 'regulatory environment'."[346] But clearly, within the framework of this regulatory environment, the role of the law is still a paramount pronouncement for the prevention of a descent into an unregulated chasm of chaos. Its coupling, with other forms of regulatory governance in the manner advanced in this chapter, would therefore be ideal in formulating a multi-faceted legal and regulatory framework, including the international dimension, for biomedical technologies.

References

Airedale NHS Trust v Bland [1993] AC 789

Aveyard H (2002) Implied consent prior to nursing care procedures. J Adv Nurs 39:201

Ayres I, Braithwaite J (1992) Responsive regulation: transcending the deregulation debate. Oxford University Press

Baldwin R, Black J (2008) Really responsive regulation. Modern Law Rev 71:59. https://doi.org/10.1111/j.1468-2230.2008.00681.x

Baldwin R, Cave M (1999) Understanding regulation: theory, strategy, and practice. Oxford University Press

Barak A (2012) Proportionality: constitutional rights and their limitations. Cambridge University Press

BBC News (14 November 2012) Abortion "Would Have Saved Wife". http://www.bbc.com/news/uk-northern-ireland-20321741

Bell AV (2016) The margins of medicalization: diversity and context through the case of infertility. Soc Sci Med 156:39

Bentham J, Burns JH, Hart HLA (1988) A fragment on government. Cambridge University Press

Beyleveld D, Brownsword R (2007) Consent in the law. Hart, Oxford

Biegel S (2001) Beyond our control?: confronting the limits of our legal system in the age of cyberspace. MIT Press

Bimber B (1990) Karl Marx and the three faces of technological determinism. Soc Stud Sci 20:333

Black J (2005) What is regulatory innovation? In: Black J, Lodge M, Thatcher M (eds) Regulatory innovation. Edward Elgar

Braithwaite J (2002) Restorative justice and responsive regulation. Oxford University Press

Braithwaite J (2017a) Types of responsiveness. In: Drahos P (ed) Regulatory theory: foundations and applications. Australian National University Press

Braithwaite J, Drahos P (2000) Global business regulation. Cambridge University Press

Braithwaite V (2017b) Closing the gap between regulation and the community. In: Drahos P (ed) Regulatory theory: foundations and applications. Australian National University Press

Brownsword R (2008) Rights, regulation, and the technological revolution. Oxford University Press

Brownsword R (2011) Why I Wrote … rights, regulation, and the technological revolution. Clin Ethics 6:207

---

[345] ibid.

[346] ibid 26.

Brownsword R, Goodwin M (2012) Law and the technologies of the twenty-first century. Cambridge University Press

Brownsword R, Yeung K (2008) Regulating technologies: legal futures, regulatory frames and technological fixes. Hart

Burns JH (2005) Happiness and utility: Jeremy Bentham's equation. Utilitas 17:46

Callahan D (1986) How technology is reframing the abortion debate. Hastings Center Rep 16:33

Cannon L (1991) President Reagan: the role of a lifetime. Public Affairs

Carroll D (2011) Genome engineering with Zinc-Finger nucleases. Genetics 188:773

Case of A, B and C v Ireland [2010] Grand Chamber 25579/05

Chaffin BC, Gosnell H, Cosens BA (2014) A decade of adaptive governance scholarship: synthesis and future directions. Ecol Soc 19:56

Clarke AC (1985) Profiles of the future, 1st edn. Warner Books

Collingridge D (1982) The social control of technology. St Martin's Press

Conseil de l'Europe (1997) Convention for the protection of human rights and dignity of the human being with regard to the application of biology and medicine: convention on human rights and biomedicine. Editions du Conseil de l'Europe. http://193.205.211.30/lawtech/images/lawtech/law/convenzioneoviedo.pdf

Cosens BA et al (2017) The role of law in adaptive governance. Ecol Soc 22:30

Crimes Amendment (Zoe's Law) Bill 2017. https://www.parliament.nsw.gov.au/bills/Pages/bill-details.aspx?pk=2936

Dafoe A (2015) On technological determinism: a typology, scope conditions, and a mechanism. Sci Technol Human Values 40:1047

Dietz T, Ostrom E, Stern PC (2003) The struggle to govern the commons. Science 302:1907

Drahos P (ed) (2017) Regulatory theory foundations and applications. Australian National University Press

Engeli I, Rothmayr CA (2016) When doctors shape policy: the impact of self-regulation on governing human biotechnology: when doctors shape policy. Regul Gov 10:248

Erdman JN (2015) The politics of global abortion rights. Brown J World Aff 39:22

European Commission JRC Science for Policy EC (2018) 'JRCF7- Knowledge Health and Consumer Safety, Overview of EU National Legislation on Genomics. European Commission. EUR29404EN. http://publications.jrc.ec.europa.eu/repository/bitstream/JRC113479/policy_report_-_review_of_eu_national_legislation_on_genomics_-_with_identifiers.pdf

European Medicines Agency (2018) Report of the EMA Expert Meeting on Genome Editing Technologies Used in Medicinal Product Development. European Medicines Agency. EMA/47066/2018

Exploring Constitutional Conflicts 'Levels of Scrutiny Under the Equal Protection Clause'. http://law2.umkc.edu/faculty/projects/ftrials/conlaw/epcscrutiny.htm

Farokhmanesh M (12 December 2016) How a Trump Administration Threatens Women's Health. The Verge. https://www.theverge.com/2016/12/12/13904032/trump-womens-reproductive-health-affordable-care-planned-parenthood

Finnis J (2015) Grounding human rights in natural law. Am J Jurisprud 60:199

Foucault M (1977) Discipline and punish: the birth of the prison. Vintage Books, Random House

Foucault M, Gordon C (1980) Power/knowledge: selected interviews and other writings, 1972–1977, 1st American edn. Pantheon Books

Francioni F (2007) Biotechnologies and international human rights. Bloomsbury Publishing

Freedland J (2 July 2015) 1984 by George Orwell, Book of a Lifetime: An Absorbing, Deeply Affecting Political Thriller. The Independent. http://www.independent.co.uk/arts-entertainment/books/reviews/1984-by-george-orwell-book-of-a-lifetime-an-absorbing-deeply-affecting-political-thriller-10360789.html

Fuller LL (1969) The morality of law. Yale University Press

Gibbs WW (2014) Biomarkers and aging: the clock-watcher. Nature 508:168

Gotsis T, Ismay L. Abortion Law: A National Perspective, Briefing Paper No. 2/2017. NSW Parliamentary Research Service. https://www.parliament.nsw.gov.au/researchpapers/Documents/Abortion%20Law.pdf

Gunningham N, Sinclair D (2017) Smart regulation. In: Drahos P (ed) Regulatory theory: foundations and applications. Australian National University Press

Gunningham N, Grabosky P, Sinclair D (1998) Smart regulation: designing environmental policy. Clarendon Press

Harari YN (2015) Sapiens: a brief history of humankind, 1st edn. Harper

Harmon SHE (2016) Modernizing biomedical regulation: foresight and values in the promotion of responsible research and innovation. J Law Biosci 3:680

Harryono M et al (2006) Thailand medical tourism cluster. Harvard Business School Microeconomics of Competitiveness

History, 'The 1960s - Facts & Summary'. (*HISTORY.com*) http://www.history.com/topics/1960s

Hobbes T. Leviathan or the Matter, Forme, & Power of a Common-Wealth Ecclesiasticall and Civill. Andrew Crooke at the Green Dragon in St Paul's Church-yard 1651. https://socialsciences.mcmaster.ca/econ/ugcm/3ll3/hobbes/Leviathan.pdf

Hume D (1896) A treatise of human nature. Oxford University Press

Inglis-Arkell E (28 April 2013) Technology isn't magic: why Clarke's third law always bugged me. io9. http://io9.gizmodo.com/technology-isnt-magic-why-clarkes-third-law-always-bug-479194151

Inhorn MC (2003) Global infertility and the globalization of new reproductive technologies: illustrations from Egypt. Soc Sci Med 56:1837

International Women's Development Agency (IWDA) (26 April 2015) Reproductive Rights, Abortion & Zoe's Law: Why Freedom of Choice Is Still Feminism's Biggest Fight. https://iwda.org.au/reproductive-rights-abortion-zoes-law-why-freedom-of-choice-is-still-feminisms-biggest-fight/

Kamel RMA (2013) Assisted reproductive technology after the birth of Louise Brown, vol 3. Gynecol Obstet, p 156

Kant I (2003) The critique of pure reason. JMD Meiklejohn tr, The Project Gutenberg

Kelly J (8 April 2016) Why are Northern Ireland's abortion laws different? BBC News. http://www.bbc.com/news/magazine-35980195

Keszthelyi C (2 May 2017) Government's "Stop Brussels" Campaign Revs Up. Budapest Bus J. http://bbj.hu/politics/governments-stop-brussels-campaign-revs-up_132259

Kleinlein T (2012) Constitutionalization of international law. Das Max-Planck-Institut für ausländisches öffentliches Recht und Völkerrecht 231:703

Laurie G, Harmon SHE, Arzuaga F (2012) Foresighting futures: law, new technologies, and the challenges of regulating for uncertainty. Law Innov Technol 4:1

Lee MYH (31 March 2016) Donald Trump's Claim He Evolved into "pro-Life" Views, like Ronald Reagan. The Washington Post. https://www.washingtonpost.com/news/fact-checker/wp/2016/03/31/donald-trumps-claim-he-evolved-into-pro-life-views-like-ronald-reagan/?noredirect=on&utm_term=.a98e5a4e62f8

Lessig L (1997) The constitution of code: limitations on choice-based critiques of cyberspace regulation. Commlaw Conspectus: J Commun Law Technol Policy 5:181

Lessig L (1999) Code and other laws of cyberspace. Basic Books

Lessig L (2001) The future of ideas: the fate of the commons in a connected world, 1st edn. Random House

Lessig L (2006) Code: Version 2.0, 2nd edn. Basic Books

Level Crossings, Law Commission Consultation Paper No. 194, Scottish Law Commission and Law Commission Discussion Paper No. 143, 'Regulatory Theory' (2010). https://www.scotlawcom.gov.uk/files/5312/8024/5698/regulatory_theory.pdf

Liebert W, Schmidt JC (2010) Collingridge's Dilemma and technoscience: an attempt to provide a clarification from the perspective of the philosophy of science. Poiesis Praxis 7:55

Mandel GN (2009) Regulating emerging technologies. Law Innov Technol 1:75

Maturo A (2012) Medicalization: current concept and future directions in a bionic society. Mens Sana Monogr 10:122

Meidinger E (1987) Regulatory culture: a theoretical outline. Law Policy 9(4):355–386

Mohd Mutalip SS (2012) Promoting Malaysia through "fertility" tourism. J Tourism Hosp Culinary Arts 4:1

Morrison M (2016) Overdiagnosis, medicalisation and social justice: commentary on Carter et al (2016) 'A definition and ethical evaluation of overdiagnosis. J Med Ethics 42:720

Müller J-W (11 February 2016) The problem with Poland. The New York Review of Books. http://www.nybooks.com/daily/2016/02/11/kaczynski-eu-problem-with-poland/

Munro K (10 March 2017) Fred Nile gives renewed push to Zoe's law to criminalise harm to a fetus. The Sydney Morning Herald. http://www.smh.com.au/nsw/fred-nile-gives-renewed-push-to-zoes-law-to-criminalise-harm-to-a-fetus-20170309-guup40.html

National Human Genome Research Institute (NHGRI) 'An Overview of the Human Genome Project'. https://www.genome.gov/12011238/an-overview-of-the-human-genome-project/

Nelson TE, Oxley ZM (1999) Issue framing effects on belief importance and opinion. J Polit 61:1040

Nemudryi AA et al (2014) TALEN and CRISPR/Cas genome editing systems: tools of discovery. Acta Naturae 6:22

Nielsen VL, Parker C (2009) Testing responsive regulation in regulatory enforcement. Regul Gov 3:376

Noonan JT (1977) Abortion in the American context. Human Life Rev 3:29

Norberg J (20 August 2016) 'Why can't we see that we're living in a golden age? The Spectator. https://www.spectator.co.uk/2016/08/why-cant-we-see-that-were-living-in-a-golden-age/

Noyes J (21 November 2013) 'On Zoe's law, and the accidental/on purpose erosion of your reproductive rights. Junkee. http://junkee.com/on-zoes-law-and-the-accidentalon-purpose-erosion-of-your-reproductive-rights/21659

Nuffield Council on Bioethics (2016) Genome editing: an ethical review. Nuffield Council on Bioethics

Nuffield Council on Bioethics (2017) Non-invasive prenatal testing: ethical issues. Nuffield Council on Bioethics

O'Connor A (28 October 2017) How the death of Savita Halappanavar changed the abortion debate. The Irish Examiner. http://www.irishexaminer.com/analysis/how-the-death-of-savita-halappanavar-changed-the-abortion-debate-461787.html

OECD (2010) Biomedicine and Health Innovation: Synthesis Report. http://www.oecd.org/health/biotech/46925602.pdf

Parens E (2013) On good and bad forms of medicalization. Bioethics 27:28

Parker C (2013) Twenty years of responsive regulation: an appreciation and appraisal: twenty years of responsive regulation. Regul Gov 7:2

Patterson D et al (2015) The dark future of constitutionalism: the cosmopolitan constitution. Const Commentary 30:667

Pennings AJ (1 July 2012) Arthur C. Clarke's three laws of innovation. Writings on Digital Strategies, ICT Economies, and Global Communications. http://apennings.com/political-economies-in-sf/arthur-c-clarkes-three-laws-of-innovation/

Peter F (2008) Pure epistemic proceduralism. Episteme: A J Soc Epistemol 5:33

Phillips DC (1995) The good, the bad, and the ugly: the many faces of constructivism. Edu Res 24:5

Poort L, van Beers B, van Klink B (2016) Introduction: symbolic dimensions of biolaw. In: Symbolic legislation theory and developments in biolaw. Springer

Posner RA (2006) The role of the judge in the twenty-first century. BUL Rev 86:1049

Powell CMH. Being human: how should we define life and personhood? Enrichment Journal. http://enrichmentjournal.ag.org/201002/201002_134_define_person.cfm

Ribeiro GL (2001) Cosmopolitanism. Int Encycl Soc Behav Sci 4:2842

Roe v. Wade, 410 U.S. 113 (1973) (*Justia Law*). https://supreme.justia.com/cases/federal/us/410/113/

Sajó A, Ryan C (2016) Judicial reasoning and new technologies: framing, newness, fundamental rights and the internet. In: Pollicino O, Romeo G (eds) The internet and constitutional law. The protection of fundamental rights and constitutional adjudication. Routledge

Sandel M (2004) The case against perfection. Atl Mon 51:293

Sanders L (2017) 40 more genes linked to intelligence. Sci News 191:14

Schloendorff v The Society of the New York Hospital (1914) 211 NY 125 (The Court of Appeals of New York)

Somek A (2014) The cosmopolitan constitution. Oxford University Press

Stock G (2005) Germinal choice technology and the human future. Reprod BioMed Online 10:27

Stockholm Resilience Centre 'Adaptive Governance' (6 December 2010). http://www.stockholm-resilience.org/research/research-streames/stewardship/adaptive-governance-.html

Strawson G (11 September 2014) Sapiens: a brief history of humankind by Yuval Noah Harari – review. The Guardian. https://www.theguardian.com/books/2014/sep/11/sapiens-brief-history-humankind-yuval-noah-harari-review

Suter SM (2002) The routinization of prenatal testing. Am J Law Med 28:233

Sweet AS, Mathews J (2008) Proportionality balancing and global constitutionalism. Columbia J Transntl Law 72:47

Taranto S (22 January 2018) How abortion became the single most important litmus test in American politics. The Washington Post. https://www.washingtonpost.com/news/made-by-history/wp/2018/01/22/how-abortion-became-the-single-most-important-litmus-test-in-american-politics/?noredirect=on&utm_term=.b4015648272b

ten Have HAMJ, Jean MS (2009) The UNESCO universal declaration on bioethics and human rights: background, principles and application. UNESCO Publishing

The Guardian (4 March 2011) Profiles of the future by Arthur C Clarke – review. The Guardian. https://www.theguardian.com/science/2011/mar/04/profiles-future-arthur-clarke-review

Trebilcock MJ, Iacobucci EM (2009) Designing competition law institutions: values, structure, and mandate. Loyola Univ Chicago Law J 41:455

Uitz R (2015) Can you tell when an illiberal democracy is in the making? An appeal to comparative constitutional scholarship from hungary. Int J Const Law 13:279

UNESCO (11 November 1997) Universal Declaration on the Human Genome and Human Rights. http://portal.unesco.org/en/ev.php-URL_ID=13177&URL_DO=DO_TOPIC&URL_SECTION=201.html

UNESCO (2006) The Universal Declaration on Bioethics and Human Rights. http://unesdoc.unesco.org/images/0014/001461/146180E.pdf

United States v. Carolene Products Co. 304 U.S. 144 (1938) (*Justia Law*) https://supreme.justia.com/cases/federal/us/304/144/case.html

van Gossum O, Arts B, Verheyen K (2010) From "smart regulation" to "regulatory arrangements". Policy Sci 43:245–261. https://doi.org/10.1007/s11077-101-9108-0

van Klink B (2016) Symbolic legislation: an essentially political concept. In: van Klink B, van Beers B, Poort L (eds) Symbolic legislation theory and developments in biolaw, vol 4. Springer

World Medical Association (19 October 2013) WMA Declaration of Helsinki - Ethical Principles for Medical Research Involving Human Subjects. https://www.wma.net/policies-post/wma-declaration-of-helsinki-ethical-principles-for-medical-research-involving-human-subjects/

Chapter 5
International Biomedical Laws in the Field of Genetic Interventions

Abstract This chapter imports an international dimension into the discourse of genetic interventions. In examining a selection of international biomedical laws in the field of genetic interventions, the international human rights dimensions that are protected in these biomedical instruments are highlighted. Of particular note is that despite the existence of these international human rights instruments, some key bioethical issues still remain unresolved, which points to a lack of international consensus and concerted will at bridging the gap that may actualize the resolution of these problems. I summarize that two of the main bioethical issues (the determination of human dignity as a grounded normative value, and germ line genetic interventions) still lead to divergent opinions in both bioethical and legal circles, and in this, demonstrates shortcomings that impact considerably on the human rights discourse.

In line with the hypothesis of the previous chapter (Chap. 4), and in expanding the scope of regulatory frameworks for biomedical technologies, this chapter begins to consider the framework of biomedical laws on an international level through an examination of selected international treaties and conventions. In fact, a recent announcement by the World Health Organization (WHO) in December 2018[1] is noteworthy following Dr. He Jiankui's claims of having gene edited two embryos using CRISPR. Following the announcement, the WHO has set up an expert advisory panel to develop global standards for the development and governance of human genome editing.[2] The panel's first meeting was on 18 to 19 March 2019. The urgency of this matter has also prompted responses from scientists and bioethists from seven countries, published in *Nature*, and to call for an 'enforceable' moratorium on germ line gene editing.[3] In contributing to the discourse on international

---

[1] GenomeWeb, 'WHO to Set Up Panel to Develop Gene-Editing Standards' *GenomeWeb* (4 December 2018) https://www.genomeweb.com/scan/who-set-panel-develop-gene-editing-standards?utm_source=addthis_shares&fbclid=IwAR320wblX33e1znwYO-8bkOLtvjc9mOrGSa Imu8ARQ22ST6k3U2xn-2injo#.XGcDwcnSMNY.facebook.

[2] World Health Organization, 'WHO Expert Advisory Committee on Developing Global Standards for Governance and Oversight of Human Genome Editing' http://www.who.int/ethics/topics/human-genome-editing/committee-members/en/.

[3] Lander et al. (2019), p. 165.

© Springer Nature Switzerland AG 2019
P. L. Lau, *Comparative Legal Frameworks for Pre-Implantation Embryonic Genetic Interventions*, https://doi.org/10.1007/978-3-030-22308-3_5

governance of genetic interventions, the aim of the initial exercise in Sect. 5.1 is to highlight the premise of Chap. 4 regarding regulatory approaches for biomedical technologies, and to demonstrate the normative regulatory framework of law, governance and values within the constitutional framework of the selected jurisdictions, which will be examined in the following Chap. 6. The fruitfulness of merging the architectural design of a quasi-legal, governance or regulatory nature envisaged in Chap. 4, with the role of international law, conventions, and guidelines, and other regional considerations (which will be analysed in this chapter) that may have an influence, whether apparent or not, on the constitutional systems in Chap. 6, will be considered. At this juncture, due to the voluminous nature and variety of legal and/ or quasi-legal instruments on a global scale, the considerations in this chapter will be limited to the exploration of the key aspects of human rights instruments which are relevant to the human genome, genetic interventions and reproductive technologies and rights. These will be examined through the lens of international biomedical law instruments.

At the international level, some examples that will be scrutinized include UNESCO's Universal Declaration on the Human Genome and Human Rights (UDHG),[4] and the Universal Declaration on Bioethics and Human Rights (UDBHR),[5] amongst others. The European system, encompassing the European Convention for the Protection of Human Rights and Fundamental Freedoms (ECHR)[6] and the Convention for the Protection of Human Rights and Dignity of the Human Being with Regard to the Application of Biology and Medicine (the Oviedo Convention)[7] is also worth considering because the region has adopted a framework on biomedical technologies and specific restrictions into medical or scientific practices that are performed on the human body and human genome. Other considerations of the European region also encompasses the European Medicines Agency (an agency of the European Union) report on genome editing technologies used in medicinal products[8] as well as the European Commission Joint Research Center Science for Policy report that encompasses an overview of EU national legislation on genomics.[9] (Collectively, I refer to all these as the International Instruments).

---

[4] UNESCO, 'Universal Declaration on the Human Genome and Human Rights: UNESCO' (11 November 1997) http://portal.unesco.org/en/ev.php-URL_ID=13177&URL_DO=DO_TOPIC&URL_SECTION=201.html.

[5] UNESCO 'The Universal Declaration on Bioethics and Human Rights' (2006) http://unesdoc.unesco.org/images/0014/001461/146180E.pdf.

[6] Warbrick (1989), p. 698.

[7] Conseil de l'Europe (1997).

[8] European Medicines Agency, 'Report of the EMA Expert Meeting on Genome Editing Technologies Used in Medicinal Product Development' (European Medicines Agency 2018) EMA/47066/2018.

[9] European Commission JRC Science for Policy, 'JRCF7- Knowledge Health and Consumer Safety, Overview of EU National Legislation on Genomics' (European Commission 2018) EUR29404EN http://publications.jrc.ec.europa.eu/repository/bitstream/JRC113479/policy_report_-_review_of_eu_national_legislation_on_genomics_-_with_identifiers.pdf.

It must be emphasized that there is deliberative selection of the International Instruments in this book. The encompassment of a majority of the instruments within the field of human rights is my paramount focus. The specific nature of the rights that are studied in this book encompass more human rights, and less broadly fundamental rights in the manner addressed by European Union (EU) treaties. The deep interest that the Council of Europe (CoE) has in biomedical issues, and its close relationship with the European Convention on Human Rights, is also one of the focal points for analysis, since the primary claim in my work is to use rights-based approaches in regulating pre-implantation genetic interventions. The exclusion of many other EU directives that also regulate, to some extent, interventions into the human genome, such as EudraLex, and specifically the EU Clinical Gene Therapy Trials Legislation [encompassing the Clinical Trials Directive [Directive 2001/20/EC), Advanced Medicinal Therapy Products (ATMPs) and Genetically Modified Organisms (GMO)], I opine, do not fit specifically within the gene editing technologies discourse in this book. In my view, the EU approach, although nodding towards broader normative (human rights) standards, are more concerned with fundamental rights, which ultimately regulates the single market. Therefore, the observation I have arrived at is that it very much posits the role of the state in regulating private relations to the extent that it enables the market to function.

Hence, the analysis that is then carried out in connection with the International Instruments (in Sect. 5.3) will contribute towards the processing of information, agenda and motivations, that I hypothesize, are made in response to global concerns that have been raised but still demonstrate a number of glaring shortcomings. By addressing two of the main current bioethical issues that form the basis of these shortcomings, I believe that this analysis allows us to view, from a bird's eye perspective, the importance in informing international programs and strategies towards the continued improvement of human rights protections. It is challenging to manage information and data that encompasses the international plane, but this task is not untenable and must be undertaken to demonstrate the possibilities of reaching a form of global biomedicine approach. This task is two-fold: first, to question if there is a universality of a shared values system that can be visualized in complex environments and as a means of beginning a conversation in how these shortcomings may impact upon the human rights discourse. Secondly, this must also arguably take into account non-Western perspectives to provide more inclusion, accountability and a negotiated universality of values.

Finally, Sect. 5.4 concludes with the expectation that an analysis of the provisions in these instruments will reveal overlapping facets in terms of the prioritization of importance, and levels of human rights protection in the different spheres of jurisdictions. It is also likely that this exercise will unveil a stark lacuna in some equally important areas that may be considered ripe for regulatory purposes, for instance, relating to concrete boundaries of permissibility, access to and costs of, various emerging technologies. What is proposed in this section therefore is the strength of unity and consensus on the international level, which may then inform a set of stable and strong bioethical and biomedical guidelines that is capable of transforming perceptions, as a complementary directory to existing legislation or regulatory measures. Doing this therefore would suggest a form of international unitary model of governance resulting from negotiated shared values that can be applicable to most constitutional, and which embodies inclusion and non-elitism.

5.1 Existing International Biomedical Laws

5.1.1 Biomedical Laws: Public Law or Private Law?

Prior to undertaking a discussion of the existing international biomedical laws, it is interesting to examine the oppositional models of constitutional human rights in the sphere of both private and public law. In which sphere specifically do biomedical laws fall under? Defining the distinctions between public and private law impacts the manner in which the scope of biomedical laws operate. In the course of writing this book, as a common-law trained lawyer, I had also come to realize that the definitional notions of public law and private law in a common-law context is somewhat less demarcated than in the manner of the civil law traditions. In terms of legal regulation (and specifically legal regulation within the realm of biomedicine), it is relevant to think about the main aspects of such regulation, which may lead to different usage of the terms public law and private law. Randy Barnett breaks this down into four specific aspects that can help us glean the distinctions. In his work, he lists the following four senses of legal regulation[10]:

(1) The kinds of substantive standards used to assess the types of conduct that may properly be subjected to legal regulation;
(2) The different status of persons or entities that may properly complain about violations of legal regulation;
(3) The different status of persons or entities that are subject to legal regulation;
(4) The different kinds of institutions that may be charged with adjudicating and enforcing legal regulations.

The Continental (civil) law traditions are much more prescriptive than the common law ones, taking its origins from Roman law, with codification and clear legal links between citizens and the State (public law) and legal links between citizens and private organizations (private law). On this basis, applying the civil law traditions to biomedicine would be straightforward. As long as legislators or regulatory authorities determine how and what biomedical technologies are authorities, used and made accessible, the public law aspect is engaged. The private law aspect would be more relevant in terms of the fundamental rights of the human person.

The peculiarity of the association of thinking from a common law aspect is that these continental aspects are, indeed, accurate; and contrary to popular belief by continental schools of thought, public law is indeed present in common law

---

[10] Barnett (1986), p. 267, 268.

jurisdictions, although not necessarily as a "body of wholly autonomous rules, entirely separate from private law."[11] Considering the scope of biomedicine, therefore, biomedical laws are capable of falling under the public law domain because it affects a community of citizens as a whole that may engage the concept of "public harm"[12]; on the other hand, some of the components of biomedical laws, for example, business aspects such as accreditation, costs, and commercialization, may also fall within the purview of private law. Therefore, considering the definitional scope could be a vital indicator of how biomedical laws may make its presence felt under a constitutional framework of rights.

Justice Aharon Barak raises the question "whether constitutional human rights merely define and delimit the relationship between citizen and state or whether such rights also apply to relations between citizens".[13] Scholarly views have been polarized as to the applicability of human rights within the field of either public law or private law, depending on the legal approaches that are being used; and Justice Barak has analysed, *vis-à-vis* the proclamation of four essential models[14] that may help to define the application of constitutional human rights, and the appropriate scope of this application, whilst tracing the roots of these models of application in the jurisprudence of various countries.

The distinction between the sphere of private law and public law is essential, especially in terms of considering some relevant human rights components that affect biomedical technologies. Legal scholarship has demonstrated, at the very least, that the dividing line between private law and public law has often become blurred, with private organizations in many jurisdictions; for example, undertaking particularized public roles and regarded as "quasi-public"[15] bodies. Governments, in the meantime, through the legislative body, enact laws that apply to private individuals and the citizens of the state.[16] Therefore, the distinction between private and public law is often softened in light of the practical realities of modern democracy, which has demonstrated that human rights do, in fact, penetrate the domain of both these legal cultures.

The suggestion made by Justice Barak in the scope of applying human rights in the realm of private law, is that an adoption of a "strengthened and augmented"[17] indirect model application[18] would most appropriately address the situation. He states succinctly "between public law (the creator of the right) and private law (the

---

[11] Harlow (1980), p. 241, quoting Vedel, *Droit Administratif* (1973) 57.

[12] Barnett (1986), p. 268.

[13] Barak (1996), p. 218.

[14] These models of application include: the direct application model, the non-application model, the indirect application model, and the application to judiciary model.

[15] Barak (1996), p. 220.

[16] ibid 221.

[17] ibid 258.

[18] ibid 225. A traditional indirect model of applying human rights to private law means that "protected human rights do not directly permeate private law", but they do, however, apply *vis-à-vis* "means of private law doctrines.

grantor of the remedy for the right) there must be "integrated tools". Where the right is recognized (in public law), the remedy must also be recognized (in private law)."[19] He notes that "the tools" to facilitate this applicability already exists in private law, referring to concepts like "good faith", "public policy" and "unconscionability".[20] There is increasing visibility in the applicability of human rights principles in private law, for example, in areas such as the law of contract,[21] tort law,[22] property law,[23] and family and inheritance law,[24] just to name a few. The widening understanding is that human rights are indeed influenced in and by the workings of private law, referred to as the "constitutionalization of private law".[25] In the United States (US), the understanding, however, is expected to be limited and scholarly work has demonstrated that the incorporation of international human rights elements into private law in the US have largely been focused on anti-discrimination laws and defamation.[26] In the case of reproductive rights, for example, the interplay between the individual and the state certainly crosses the boundaries of private and public law. By way of example, abortion legislation that restrictions abortion coverage in medical insurance made by a particular state, targeted towards private citizens, can certainly be deemed to present itself into the realm of private law, between said individual citizen and insurance companies (traditionally deemed to be a contractual arrangement between these parties) in terms of private insurance coverage in private health care plans. In the Netherlands, interesting perspectives have been presented *vis-à-vis* the application of fundamental rights in contract law,[27] where "a patient's constitutional right to bodily integrity was invoked as a reason to undergo AIDS testing."[28]

Regulatory approaches can reveal that the intersectional relationship between human rights in private and public law is a contemporary revelry in democratic maturity, and for good reason. Within the realm of pre-implantation genetic interventions, the validation becomes much more acute through the consideration of human rights. From the consideration of how assisted reproductive technologies (ARTs) and more particularly, in-vitro fertilization (IVF) procedures have shaped the landscape that ushered in the era of Pre-Implantation Genetic Diagnosis (PGD) laws (or quasi-laws) in the various selected jurisdictions (in Chap. 3), to the drawing of parallel discourses and laws in the abortion debates and earlier prenatal testing

---

[19] ibid 261.

[20] ibid 271–277. According to Barak, these concepts are consistent with "the guiding impulses that drive human rights". These include respect for equality, dignity and individual autonomy.

[21] Trstenjak and Weingerl (2016).

[22] ibid 22.

[23] ibid 28.

[24] ibid 35.

[25] ibid 7.

[26] Miller (2014), p. 133.

[27] Trstenjak and Weingerl (2016), p. 15.

[28] ibid. See also the Dutch Hoge Raad (Supreme Court) case: HR 12 December 2003, NJ 2004, 117 (*Aidstest II*).

technologies, and their underlying regulatory design framework and values as equivalent models of consideration for PGD and genetic interventions (in Chap. 4), the examination of human rights in pre-implantation genetic interventions within the constitutional frameworks of the selected jurisdictions, is now an appropriate enterprise.

Insofar as values in regulation, and its influence on constitutional frameworks are concerned, the relationship of these regulatory values within a human rights network (birthed within the constitutional frameworks of the jurisdictions' legal systems), are synergetic and reciprocal. Mary Robinson appropriately sums the relationship in this manner: "human rights are the closest thing we have to a shared values system for the world. We should take every opportunity to see them not simply as shared goals, but as legal obligations and policy-making tools that can assist those charged with making complex decisions."[29] Diana Boer and Klaus Boehnke state, in respect of values, "Variability across cultures and the change of value preferences across time are important aspects to consider beyond individual value preferences."[30] Ethical values and moral perspectives especially in the field of PGD and genetic interventions are also rudimentary parts of the discourse; in relative consideration to perspectives such as moral conservatism, welfare and risk, moral norms and human rights, the moral enterprise of science and technologies, and social justice.[31] Moral considerations are therefore equally as momentous in a shared values system. The shared oneness of values within national, constitutional systems is apparent from the various fields of the abortion debates and earlier prenatal testing technologies highlighted in Chap. 4, and even from beyond its scope, venturing into industries such as competition law, cyberspace regulation, and the management of socio-ecological systems and environmental law; all of which demonstrate a unification of some type of universal acceptance.

5.1.2 Principles of International Biomedical Laws

Is there an internationalized form of biomedical laws thus far? In his work, Roberto Andorno credits the UNESCO and the Council of Europe for an emergence of core principles relating to biomedicine since the late 1990s period.[32] To identify these core principles, we look to an engagement with the limited number of international instruments that deal specifically with the human genome and the interventions that may be performed on the human genome. The selection of instruments here include UNESCO's Universal Declaration on the Human Genome and Human Rights of 1997, (UDHG),[33] the International Declaration on Human Genetic Data of 2003

---

[29] Robinson (2007), p. 241, 242.

[30] Boer and Boehnke (2015), p. 129.

[31] Nuffield Council on Bioethics (2016).

[32] Andorno (2013a).

[33] 'Universal Declaration on the Human Genome and Human Rights: UNESCO' (n 4).

(IDHGD),[34] and the Universal Declaration on Bioethics and Human Rights of 2005 (UDBHR)[35]; (hereinafter collectively referred to as the UNESCO Declarations). We also look at the Council of Europe's Convention on the Protection of Human Rights and Dignity of the Human Being with regard to the Application of Biology and Medicine (the Oviedo Convention),[36] the European Medicines Agency report on genome editing technologies used in medicinal products[37] as well as the European Commission Joint Research Center Science for Policy report that encompasses an overview of EU national legislation on genomics.[38] The World Medical Association's non-binding Declaration of Helsinki: Ethical Principles for Medical Research involving Human Subjects (Declaration of Helsinki)[39] also imparts an ethical component to the consideration of human rights within these fields, and targets the primary concerns related to physicians and medical. (These international instruments are hereinafter collectively referred to as the International Instruments).

The claims put forward by Andorno is premised on the fact that the approval of the UNESCO Declarations particularly, already indicates a universal and international acceptance of the fundamental principles in bioethical practice and research around the world.[40] He identifies the key purpose of each of the International Instruments and three main features of international biomedical law.[41] These are: firstly, that human dignity is an "overarching principle",[42] a similar position that I also assert in Sect. 5.2 below; secondly, that the human rights framework is a suitable one to be used in the field of biomedicine[43]; and thirdly, that broad principles in the manner of "norms included in international bio-legal instruments"[44] are adopted to allow a proper procedural transformation and transition into national legal systems. He translates this into a set of comprehensive guiding principles evolved as norms in the international arena, what he terms the "core principles of international

---

[34] UNESCO, 'International Declaration on Human Genetic Data' (16 October 2003) http://unesdoc.unesco.org/images/0013/001361/136112e.pdf.

[35] 'The Universal Declaration on Bioethics and Human Rights' (2006).

[36] Conseil de l'Europe (1997).

[37] European Medicines Agency (2018).

[38] JRC Science for Policy (2018).

[39] World Medical Association, 'WMA Declaration of Helsinki - Ethical Principles for Medical Research Involving Human Subjects' (19 October 2013) https://www.wma.net/policies-post/wma-declaration-of-helsinki-ethical-principles-for-medical-research-involving-human-subjects/.

[40] Andorno (2013a), p. 14.

[41] ibid 15.

[42] ibid 16–17.

[43] ibid 17.

[44] ibid 18.

biolaw."[45] In his view, even though the development of a universally accepted framework of biomedical laws is challenging, it is nevertheless a feasible endeavour.[46]

As optimistic as it may seem to will the acceptance of some kind of internationalized biomedical stand, the reality of a universal system is much more bristling in nature. One of the challenges correctly raised by Atina Krajewska is that despite the increasing recognition that biomedical technologies need to be more firmly regulated on the international level, "little scholarly consideration is paid to the institutional basis and the normative status of international biomedical law."[47] In particular, she addresses this proliferation of biomedical law on the international level *vis-à-vis* the theories of "fragmentation" and "global legal pluralism".[48] For example, Krajewska highlights that as medical and scientific technologies increase in importance, so too does the emergence of an international health framework. This has inevitably resulted in "numerous global health actors",[49] encompassing different functions that relate to business and commercialization, standards, and operational policy functions. The drawback in this consequence is "the difficulty of achieving coherence and coordination in the social, economic and humanitarian fields."[50] She quotes Alan Taylor[51] who states that "dramatic advances in the field of biomedical science have recently triggered numerous, uncoordinated regional and global initiatives, which, while undertaken without meaningful consultation, coordination or planning, obscure rather than rationalized the global legal framework."[52] As such, the points raised by Krajewska are indeed valid ones.

Moving one step further, I would also raise critique, that insofar as the principles highlighted by Andorno are universal in nature, the Western-centric nature of the International Instruments to a very large extent, may contribute to the fragmentation and legal pluralism issues raised by Krajewska. In a powerfully scathing piece of work, Makau Mutua criticizes the Western-centric nature of the human rights corpus[53] that fails to take into reasonable account the validity of non-Western cultural plurality and determinism, and for projecting a disillusioned metaphor of the

---

[45] ibid 19–34. Some of these principles draw on personhood (Principle 1, the "recognition of human dignity as an overarching principle" and Principle 2, "the primacy of the human individual over the sole interest of science and society"), to equitable health care (Principle 5, "equitable access to health care"), to autonomy and privacy (Principle 4, "respect for the autonomy of patients and research subjects", and Principle 8, "right to know and right not to know one's health (especially genetic) information"), and several others.

[46] ibid 35.

[47] Krajewska (2012), p. 121, 124.

[48] ibid. Krajewska refers to the following references in her work: in respect of the fragmentation theory; Koskenniemi and Leino (2002), pp. 553–579; and in respect of global legal pluralism, to Fischer-Lescano and Teubner (2004), pp. 99–1045.

[49] ibid 127.

[50] ibid 128. Krajewska further refers to de Anarclens (2007), pp. 8–35.

[51] ibid 130.

[52] Taylor (2004), p. 500, 504.

[53] Mutua (2001), p. 201.

"savages-victims-saviours" construction,[54] which he claims is put forward by the United Nations, Western states, international non-governmental organizations and senior Western academics.[55] The crux of Mutua's article criticizes this narrative as a flawed one because it fails to capture, amongst others, the historical context of human rights development within non-Western or "European tyranny and imperialism."[56] Having mentioned this, and although recognizing that there is some truth in Mutua's discourse, the premise that I put forth is more neutralized. As important as historical and cultural facets are in establishment the foundations of human rights discourse in a particular locality, the so-called "European ideal"[57] of human rights that Mutua claims is projected, cannot be a bad one. The ideals, for example, relating to human dignity and personhood, equality and non-discrimination, and many such more, are ideals that should be universal in nature simply because they contribute to the betterment of all sections of society generally, regardless of legal, cultural or historical pluralism. These ideals, I claim, however, can only be deemed to be shared values through a process of thorough understanding and negotiation by state parties, which would then throw off the 'imperialist' claims of Mutua. Insofar as reaching a desirable outcome is concerned, these negotiated share values therefore would present an important starting point in reaching an international consensus on biomedical technologies generally, and in the future, genetic interventions specifically.

It is, however, recognized that the manner in reaching the desirable outcome is much more challenging than a theoretical discussion of the circumstances. Inevitably the problems of fragmentation of international biomedical laws and the multi-normative involvement of different global health stakeholder groups also impacts on the interpretation and binding reach of public international law. In his work, Thomas Kleinlein raises the interesting feasibility of "constitutionalization in public international law."[58] In essence, the recognition of objectivity in an international dimension, where the "common interest of humanity [transcends] state interests",[59] can be thus reflected in a national constitutional system by transposing this objectivity in a constitutional order. He identifies that "adherence to human rights is an essential element in a framework for the justification of the exercise of authority in public international law, and this framework is obligatory for a constitutional perspective."[60] Whether Kleinlein's constitutionalization dissertation would be workable, however, from the international biomedical law perspective would therefore require an engagement with not only public international law, but also the institutional and normative understanding of aspects that may be peculiar to biomedicine in general.

---

[54] ibid 201.

[55] ibid 202.

[56] ibid 205.

[57] ibid 206.

[58] Kleinlein (2012), p. 703.

[59] ibid 703.

[60] ibid 704.

On raising these premises and translating them into the field of biomedical discourse, it is my position that constitutional frameworks are similarly infused, to some extent, with some of these shared values system from the international dimension. By way of example, the US Constitution,[61] deriving its birth from the Declaration of Independence,[62] expresses core democratic values as its guiding light. These core democratic values encompass fundamental beliefs such as life, liberty, justice, equality, diversity, truth, and popular sovereignty, amongst other beliefs.[63] Conversely, the Chinese Constitution,[64] on the other hand, employs a combination of key values that emanate from its "tripartite culture of Confucianism-Buddhism-Taoism"[65] and modernized "Western" values after its revolution in 1840,[66] comprising key values such as progress,[67] affluence,[68] peace and safety,[69] and harmony.[70] In the Constitution of Kenya Review Commission,[71] the Living Values Statements encompass freedom, peace, respect, love, happiness, honesty, humility, responsibility, simplicity, cooperation, tolerance, and unity. It was emphasized in this review that ethics and 'living values' play an important role in a constitution, because it reaffirms "the acceptance of the dignity and worth of the individual and sacredness or divine nature of human life."[72] Australia uses "values and principles" as part of the discourse on constitutionalism, referring to these as "the civic values of [our] community."[73] These civic values are no different from the other values illustrated in the previous examples: "the rule of law, the democratic principles of government, acceptance of cultural diversity (tolerance), equality of sexes and ethnic backgrounds, equality of opportunity",[74] just to name a few. At the heart of the matter, these illustrations on the significance of values in different constitutions, demonstrate that the diversity and plurality of constitutional values due to geographical, historical, socio-economic conditions, however, do not run very far from each other, and work towards common goals. Hence, building on the analysis conducted in this chapter, the common denominator of a negotiated shared values system on an international level, will diversify in Chap. 6, into a much more detailed

---

[61] Sholley (1951).

[62] Jefferson (2002).

[63] Stone et al. (2005).

[64] Chen (1994), p. 77.

[65] Ma et al. (2015), p. 75.

[66] ibid.

[67] ibid 77.

[68] ibid 78.

[69] ibid.

[70] ibid 79.

[71] Owigar (2002), p. 4.

[72] ibid.

[73] Department of Social Services, Australian Government, 'Values and Law' https://www.dss.gov.au/our-responsibilities/settlement-and-multicultural-affairs/programs-policy/taking-the-initiative/resources/values-and-law.

[74] ibid.

examination of the underlying values of the constitutional systems in each of the jurisdictions, and how these values can then saturate the legitimacy, operability and enforcement of human rights in the selected jurisdictions, in pre-implantation genetic interventions.

5.2 Selected International Human Rights Instruments in Biomedicine

In this section, the underlying aims and principles of the International Instruments is illustrated in connection with interventions into the human genome. The expected hypothesis from this analysis is to reveal common facets that form the basis of protection for a community of persons within the scope of biomedical genetic interventions. It is recognized that a human rights framework is only one of many possible considerations in regulating biomedical technologies on an international level; but it is perhaps one of most fundamental components in considering an over-arching regulatory international biomedical law framework.

At this juncture, the first very clear human rights component that stands out in ALL these instruments is the respect for human dignity and a variation of the recognition of the primacy of the human being *vis-à-vis* this framework of dignity recognition.[75] This is not surprising within the predominantly European framework of these instruments, where the trajectory of recognizing human dignity has expanded into the field of biomedicine and bioethical discourse. In characterizing the relationship between bioethics and law, Andorno states that if we employ a broad meaning of bioethics, this also means it is an "extension of international human rights law into the field of biomedicine".[76] Therefore, the understanding of human dignity and human rights can be viewed as a "common ground for a global bioethics."[77] The increased emphasis on human dignity as a fundamental component in international human rights discourse is due to "biomedical practice [being] closely related to the most basic human rights, namely the rights to life, to physical integrity, to privacy, to access to basic health case, among others."[78] In addition to this, Andorno views human dignity as "the last barrier against the alteration of some basic features of the

---

[75] See Article 10 of the UDHGHR ("Research on human genome shall not prevail over human rights, fundamental freedoms and human dignity"); Article 3 of the UDBHR ("Human dignity, human rights and fundamental freedoms are to be respected"); Article 1 of the Oviedo Convention ("Parties to this Convention shall protect the dignity and identity of all human beings and guarantee everyone, without discrimination, respect for their integrity and other rights and fundamental freedoms with regard to the application of biology and medicine"); and Article 9 of the Declaration of Helsinki ("It is the duty of physicians who are involved in medical research to protect the life, health, dignity, integrity, right to self-determination, privacy, and confidentiality of personal information of research subjects.").

[76] Andorno (2009), p. 223, 225.

[77] Andorno (2009).

[78] ibid 227.

human species that might result from practices such as reproductive cloning or germ-line interventions."[79]

This is in fact consistent with the view of Justice Barak insofar as the human dignity discourse has developed in international conventions.[80] In recognizing that human dignity is one of the main tenets recognized as a general principle in international law, and very much particularly so after the Second World War, Justice Barak goes on to highlight the length and breadth of international conventions that has proclaimed the importance of human dignity in some form or other. Some of these more commonly include the United Nations human rights conventions, the International Covenant on Civil and Political Rights (ICCPR) and the International Covenant on Economic, Social and Cultural Rights (ICESCR),[81] but also encompass other conventions that focuses on very particularized issues, such as the International Convention on the Elimination of All Forms of Racial Discrimination,[82] the Convention on the Elimination of All Forms of Discrimination Against Women (CEDAW),[83,84] the International Convention for the Protection of All Persons from Enforced Disappearances,[85] and the Convention on the Rights of Persons with Disabilities (CRPD),[86] amongst many others. In addition to these international conventions, the regional sphere is also not exempt from its interpretation of the role of human dignity, making the concept felt in regional conventions in Europe (the European Convention on Human Rights and the Charter of Fundamental Rights for the European Union),[87] the Americas (the American Declaration on the Rights and Duties of Man),[88] and in Africa (the African Charter on Human and Peoples Rights).[89]

In fact, such is the importance attributed to human dignity that in some jurisdictions, it has been accorded as a constitutional value that plays a "central normative role"[90] in the unification of human rights discourse. Justice Barak goes on to say that it serves three main roles: first, "as a normative basis for constitutional rights set out in the constitution"[91]; secondly, "as an interpretive principle for determining the

---

[79] ibid 228.

[80] Barak (2015), p. 37.

[81] ibid 38.

[82] ibid 39.

[83] ibid.

[84] UN General Assembly, 'Convention on the Elimination of All Forms of Discrimination against Women' (1979) 20 Retrieved April 2006.

[85] Barak (2015), p. 40.

[86] ibid 41.

[87] Warbrick (1989).

[88] Ruiz-Chiriboga (2013), p. 159.

[89] 'African Charter on Human and Peoples Rights' (21 October 1986) http://www.humanrights.se/wp-content/uploads/2012/01/African-Charter-on-Human-and-Peoples-Rights.pdf.

[90] Barak (2015), p. 103.

[91] ibid.

scope of constitutional rights, including the right to human dignity"[92]; and finally, "in determining the proportionality of a statute limiting a constitutional right."[93] Flowing from human dignity, he opines are further rights that reflect "different levels of generality, the various civil and social aspects of human dignity…different categories of acts or omissions that can affect human dignity."[94] Within the framework of comparative constitutional law, the exposure of the nature of human dignity and whether or not it occupies a central space in the constitutional framework of a country's legal system is an interesting dimension to explore, and is not without its provocations.[95]

The second human rights component that makes it presence immensely felt in these International Instruments is the right to privacy.[96,97] In the framework of human rights protection, the right to privacy covers a wide range of issues, such as reproductive freedom,[98,99] privacy in the context of digitalization,[100] the protection of personal information and other forms of data protection,[101] the sphere of family and personal relationships,[102] and much more. Very closely relative to the right to privacy, which is ever prevalent in many constitutional systems, is the respect for the principle of autonomy and individual decision-making,[103,104] which underlies the

---

[92] ibid 104.

[93] ibid.

[94] ibid 160.

[95] Notwithstanding the apparent positivity of the recognition of and respect for human dignity as a fundamental underlying value from the international perspective, there may still be some disparity on the interpretation of "human dignity" itself. Additionally, the same may not be so easily transposed into a constitutional sphere. The exploration of the concept and principles of human dignity and its normative status in international law will be detailed and critiqued more fully in Sect. 5.3.1 of this chapter.

[96] Sharp (2013).

[97] See: Article 9 of the UDHGHR ("In order to protect human rights and fundamental freedoms, limitations to the principles of consent and confidentiality may only be prescribed by law, for compelling reasons within the bounds of public international law and the international law of human rights."); Article 5 of the UDBHR ("The autonomy of persons to make decisions, while taking responsibility for those decisions and respecting the autonomy of others, is to be respected.") and Article 9 of the UDBHR ("The privacy of the persons concerned and the confidentiality of their personal information should be respected."); Article 10 of the Oviedo Convention ("Everyone has the right to respect for private life in relation to information about his or her health."); and Article 9 of the Declaration of Helsinki ("It is the duty of physicians who are involved in medical research to protect the life, health, dignity, integrity, right to self-determination, privacy, and confidentiality of personal information of research subjects.").

[98] See example: Roe v Wade, 410 U.S. 113 (1973).

[99] Hallich (2017), p. 107.

[100] Geller (2016), p. 21.

[101] Nunan and Di Domenico (2017), p. 481.

[102] See example: Douglas v Hello! Ltd [2005] EWCA Civ 595.

[103] Gillon (1985), p. 1806.

[104] Coggon and Miola (2011), p. 523.

exercise of a right to privacy.[105] In the context of reproductive liberty, and there being a lacuna regarding interventions in the human genome, which have yet to be clarified by judicial or other forms of interpretive mechanisms, the realm of privacy bears immense value because "the fight for constitutionality starts with the human's fight for conquest of the personal rights."[106] In the European sphere, as is the case illustrated by the International Instruments in this section, the right to privacy bears "a long tradition of constitutional acknowledgment".[107] In several European countries, certainly those that are also party to these international instruments, the right to privacy is conscientiously guarded via both explicit and implicit mechanisms, whether contained within the expressly mentioned constitution of such countries, or given protection by way of other means and derivatives through other constitutional provisions that guard the "inviolability of the person, equality in freedoms and rights, as well as in the provisions for the specific protection of marriage, family, parental rights, etc."[108]

Finally, the third componential human rights value as a commonality of these International Instruments is the right to equality and non-discrimination.[109] In the context of globalization and emerging economies, international migration and movement, and particularly through the sophistication of automation and technologies, the right to equality has weathered some one-dimensional opposition through the forces of anti-liberalism. The realization of true equality and non-discrimination, however, is much more challenging: current foreign policies in countries such as the US[110] and the UK,[111] and within the context of refugee asylum and migration within

---

[105] Sharp, June 12 and ET (2013).

[106] Karovska-Andonovska (2014), p. 127, 128.

[107] ibid 132.

[108] ibid.

[109] See: the General Principles of the UDHGHR rejects the doctrine of inequality of men and races, and embraces "democratic principles of the dignity, equality and mutual respect of men." Article 6 of this instrument also states that "No one shall be subjected to discrimination based on genetic characteristics that is intended to infringe or has the effect of infringing human rights, fundamental freedoms and human dignity." In the UDBHR, its stated aims include "promoting equitable access to medical, scientific and technological developments" and further safeguards equality and non-discrimination via Article 10 ("The fundamental equality of all human beings in dignity and rights is to be respected so that they are treated justly and equitably.") and Article 11 ("No individual or group should be discriminated against or stigmatized on any grounds, in violation of human dignity, human rights and fundamental freedoms."). The Oviedo Convention also protects this right in Article 1 ("Parties to this Convention shall protect the dignity and identity of all human beings and guarantee everyone, without discrimination, respect for their integrity and other rights and fundamental freedoms with regard to the application of biology and medicine."), Article 3 ("Parties, taking into account health needs and available resources, shall take appropriate measures with a view to providing, within their jurisdiction, equitable access to health care of appropriate quality.") and Article 11 ("Any form of discrimination against a person on grounds of his or her genetic heritage is prohibited."). Within the purview of the Declaration of Helsinki, Article 1-A contains a statement of ethical principles which act as a form of guidance to physicians and other participants in medical research involving human subjects.

[110] Jacobs (2017).

[111] Powell (2017).

Central and Eastern Europe,[112] indicate a shift in equality 'trends' that are frighteningly disturbing. Under the scope of international human rights law such as the UDHR, ICCPR and ICESCR, for example, scrutiny has been focused on enforcement and state obligations in the protection of the right to equality.[113] The acceptance of the UDHR as *jus cogens* notwithstanding, has not made this an easy task; scholars have viewed the principles as being "too vague to be enforceable, and are therefore opposed to undertaking international obligations which would supersede domestic jurisdictions with explicit, enforceable provisions."[114] The containment of the right to equality in the international context, is no doubt a challenging one because of the prioritization of issues that confront equality and non-discrimination; it is expected that justifications relating to biomedical interventions, let alone PGD and genetic interventions may not feature within the wider scheme of discourse just yet and may remain grounded on a theoretical basis for now (at the time of writing this).

Notwithstanding, one of the saving graces is that the applications of PGD and its boundaries of permissibility may, in some aspects, be regulated under the purview of laws that relate to reproduction and all other forms of assisted reproductive technologies. As indicated in Chap. 3, there is at least some form of governance or regulatory framework for PGD in the selected jurisdictions, which may require varying degrees of clarification or reforms. With regards to interventions into the human genome, however, other than a worldwide partial moratorium on human germ-line genetic modifications,[115] the situation is much less clear, although some recent developments have indicated that experts have begun the process to develop international guidelines and standards for gene editing.

For example, the European Medicines Agency (EMA) report,[116] issued in September 2018, although largely concerned with site-specific targeting of gene editing technologies, and the relevance of non-clinical critical findings about the technologies that still need to be further investigated,[117] has stressed upon the need to have early dialogues with regulatory authorities. Bernd Gänsbacher, a member of the Committee for Advanced Therapies (CAT) highlighted the need for these dialogues, which would allow "progress into clinical trials and further marketing authorities and access to patients at the EU and international level."[118] Industry stakeholder, Intellia Therapeutics, encouraged the community

.....to maintain open-mindedness in the use of different analytic tools, and to take a pragmatic approach to the use of preclinical models. In addition, the audience was asked to consider advanced clinical trial designs including umbrella approaches and extrapolation, conditional marketing approval, incorporation of real-world evidence, and the development

---

[112] Tisdall (2017).

[113] Lie (2004).

[114] ibid 19.

[115] Guttinger (2017).

[116] European Medicines Agency (2018).

[117] ibid 8.

[118] ibid 4.

of guidelines to specifically address genome editing therapies. Going forward, the need for continued constructive dialogue on the therapeutic application of genome editing was considered paramount.[119]

The recent publication of the European Commission's Joint Research Centre (JRC) on a review of EU national legislation on genomics[120] is also an important instrument to consider. The aim of the report, to map existing national legislations linked to genomics, "can be used as a baseline for the analyses of possible consequences for EU policies already in place, and to forecast policy gaps and eventual interventions."[121] The conclusions drawn by the report, in fact, highlights the key issues that have been pervasive throughout the entire discourse of pre-implantation genetic interventions in this book. Primarily, the JRC report emphasized the importance of evaluating laws relating to human germ line modifications, highlighting that if it were to be permissible, then there must be "limits on permitted modifications and the purposes for which modifications are allowed should be outlined in the legislative instruments."[122] The report also highlighted the applications of gene and genome editing technologies, (besides human subjects), to plant and animals, "especially on species relevant to the food and feed industry."[123]

In the meantime, the evidentiary provisions of the International Instruments that relates to interventions into the human genome is presented in Table 5.1, and supports the analysis made in this Sect. 5.2 of this chapter.

5.3 Shortcomings of the International Instruments in Current Bioethical Issues

The analysis of the International Instruments above reveals two important points. Firstly, that concerns in biomedical laws have propelled the international legal and scientific community to respond in appropriate ways via suitable legislation. Secondly, however, the inability of the full strength of the provisions of the International Instruments to effectively combat and solve the disparities relating to lasting bioethical issues, still persists. In a similar manner that the abortion debates have been polarized to divergent opinions in the entirety of its history, the same may be said of a plethora of as yet unresolved current bioethical issues. For the purposes of discussion in this book, however, the current bioethical issues that will be highlighted in this section are firstly; the ones relating to the concept of human dignity in biomedical discourse and whether it has or will be more consistently granted with normative and definitive legal status in the international community. Secondly, the

---

[119] ibid 6.

[120] JRC Science for Policy (2018).

[121] ibid 4.

[122] ibid 75.

[123] ibid.

Table 5.1 Illustration of International Instruments Relating to Human Genetic Interventions and the Human Genome

Organization	Instrument	Key provisions of the instruments			
		Aims	General principles	Rights of persons	Positive obligations of the state
UNESCO	Universal Declaration on the Human Genome and Human Rights		• 'Democratic principles of the dignity, equality and mutual respect of men' • Rejection of 'doctrine of inequality of men and races' • Universal principles of human rights • Art. 1: 'inherent dignity and diversity' and 'respect of uniqueness and diversity'	• Art. 5- research/treatment risks and safety; informed consent; international standards, protocols and guidelines • Art. 6- non-discrimination • Art. 10- research on human genome shall not prevail over human rights, fundamental freedoms and human dignity • Art. 11- prohibition against reproductive cloning	• Art. 15- framework for free exercise of research • Art. 17- solidarity and international cooperation • Art. 18- international dissemination of scientific knowledge • Art. 19- encouraging measures to comply and implement principles • Art. 24- International Bioethics Committee and the dissemination of principles
	Universal Declaration on Bioethics and Human Rights	• Universal framework of principles and procedures • Respect for human dignity and protection of human rights • Recognizing freedom of scientific research • Fostering multidisciplinary and pluralistic dialogue about bioethical issues • Promoting equitable access to medical, scientific and technological developments • Safeguarding and promoting interests of present and future generations • Importance of biodiversity	• Art. 16- protection of future generations • Art. 17- protection of the environment, biosphere and biodiversity • Art. 18- applying the principles and • Art. 20- risk management and assessment • Art. 26- complementarity • Art. 28- denial of acts contrary to human rights, fundamental freedoms and human dignity	• Art. 3- human dignity and human rights • Art. 4- benefits and harm • Art. 5- Autonomy and individual responsibility • Art.6- consent • Art. 9- privacy and confidentiality • Art. 10- Equality and justice • Art. 11- non-discrimination and non-stigmatization • Art. 12- cultural diversity and pluralism • Art. 14- social responsibility and health	• Art. 22-Role of the states • Art. 23- bioethics education, training and information • Art.24- international cooperation • Art. 25- follow-up action by UNESCO

The Council of Europe	Convention on the Protection of Human Rights and Dignity of the Human Being with regard to the Application of Biology and Medicine: Convention on Human Rights and Biomedicine (the Oviedo Convention)	• Art. 1- Protect the dignity and identity of all human beings and guarantees (without discrimination) respect for integrity and other fundamental rights and freedoms	• Art. 2- primacy of the human beings • Art. 3 - equitable access to healthcare	• Art.5- consent • Art. 10- privacy and right to information • Art. 11- non-discrimination • Art. 12- predictive genetic testing • Art. 13- interventions into the human genome • Art. 14- non selection of sex • Art. 16- protection of persons undergoing research • Art. 18- research on embryos in vitro • Art. 21- prohibition of financial gain for human body and its parts	• Art. 23- appropriate judicial protection to prevent unlawful infringement of rights and principles • Art. 25- sanctions in the event of infringement • Art. 28- public debate • Art. 29- interpretation of the Convention by the European Court of Human Rights • Art. 33- Signature, ratification and entry into force
World Medical Association (Non-Binding)	Declaration of Helsinki: Ethical Principles for Medical Research involving Human Subjects	• Art. 1- A statement of ethical principles to provide guidance to physicians and other participants in medical research involving human subjects • Art. 6- Primary purpose of medical research to improve prophylactic, diagnostic and therapeutic procedures and understanding of aetiology and pathogenesis of disease	• Art. 9- duty of physician in medical research to protect life, health, privacy and dignity of the human subject • Art. 13- formulation of experimental procedures in research protocols • Art. 15- qualification of clinically competent medical person	• Art. 16- Assessment of risks and burdens vs. comparable benefits • Art. 20- voluntariness / informed participants • Art. 21- safeguard of integrity and respect of privacy and confidentiality • Art. 22- informed consent	Not applicable

divided international discourse between scientists and academics relating to germ-line genetic interventions, particularly because of the advent of the CRISPR/Cas-9 gene editing tool. The impact of both of these current matters, if considered from juridical positioning, in the event they continue to be unresolved, is acute enough to affect scientific and medical research and development in the long term.[124] Finally, the contributory effect of these twin problems are also magnified when we consider that notions of human dignity and germ line genetic interventions are concepts that do not hold a similar moral weight in Southeastern societies, as they may do in Western communities.

5.3.1   Human Dignity as a Grounded Normative Value

The role of the consideration of human dignity has a firm place in the bioethical circles; and has made its presence felt within the context of contemporary laws. Charles Foster argues that he centrality of any form of bioethical or biomedical discourse is that "dignity is the only sustainable Theory of Everything[125] in bioethics."[126] In this section, I argue that the concept of human dignity, as vast as its permutations have featured in contemporary laws and constitutional systems, has been subject to varied juridical reasoning and interpretations. I argue that this is largely due to the fact that human dignity has generally been accepted as a grounded normative value in Western and European communities, but the reasoning behind its recognition as a normative value are inconsistent, and may, in some jurisdictions, operates as "window-dressing"[127] for a complex dilemma that legislators and the judiciary alike are unwilling to address more specifically. Lending support to the misunderstanding of the concept of human dignity, Roberto Andorno claims that a lack of distinction in the dual role played by human dignity contributes to a misun-derstanding of its true value.[128]

The repository of legal cases, by way of example, from the ECtHR, on the con-siderations of the fulfilment of state obligations in human dignity, are varied and very much contextual to the presented circumstances.[129] Indeed, Christopher McCrudden has expounded on human dignity in the judicial interpretation of human rights, providing us with a rich background of the historical concepts of human dignity,[130] and the trove of references to human dignity in various national texts,

---

[124] The impact of the shortcomings in the International Instruments in being able to resolve the cur-rent bioethical dilemmas is raised in this Sect. 5.3.3 in greater detail.

[125] Hawking (2005).

[126] Foster (2011).

[127] ibid.

[128] Andorno (2013b), p. 967.

[129] Groll and Lott (2015), p. 623.

[130] McCrudden (2008), p. 655, 656–663.

constitutional documents, and international and regional texts.[131] As a concept, human dignity appears to have been universally accepted in most jurisdictions around the world, and its tantamount ability to the nature of humankind has found support in a variety of legal cases and circumstances that range from equality and non-discrimination, to autonomy and liberty, and to circumstances in the workforce and employment, amongst many others. As rich as the definitions may be as to what encompasses a violation of human dignity, the scope of these definitions are also flung far and wide between because of the judicial application of human dignity in various scenarios.

The premise of my work is not to criticize or fundamentally question the value of the concept of human dignity,[132] but to raise questions whether human dignity has truly achieved a grounded normative value that applies irrevocably and universally around the world (to which my answer is in the negative); and therefore, what more can be done to elevate human dignity to a truly internationally universal approach. One of the first questions that must be raised is with regards to the diversified but eloquent judicial interpretation of human dignity; and the second question regards the reality of the exercise (or lack thereof) of dignity recognition in some jurisdictions around the world, particularly in the selected jurisdictions of Malaysia and Thailand.

In the judicial interpretation of human dignity, the ECtHR has interestingly found a violation of human dignity in a number of cases, despite the fact that human dignity has not been expressly mentioned in the European Convention on Human Rights (ECHR). This express omission of such words may, indeed, come as a surprise,; but if we take into account the intended nature of the convention, that is, that of a "living instrument", this omission may begin to make sense. In any event, the rich jurisprudence of the ECtHR case law is proof that this omission has little bearing on the substantive application of human dignity. In fact, such is its value that the ECtHR has deemed it to be the "very essence" of the ECHR.[133] The diversity of the type of cases brought to the ECtHR often cite various violations of the provisions in the ECHR that then impact upon the respect for human dignity in those circumstances. A number of cases in the ECtHR's jurisprudence concern offences or violations of Articles 2, 3 and 4 of the ECHR,[134] particularly those that deal with "disproportionate use of physical force against people in vulnerable situations".[135] Some of these cases include *Siliadin v France*,[136] concerning a serious issue of modern day slavery where a young girl was made to work in a household, amounting to forced servitude where she was treated badly and denied from wages and nutrition. This position was further reinforced in another very serious case, in *Rantsev v*

---

[131] ibid 664–673.

[132] O'Mahony (2012), p. 551.

[133] ibid 553.

[134] Conseil de l'Europe (1953).

[135] Costa (2013), p. 396.

[136] *Siliadin v France (Application No 73316/01)* [2005] HUDOC (Second Section, European Court of Human Rights).

Cyprus & Russia,[137] where a young girl was trafficked from Russia to Cyprus under a specialized working visa to become a prostitute. She subsequently died under mysterious circumstances. In both these cases, notwithstanding the outcome in the respective judgments, the ECtHR found that the human dignity of the individuals concerned had been violated in those circumstances.

The ECtHR was also willing to find that the human dignity and human freedom had been violated in other cases brought under Article 8 (concerning the right to private and family life, usually interpreted as a right to privacy too). In *Christine Goodwin v the United Kingdom*,[138] commonly known as the transsexual case, the court found that the state's refusal to grant two post-operative transsexuals the right to be recognized with their new gender and marry individuals of the opposite sex[139] amounted to a violation of Article 8 which did not respect the applicant's human dignity. In addition to cases of this nature, the human dignity recognition by the courts have also been tested in the domain of bioethics, and this would be pertinent to the underlying premises made in this work. In considerations that touch on issues of assisted suicide in *Pretty v the United Kingdom*,[140] to general issues of consent in medical procedures in *Jalloh v Germany*,[141] and especially in cases regarding the legality of abortions in *A, B and C v Ireland*[142] and *Tysiac v Poland*,[143] amongst many others, it may be said that the concept of human dignity that features in any bioethical or biomedical discourse has always been accepted as one of its foundational principles.[144] Even though human dignity has not been expressly recognized as the basis for all violations of human rights within the jurisprudence of the ECtHR, an analysis does reveal however that "only the most fundamental, vital rights are protected by human dignity,"[145] where the violations perpetrated against an individual is so disgraceful or outrageous that it warrants the intervention of human dignity concepts.[146]

The recourse to human dignity as a fundamental principle within the protection of rights for individuals have also been famously pronounced in France, in the

---

[137] *Rantsev v Cyprus & Russia (Application No 25965/04)* HUDOC (First Section, European Court of Human Rights).

[138] *Christine Goodwin v the United Kingdom (Application No 28957/95)* [2002] HUDOC (Grand Chamber, European Court of Human Rights).

[139] Costa (2013), p. 399.

[140] *Pretty v the United Kingdom (Application No 2346/02)* [2002] HUDOC (Fourth Section, European Court of Human Rights).

[141] *Jalloh v Germany (Application 54810/00)* [2006] HUDOC (Grand Chamber, European Court of Human Rights).

[142] *A, B and C v Ireland (Application No 25575/05)* [2010] HUDOC (Grand Chamber, European Court of Human Rights).

[143] *Tysiac v Poland (Application No 5410/03)* [2007] HUDOC (Fourth Section, European Court of Human Rights).

[144] Research and Library Division Department of Jurisconsult, 'Bioethics and the Case-Law of the Court' (Council of Europe/European Court of Human Rights 2016) Research Report.

[145] Costa (2013), p. 402.

[146] ibid.

Conseil d'Etat's decision in *Commune de Morsang-sur-Orge v Societe Fun Production et M. Wackenheim*,[147] where it recognized a violation of human dignity in the activity of dwarf-tossing that outweighs freedom of livelihood and commerce in that instance. In the German *Grundgesetz* (Constitution, Basic Law), the inviolability of human dignity is a fundamental constitutional right,[148] leading to the German *Bundesverfassungsgericht* (Federal Constitution Court) to find a violation of the principle of human dignity[149] against a satirical magazine that had published very demeaning depictions of Franz Josef-Strauß as a pig.[150] Denmark has equally pronounced its commitment to human dignity in its candidature for a seat at the United Nations Human Rights Council, proclaiming its commitment to human rights based on three (3) Ds: Dignity, Dialogue and Development.[151]

Within the jurisprudence of the US, the importance of the role of human dignity is less pronounced, although in the field of bioethics, the President's Council of Bioethics in 2008 produced a report on the operation of human dignity in bioethics. Even in Australia, although there is no federal legislation that protects human dignity on a constitutional level, a recent Australian Bill of Rights 2017 is being proposed at the Parliament,[152] which would specifically spell out the inherent human dignity of all persons and constitutionalize such a recognition.[153]

In the context of Asian jurisprudence however, the importance accorded to the principle of human dignity appears to be very different. Not only is human dignity absent from the umbrella of constitutional protections and liberties, its interpretation is markedly different from its Western counterparts, focusing more on aspects such as respect and honour, as opposed to dignity engrained as part of individual personhood. This is not to say that it is not respected at all, but the contextual interpretation of a dignity of a human person within Asian societies, like Malaysia and Thailand as examples, will certainly differ radically. By way of example, if we look at the arguments relating to the status of the human embryo, the concept of dignity has often been cited and attributed to the embryo in several European jurisdictions as a life worthy of protection. In Malaysia and Thailand, the protected status of the human embryo may not hinge on human dignity, but instead on religious and cultural interpretation of that embryo's status. Indeed, it has been recognized that in view of alternative political and legislative systems in the Asian context, the recog-

---

[147] *Commune de Morsang-sur-Orge v Societe Fun Production et MWackenheim* [1995] Conseil d'Etat 136727, Cons Etat.

[148] Rosen (2012), p. 77.

[149] *BVerfGE 75* [1987] Bundesverfassungsgericht 369 1 BvR 313/85.

[150] Rosen (2012), p. 76.

[151] Ministry of Foreign Affairs of Denmark, 'Denmark for the UN Human Rights Council' (*Ministry of Foreign Affairs of Denmark*) http://um.dk/en/foreign-policy/denmark-for-the-un-human-rights-council/ accessed 18 June 2018.

[152] House of Representatives, The Parliament of the Commonwealth of Australia, 'Australian Bill of Rights Bill 2017' https://www.legislation.gov.au/Details/C2017B00161/Html/Text.

[153] Mr Wilkie, 'Australian Bill of Rights Bill 2017 No. 2017' 38.

nition of human dignity is not entirely absent, but is interpreted in ways that are different to the Western context.[154]

In Malaysia, for example, a plethora of cases that have been brought before its courts demonstrates how human dignity has been interpreted to mean a damage or tarnishing of reputation that invites constitutional scrutiny *vis-à-vis* Article 5 of the Malaysian Federal Constitution.[155] This is very interesting because Articles 5 to 13 of the Malaysian Federal Constitution protects the fundamental liberties of its citizens. Particularly, Article 5(1)[156] has been interpreted in a manner where a person's "life" extends to his or her right to livelihood; cases such as *Mohamed Bin Senik v Public Prosecutor*,[157] *Lembaga Tatatertib Perkhidmatan Awam, Hospital Besar Pulau Pinang v Utra Badi A/L K Perumal*,[158] and *Tan Tek Seng v Suruhanjaya Perkhidmatan Pendidikan & Anor*,[159] are all illustrative of the cognizance placed on the value of human dignity and its connection to the constitutional right to life and liberty.[160]

In the field of biomedical technologies, bioethics and interventions into the human body specifically, the role of human dignity therefore imparts more value within a Western ethnocentric interpretation than it does within the Asian context. The linguistic differences of what human dignity means is one way in which the grounded normative value that it provides to any human rights discourse should be carefully considered. Indeed, in the wake of the fact that human dignity does indeed demonstrate an overlapping of consensus in various constitutional systems,[161] the challenge now would be to find a meaningful definition in which its true normative value can be realized in the biomedical discourse without exclusion of other forms of dignity in the Asian context. We are, here, not attempting to shatter the value of human dignity as a concept in biomedical technologies.[162] Instead, to recognize how it may be more effectively perpetrated through a pluralistic and concerted understand of its multi-layered meanings, taking into account structural inequalities that exist in different societies,[163] and feminist interpretations,[164] (especially where

---

[154] Lee (2008), p. 1.

[155] Federal Constitution of Malaysia 1957. Article 5(1) of the Malaysian Federal Constitution reads: "No person shall be deprived of his life or personal liberty save in accordance with law."

[156] ibid.

[157] *Mohamad Bin Senik v Public Prosecutor* (2005) 4 Malayan Law Journal 164 (Kuala Lumpur Court of Appeal).

[158] *Lembaga Tatatertib Perkhidmatan Awam, Hospital Besar Pulau Pinang v Utra Badi A/L K Perumal* (2000) 3 Malayan Law Journal 281 (Kuala Lumpur Court of Appeal).

[159] *Tan Tek Seng v Suruhanjaya Perkhidmatan Pendidikan & Anor* (1996) 1 Malayan Law Journal 261 (Kuala Lumpur Court of Appeal).

[160] Lee (2008), p. 24.

[161] Addis (2015), p. 1, 16.

[162] Macklin (2003), p. 1419.

[163] Rawlinson and Donchin (2005), p. 258, 262.

[164] ibid 259.

reproduction may be concerned) may actually contribute to achieving a true sense of universality on what human dignity entails.

5.3.2 Germ-Line Genetic Interventions

In recent years, the development and initial successes of CRISPR/Cas-9 as a revolutionary gene-editing tool has provoked mixed responses. What cannot be denied, however, is the wealth of potential in its future applications, and the manner in which these applications will dramatically change the landscape of medical and scientific treatments, particularly where illnesses or conditions involve the human genes. In a short period of time spanning the last 5 to 6 years, although gene editing tools are no strangers in biomedical advancements, CRISPR however has revealed its enormous potential vis-à-vis successful experimentation for a variety of treatment of human conditions, as well as promising sustainability of biofuel[165] outside of the field of medicine. CRISPR's allure lies in its precision, cost and flexibility in comparison to earlier gene editing tools.

Within the field of medical treatment, 2017 alone was a year in which CRISPR proved itself formidable in moving forward the discourse on the future of humanity through biomedical technologies. In cancer treatment, CRISPR was able to create a cancer-destruction gene, called a "suicide gene" that is able to shrink aggressive tumours[166]; another cancer treatment study demonstrates how the growth of cancer cells may be slowed down by tweaking with the Tudor-SN cell replication protein.[167] Most promising as well was a recent research that discovered two specific proteins that would contribute to a CRISPR "off-switch" which would enable better control for gene editing in human subjects.[168] These are indeed truly revealing and revolutionary potential treatments in medical science. The situation however may evoke less than a positive outlook when we consider CRISPR's interventions in pre-implantation human embryos. In the United States for the first time in 2017, scientists were successful in editing the genes of a human embryo by correcting the DNA and removing what is called the MYBPC3 gene. This gene specifically causes a heritable disease called hypertrophic cardiomyopathy. Post-editing of the MYBPC3 gene, the embryos continued to divide and grow, repairing themselves with 72% of the cells in the embryo having been corrected.[169] Although this is generally to be deemed positive news, the understandable concerns are often accompanied by images and possibilities of a genetically engineered future that no longer chances its fates on the genetic lottery.

---

[165] Schwartz et al. (2016), p. 356.

[166] Chen et al. (2017), p. 543.

[167] Elbarbary et al. (2017), p. 859.

[168] Rauch et al. (2017), p. 150.

[169] Hong et al. (2017), p. 413.

In early 2015, Chinese scientists, for the very first time in the world, edited the genes of a non-viable human embryo,[170] and inevitably created a scandal in the international community of medical professionals, scientists and ethicists particularly. It certainly was not only limited to the fact of the magnanimity of CRISPR's potential applications, but also because the almost 'sacred' acceptability that science would not seek to intervene into the inheritable characteristics of the human genome. In December 2015, the International Summit on Human Gene Editing (also referred to as the Washington Summit) spear-headed by the National Academies of Sciences, Engineering and Medicine in the United States, the Chinese Academy of Sciences and the Royal Society of London put forth a call for a global ban or moratorium on inheritable gene editing outcomes that would affect the human germ-line.[171] The reasons cited for this global ban (which is intended to be temporary in nature until such future time to be determined) is that it would be "irresponsible" in view of on-going issues of safety and standards about CRISPR and the fact that societal consensus could not be definitely determined regarding its use (whether therapeutic or non-therapeutic).[172] As the Washington Summit brought together a group of biologists, bioethicists, medical professionals and scientists, it is not surprising that the reason for the ban appears to be one that is more practical in nature and embedded in a predominantly scientific reasoning. This concern, however, may not extend to other grounds that are founded in religious, cultural, social and ethical sensitivities, which is why I propose a more formalized negotiated system of shared values that would overcome these barriers of sensitivities.

It appeared, however, that this so-called global moratorium did little to dissuade further research actions into germ-line gene editing experimentation. The case of Dr. He Jian Kui, earlier highlighted in this book, is evidence that such a moratorium lacking the force of law, does little to bridge the germ-line concern, and actual practical experimentation with clinical applications. Thus, the scramble to now put forward forms of regulation or governance for gene editing in the future, has become much more urgent.

Barring the on-going concerns about the safety of any scientific or medical technologies, the issue of germ-line genetic interventions has failed to exhibit a cohesive and comprehensive consensus at the societal level, which is a fundamental consideration in the framing of appropriate uses and applications and a manner in which to govern such use and application. As pointed out by Robert Andorno, the reflections on germ line interventions are very divisively clear: either they are strongly encouraged or strongly prohibited.[173] For example, in the US, in 2016, the Harvard School of Public Health's poll findings on the public perception towards genetic editing, testing and therapy yielded very mixed responses, but with a leaning

---

[170] Cyranoski and Reardon (2015).

[171] Wade (2018).

[172] Travis (2015).

[173] Andorno (2002), p. 5, 961.

towards either no knowledge of, or objection to gene editing.[174] In another study funded by the National Institutes of Health in 2017, the outcome of the polls were slightly more positive with less reticence towards human gene editing.[175] In Japan, studies indicate that the public perception and attitudes are generally receptive in nature to gene editing although the risks were understood, and even where it may result in heritability for future generations.[176] In the meantime, China has proven itself unafraid to challenge internationally accepted norms, and with its wealth of advancements in medical and scientific research, and pool of scientific talents, it has been poised to become the "petri dish" of future genetic enhancements.[177]

I point out at this juncture that the discourse of human germ-line interventions has generally been opposed in most parts of Europe and the US, which explains the prevalence of rich literature from these regions that continue to harp upon the ban of germ-line genetic interventions on ethical grounds. But within the scope of this rich saturation, however, I posit that two essential but missing links has perhaps also indirectly contributed to the conflagration of such discourse. Firstly, the lack of peremptory force or lack of resources on an international level in pushing forward a formal legalized ban on germ-line interventions has created an unnecessary likelihood of panicked attitudes, in much the same way the responses were recorded when Dolly the Sheep was cloned.[178] Secondly, by underestimating China's power in scientific research and advancements and its bold attempts at pushing the boundaries of the genetic scientific revolution, as well as the willingness of Asian countries like India, Malaysia and Thailand to move beyond ethical proprieties of genetic interventions, serves to highlight more acutely that a more well-rounded and inclusive discourse is necessary.[179]

Another aspect of the human germ-line bioethical dilemma, which requires further and deepened conversation, is precisely the use of CRISPR or other genetic intervention technologies in reproductive medicine, such as PGD. Human germ-line editing, on its own broad considerations, has inspired a rich trove of academic and scholarly research and literature, as it tends to focus on the broader aspects of ethical proprieties. These ethical questions are very broad in nature, touching on issues of human dignity, enhancement of physical or intellectual capabilities, the 'consent'

---

[174] Harvard T.H. Chan School of Public Health and STAT, 'The Public and Genetic Editing, Testing and Therapy' (Harvard TH Chan School of Public Health 2016) https://cdr1.sph.harvard.edu/wp-content/uploads/sites/94/2016/01/STAT-Harvard-Poll-Jan-2016-Genetic-Technology.pdf Accessed 30 May 2018.

[175] Weisberg et al. (2017).

[176] Uchiyama et al. (2018), p. 745.

[177] Schaefer (2016).

[178] Gabbatis (2018).

[179] In Sect. 5.4 of this chapter, I expand on a suggestion for an inclusive model for meaningful international discourse and global governance of biomedical technologies that take into account human rights components. I do this with the explicit recognition that it will be a very difficult task to engage all relevant stakeholders, but I believe, however, that in light of technological development and the sophistication propelled by our community of citizens, that is still a worthwhile task to pursue.

of the future child, parental decision-making and autonomy, and the like. With the exception of the dystopian "designer babies" concerns,[180,181,182] however, the human germ-line editing discourse has been less grounded in the factual realities of what the technology itself may be able to feasibly achieve.[183] Indeed, what has been pointed out is the fact that not enough concern has been attributed to inter-generational monitoring, within the clinical setting.[184] In ensuring safe and long-term use of gene editing technologies, Bryan Cwik proposes that a strategy for inter-generational monitoring[185] is crucial; this would be helpful in properly assessing the fundamental aspects of safety and effects of gene editing which, in the future, may become an integral part of how safety and quality standards of such technologies are measured.[186] Although the Nuffield Council of Bioethics in the UK has suggested the creation of a "centralized register of trials"[187] which should be made available to researchers for an extended period of time, this suggestion was made particularly in terms of mitochondrial replacement therapy. The additional question is also whether there would be a possibility of opening up the creation of such registry of trials for other forms of gene editing outcomes that need to be monitored on a long-term basis.

From a legal perspective, the human germ line discourse has yet to find a constructive juridical positioning. The discourse of CRISPR's application, however, to plant and animal life forms have found resonance in the opinion of the Advocate General of the Court of Justice of the European Union (CJEU)[188] regarding an enquiry from the Conseil d'Etat in France. The question was concerned with the interpretation within the framework of Directive 2001/18/EC (the GMO Directive) as to mutagenesis in gene editing processes. The GMO Directive made it incumbent that genetically modified organisms (GMO) must be subject to an environmental risk assessment, as well as being subject to other monitoring obligations before being authorized and placed within the EU market. The GMO Directive, however, exempted "organisms obtained through certain techniques of genetic modification, such as mutagenesis."[189] The main issue was whether the herbicide resistant seeds developed by the applicants, and represented by the Confederation Paysannes, were indeed such organisms exempt from the GMO Directive. The CJEU delivered its

---

[180] Fox (2010), p. 170.

[181] Turriziani (2014), p. 595 http://scholarship.shu.edu/cgi/viewcontent.cgi?article=1595&context =student_scholarship.

[182] Bawden (2017).

[183] Yong (2017).

[184] Cwik (2017), p. 1911.

[185] ibid 1912. By reference to "intergenerational monitoring", Cwik refers to "follow-up monitoring of not just participants in the original trial, but also their children and even grandchildren."

[186] ibid.

[187] ibid.

[188] *C-528/16 Fédération Nature & Progrès v Premier Ministre, Ministre de l'agriculture, de l'agroalimentaire et de la forêt* (2018) Court of Justice of the European Union C-528/16, Curia.

[189] Ibid.

opinion on 25 July 2018, stating that organisms obtained by mutagenesis are subject to the GMO Directive. The juridical positioning of the germ line discourse relating to human beings is still, as yet, not addressed.

5.3.3   The Impact of the Shortcomings on Human Rights Discourse

The abovementioned bioethical questions are but part of a wider network of the shortcomings of the International Instruments. As the only significant instruments that currently exist to regulate biomedical technologies and loosely, genetic interventions (on an international level), much must be expected from these International Instruments as a way to effectively manage the course of technologies in the near future. This does not detract from the fact that these International Instruments currently do exist as a form of buffer in terms of determining what activities may or may not be carried out. However, the hypothesis reached at this juncture is simply that the formation of the International Instruments as part of the scope of an international body of biomedical laws, does not really demonstrate a significantly universal legal standard that relates human rights to biomedical approaches.

When shortcomings of this nature are observable at this level, what it does serve to illustrate on a deeper level are two things in particular: one, that both societal and structural inequalities may continue to be perpetuated, especially in the field of reproductive biomedical technologies and the framing of the discourse that is absent the perspective of feminist and gender dimensions; and two, that the boundaries of what may or may not be permissible is blurred because pronouncements via moratoriums, for example, are not radically enforced.

In terms of the impact of societal and structural inequalities, Rawlinson and Donchin allegorize that the UDBHR relies on "abstract concepts of equality that obscure the real inequities that characterize contemporary ethical urgencies",[190] and fails to "articulate a sense of universality ample enough to address the actual inequalities of power and resources that prevail across the globe."[191] Flowing from this allegation, Rawlinson and Donchin call for a necessary formulation of truly universal principles by not only considering shared ethical values (as I have posited in this book), but also "differences in ethical values that obtain across cultures."[192] I do not disagree with this notion but I theorize that it is significantly more challenging to find truly shared values, than it is to find the differences in ethical values between transnational boundaries. In this respect, the proposition made is to focus on negotiating these shared values, and focusing on the protective aspects of pre-

---

[190] Rawlinson and Donchin (2005), p. 260.

[191] ibid.

[192] ibid 261.

implantation genetic interventions that necessitates the robust input of feminist legal theories and perspectives.[193]

On the issue of boundaries of permissibility, the International Instruments has revealed more measures of good practices, ethics, accountability and responsibility in the manner of principles, much more than it has in express prohibition of certain activities. With the exception of an express ban on reproductive human cloning,[194] other margins of medical genetic interventions are less apparent. For example, with reference to the global moratorium on human germ line gene editing[195] not only at the Washington Summit, but also by UNESCO,[196] it appeared that a cohesive unanimous view was reached on the issue of reproductive gene editing until such further time to be discussed again. However, the part of the problem posed here was likely the clarity of what such moratorium would entail, and how a violation of the moratorium would be dealt with. Often touted to be rising powers in biomedical innovations, India and particularly China has demonstrated themselves to be forces to be reckoned with. Indeed, part of the recognition regarding the ban was that regulation on gene editing is lax in other countries, and therefore, "scientists should avoid even attempting, in lax jurisdictions, germline genome modification for clinical applications in humans"[197] until the full implications "are discussed among scientific and governmental organizations."[198] Because the moratorium is not legally enforceable, and it does have the strength of guidelines that many countries follow, the question still remains how the international community of laws can respond in the event of a violation. If this is the case, then although the guidelines have served a purpose, its outcomes in ensuring accountability and full responsibility by all those who subscribe to them, is meaningless.

At this juncture, issues relating to the possibilities of human gene editing are no longer completely new, and although conversations have improved as to how it may be governed, this negotiation on how an international community of vested stakeholders can deal with the advancements of science and technology has reached a stalemate and need to be revisited. From as early as the late 1990s, Julia Black has provided an indication as to how we may manoeuvre our impending genetic revolu-

---

[193] Please see Chap. 3. I am also of the opinion that taking into account feminist legal jurisprudence can contribute to changing mindsets of patriarchy in legislation, both on a societal and political level. Because pre-implantation genetic interventions form part of reproductive technologies services in biomedical discourse, the framing of the appropriate mechanisms in reproduction becomes more important in determining how the human rights that relate to gender dimensions can be better fulfilled, and dissuade, or at least, mitigate, the inequalities that may exist.

[194] United Nations General Assembly, Report of the Sixth Committee 'International Convention against the Reproductive Cloning of Human Beings (A/59/516/Add.1)' https://digitallibrary.un. org/record/542699/files/A_59_516_Add.1-EN.pdf.

[195] Wade (2017).

[196] UNESCO, 'UNESCO Panel of Experts Calls for Ban on "Editing" of Human DNA to Avoid Unethical Tampering with Hereditary Traits' (5 October 2015) https://en.unesco.org/news/ unesco-panel-experts-calls-ban-editing-human-dna-avoid-unethical-tampering-hereditary-traits.

[197] Wade (2018).

[198] Wade (2017).

tion[199]; I posit that the genetic revolution has indeed transcended upon our communities in more significant ways than it had before. Negotiating our legal, socio-ethical and moral pathways in the science of this revolution therefore necessitates a vigorous facilitation on an inclusive international level.

5.4 Guidelines/Recommendations for Genetic Interventions: Are They Enough?

The amicability of earlier conversations on the impact of CRISPR/Cas9 on manipulating the human genome, and more particularly, editing the human germ-line,[200] for example, is to be lauded, and some suggestions made by a group of researchers, including Jennifer Doudna and David Baltimore, appear to be workable. From the international perspective, the reliance thus far, other than the International Instruments, are recommendations made following the Washington Summit. These recommendations include: firstly, more intensive basic and pre-clinical research[201]; secondly, more rigorous study on somatic clinical use[202]; thirdly, that clinical use of germ line editing should not proceed and should continue to be discussed[203]; and finally, the need for a continuous and ongoing forum of this nature.[204] In particular, the fourth and final recommendation or finding post-summit is the likely to constitute one of the most crucial ones insofar as the international dimension is concerned. The statement made after the Washington Summit regarding the ongoing forum puts forward its promotion for inclusion among nations, and to engage "a wide range of perspectives and expertise- including from biomedical scientists, social scientists, ethicists, health care providers, patients and their families, people with disabilities, policymakers, regulators, research funders, faith leaders, public interest advocates, industry representatives, and members of the general public."[205] In this, the direction of the Washington Summit is a powerful indicator that the possibilities of regulatory unity on an international may be possible, despite the magnitude of this task. It also revealed telling signs that regulating a phenomenal scientific technology such as CRISPR or gene editing of various forms was subject to divergent opinions. But generally reports from the summit are positive indicators that a summit of this nature is a first step in discussing regulatory pathways for genetic interventions. However, and as mentioned in this chapter, a more diversified and inclusive

---

[199] Black (1998), p. 621.

[200] Baltimore et al. (2015), p. 36.

[201] Committee on Science, Technology, and Law, Policy and Global Affairs and National Academies of Sciences, Engineering, and Medicine, *International Summit on Human Gene Editing: A Global Discussion* (Steven Olson ed, National Academies Press 2016) 6.

[202] ibid.

[203] ibid 6–7.

[204] ibid 7.

[205] ibid 7–8.

discussion is needed at this level: many more countries beyond the US, the UK and China need to be represented, and vested interest groups such as those with disabilities or genetic conditions, the 'lay' persons, are also under-represented.[206]

One suggestion, in particular, calls for a "globally representative group of developers and users of genome engineering technology and experts in genetics, law, and bioethics, as well as members of the scientific community, the public, and relevant government agencies and interest groups"[207] because a proper legal framework on how to deal with CRISPR and other gene editing technologies are very urgently needed. The gargantuan problem that is likely to be faced here is the magnitude in the convening of such a working group. For example, who would monitor the processes and meetings of the working group? Who would determine the composition of the working group? How would global representatives be selected? How would this working group be established? Would it be inter-organizational or inter-governmental, under the auspices of an existing international or regional supervisory organization? What are the representation of countries to be included in the working group? Would this be duplicitous of the role of the United Nations International Bioethics Committee (IBC), or could this perhaps be incorporated as a special working group focused on issues of human germ line gene editing under the IBC? On the assumption that such a working group could effectively be established, then other concerns relating to the International Instruments also need to be undertaken; this would be a new dimension of legal and regulatory challenges in providing clarity to existing legislation, or to propose the development of new treaties, conventions or legislation in a manner that could fluidly deal with the impetuosity of biomedical technologies, as well as the very important issues that relate to enforcement, transparency and accountability. It is hoped that the World Health Organization's (WHO) recent initiative in setting up an expert panel to develop guidelines and standards for gene editing will be successful.[208] Besides the issues that have been highlighted in this book, it is also imperative to consider the expert views of the scientists who have published a call to regulate gene editing, *vis-à-vis* an article in *Nature*.[209]

Guidelines, or soft law, on its own basis, can contribute to the changing dynamics on how we view the power structures that exist within the legal system and its relationship with laws and enforceability. Krajewska accurately points out that soft law do not have in common "a uniform standard of intensity as far as their legal scope is concerned, but they do share a desire to influence the practice of states, international organizations, and individuals."[210] The scope of international biomedical laws at present is not well saturated enough that it may be considered a "self-contained regime", but it is also true that biomedical technologies are very fluid and dynamic

---

[206] Reardon (2015), p. 173.

[207] Baltimore et al. (2015), p. 3.

[208] 'WHO | WHO Expert Advisory Committee on Developing Global Standards for Governance and Oversight of Human Genome Editing' (n 2).

[209] Wade (2018).

[210] Krajewska (2012), p. 17.

subject matters in themselves, and even reaching an international level of "self-contained regime" is likely to be as challenging as convening an international summit in the nature of the Washington Summit. At the heart of it, the recommendations made by the Washington Summit are, in essence, soft law, and do not have the force of law, not in the manner that the International Instruments do have. But it does not mean that the soft law effect of the Washington Summit guidelines, or any other international guidelines that pertain to biomedical technologies and genetic interventions are baseless.

In this respect, I return to my underlying hypothesis in Chap. 4, where I proposed a combination of regulatory approaches that I believe would be useful in aiding the governance and resolution of problematic circumstances relating to biomedical technologies. These regulatory approaches are governance or framework approaches, certainly not in the same league as explicit legal rules or legislation, but as some form of soft law that would complement a legislative framework, indeed, even those of the International Instruments. Perhaps this may be viewed as a stop-gap measure or a temporary deputization of a soft law or guidelines framework; in no means is this a recommendation that the role of guidelines or soft law should trump an express legislative pronouncement. However, bearing in mind that "treaties and other formal international agreements require large commitments of resources, time, and political capital and often pose enforcement challenges",[211] these guidelines are capable of filling existing lacunas until such time express legislative pronouncements can be effectively made. These guidelines may be adequate for the time being, but they may not be for long. The urgency in filling in regulatory gaps, addressing diversified opinions, finding negotiated shared values and streamlining a unified acceptable and responsible global biomedical approach is near.

References

A, B and C v Ireland (Application No 25575/05) [2010] HUDOC (Grand Chamber, European Court of Human Rights)

Addis A (2015) Human dignity in comparative constitutional context: in search of an overlapping consensus. J Int Comp Law 2:1

African Charter on Human and Peoples Rights (21 October 1986). http://www.humanrights.se/wp-content/uploads/2012/01/African-Charter-on-Human-and-Peoples-Rights.pdf

Andorno R (2002) Biomedicine and international human rights law: in search of a global consensus. Bull World Health Org 5

Andorno R (2009) Human dignity and human rights as a common ground for a global bioethics. J Med Philos 34:223

Andorno R (2013a) Principles of international biomedical law. In: Principles of international biolaw: seeking common ground at the intersection of bioethics and human rights. Editions Bruylant

Andorno R (2013b) The dual role of human dignity in bioethics. Med Health Care Philos 16:967

---

[211] *Committee on Science, Technology, and Law, Policy and Global Affairs and National Academies of Sciences, Engineering, and Medicine* (2016), p. 5.

Assembly UG (1979) Convention on the elimination of all forms of discrimination against Women. 20 Retrieved April 2006

Baltimore D et al (2015) A prudent path forward for genomic engineering and germline gene modification. Science 348:36

Barak A (1996) Constitutional human rights and private law. Rev Const Stud III:218

Barak A (2015) Human dignity: the constitutional value and the constitutional right. Cambridge University Press

Barnett RE (1986) Four senses of the public law-private law distinction. Harv J Law Public Policy 9:267

Bawden T (27 July 2017) Scientists call for new rules on GM designer babies. iNews. https:// inews.co.uk/news/health/new-rules-will-be-needed-to-exploit-designer-baby-breakthrough-in-britain. Accessed 27 Jan 2018

Black J (1998) Regulation as facilitation: negotiating the genetic revolution. Mod Law Rev 61:621

Boer D, Boehnke K (2015) What are values? Where do they come from? A developmental perspective. In: Brosch T et al (eds) Handbook of value: perspectives from economics, neuroscience, philosophy, psychology and sociology. Oxford University Press

BVerfGE 75 [1987] Verfassungsgericht 369 1 BvR 313/85

C-528/16 Fédération Nature & Progrès v Premier Ministre, Ministre de l'agriculture, de l'agroalimentaire et de la forêt [2018] Court of Justice of the European Union C-528/16, Curia

Chen QB (1994) Chinese Constitutional Law. BLJ 26:77

Chen Z-H et al (2017) Targeting genomic rearrangements in tumor cells through Cas9-mediated insertion of a suicide gene. Nature Biotechnol 35:543

Christine Goodwin v the United Kingdom (Application No 28957/95) [2002] HUDOC (Grand Chamber, European Court of Human Rights)

Coggon J, Miola J (2011) Autonomy, liberty, and medical decision-making. Camb Law J 70:523

Committee on Science, Technology, and Law, Policy and Global Affairs and National Academies of Sciences, Engineering, and Medicine (2016) Olson S (ed) International summit on human gene editing: a global discussion. National Academies Press

Commune de Morsang-sur-Orge v Societe Fun Production et MWackenheim [1995] Conseil d'Etat 136727, Cons Etat

Conseil de l'Europe (1953) Convention for the protection of human rights and fundamental freedoms

Conseil de l'Europe (1997) Convention for the Protection of Human Rights and Dignity of the Human Being with Regard to the Application of Biology and Medicine: Convention on Human Rights and Biomedicine. Editions du Conseil de l'Europe. http://193.205.211.30/lawtech/ images/lawtech/law/convenzioneoviedo.pdf

Costa J-P (2013) Human Digity in the jurisprudence of the European Court of Human Rights. In: McCrudden C (ed) Understanding human dignity. The British Academy

Cwik B (2017, 1911) Designing ethical trials of germline gene editing. New Engl J Med 377

Cyranoski D, Reardon S (22 April 2015) Chinese Scientists Genetically Modify Human Embryos. Nature News. http://www.nature.com/news/ chinese-scientists-genetically-modify-human-embryos-1.17378

de Anarclens P (2007) The United Nations as a social and economic regulator. In: de Sanarclens P, Kazancigli A (eds) Regulating globalization, critical approaches to global governance. United Nations University, pp 8–35

Department of Jurisconsult R and LD, 'Bioethics and the Case-Law of the Court' (Council of Europe / European Court of Human Rights 2016) Research Report

Department of Social Services, Australian Government, 'Values and Law'. https://www.dss.gov.au/ our-responsibilities/settlement-and-multicultural-affairs/programs-policy/taking-the-initiative/ resources/values-and-law

Douglas v Hello! Ltd [2005] EWCA Civ 595

Elbarbary RA et al (2017) Tudor-SN–mediated endonucleolytic decay of human cell MicroRNAs promotes G $_1$ /S phase transition. Science 356:859

European Commission JRC Science for Policy (2018) JRCF7- Knowledge Health and Consumer Safety, Overview of EU National Legislation on Genomics. European Commission. EUR29404EN http://publications.jrc.ec.europa.eu/repository/bitstream/JRC113479/policy_ report_-_review_of_eu_national_legislation_on_genomics_-_with_identifiers.pdf

European Medicines Agency (2018) Report of the EMA Expert Meeting on Genome Editing Technologies Used in Medicinal Product Development. European Medicines Agency. EMA/47066/2018

Federal Constitution of Malaysia 1957

Fischer-Lescano A, Teubner G (2004) Regime collisions: the vain search for legal unity in the fragmentation of global law. Mich J Int Law 25:99–1045

Foster C (2011) Human dignity in bioethics and law. Hart

Fox D (2010) Retracing liberalism and remaking nature: designer children. research embryos, and featherless chickens. Bioethics 24:170

Gabbatis J (14 February 2018) Dolly the Sheep: 15 years after her death, cloning still has the power to shock. The Independent. https://www.independent.co.uk/news/science/dolly-the-sheep-cloning-15-years-death-future-humans-monkeys-what-next-a8208896.html

Geller T (2016) In privacy law, it's the U.S. vs. the world. Commun ACM 59:21

GenomeWeb (4 December 2018) WHO to set up panel to develop gene-editing standards. GenomeWeb. https://www.genomeweb.com/scan/who-set-panel-develop-gene-editing-standards?utm_source=addthis_shares&fbclid=IwAR320wblX33e1znwYOf8bkOLtvjc9mOr GSaImu8ARQ22ST6k3U2xn-2injo#.XGcDwcnSMNY.facebook

Gillon R (1985) Autonomy and the principle of respect for autonomy. Br Med J (Clinical research ed.) 290:1806

Groll D, Lott M (2015) Is there a role for "Human Nature" in debates about human enhancement? Philosophy 90:623

Guttinger S (2017) Trust in science: CRISPR–Cas9 and the Ban on human Germline editing. Sci Eng Ethics

Hallich O (2017) Sperm donation and the right to privacy. New Bioethics 23:107

Harlow C (1980) "Public" and "Private" law: definition without distinction. Mod Law Rev 43:241

Harvard TH (2016) Chan School of Public Health and STAT, 'The Public and Genetic Editing, Testing and Therapy'. Harvard TH Chan School of Public Health. https://cdn1.sph.harvard. edu/wp-content/uploads/sites/94/2016/01/STAT-Harvard-Poll-Jan-2015-Genetic-Technology. pdf

Hawking SW (2005) The theory of everything: the origin and fate of the universe. Phoenix Books

Hong M et al (2017) Correction of a pathogenic gene mutation in human embryos. Nature 548:413

House of Representatives, The Parliament of the Commonwealth of Australia, 'Australian Bill of Rights Bill 2017'. https://www.legislation.gov.au/Details/C2017B00161/Html/Text

Jacobs B (2 August 2017) Donald Trump proposes law to cut immigration numbers by half in 10 years. The Guardian. http://www.theguardian.com/us-news/2017/aug/02/ trump-immigration-law-reduction-10-years

Jalloh v Germany (Application 54810/00) [2006] HUDOC (Grand Chamber, European Court of Human Rights)

Jefferson T (2002) The Declaration of Independence. Scholastic Inc

Karovska-Andonovska B (2014) Right to privacy in the constitutions of the European Countries and the US constitution. Vizione 22:127

Kleinlein T (2012) Constitutionalization of international law. Das Max-Planck-Institut für auslän-disches öffentliches Recht und Völkerrecht 231:703

Koskenniemi M, Leino P (2002) Fragmentation of international law? Postmodern anxieties. Leiden J Int Law 15(3):553–579

Krajewska A (2012) International biomedical law in search for its normative status. Revista De Derecho Y Genoma Humano = Law and the Human Genome Review 36:121

Lander ES et al (2019) Adopt a moratorium on heritable genome editing. Nature 567:165

Lee MYK (2008) Universal human dignity: some reflections in the Asian context. Asian J Comp Law 3:1

Lembaga Tatatertib Perkhidmatan Awam, Hospital Besar Pulau Pinang v Utra Badi A/L K Perumal (2000) Malayan Law J 3:281 (Kuala Lumpur Court of Appeal)

Lie W (2004) Equality and non-discrimination under international human rights law. Research Notes, Norwegian Centre for Human Rights: University of Oslo

Ma Y, Zhao Y, Liao M (2015) The values demonstrated in the constitution of the People's Republic of China. In: Ladikas M et al (eds) Science and technology governance and ethics. Springer

Macklin R (2003) Dignity is a useless concept. BMJ 327:1419

McCrudden C (2008) Human dignity and judicial interpretation of human rights. Eur J Int Law 19:655

Miller J (2014) The influence of human rights and basic rights in private law in the United States. Am J Comp Law 62:133

Ministry of Foreign Affairs of Denmark, Denmark for the UN Human Rights Council. http://um.dk/en/foreign-policy/denmark-for-the-un-human-rights-council/

Mohamad Bin Senik v Public Prosecutor (2005) Malay Law J 4:164 (Kuala Lumpur Court of Appeal)

Mutua M (2001) Savages, victims, and saviors: the metaphor of human rights. Harv Int Law J 42:201

Nuffield Council on Bioethics (2016) Genome editing: an ethical review. Nuffield Council on Bioethics

Nunan D, Di Domenico M (2017) Big data: a normal accident waiting to happen? J Bus Ethics 145:481

O'Mahony C (2012) There is no such thing as a right to dignity. Int J Const Law 10:551

Owigar JWB (2002) Ethics and living values in constitution. Const Kenya Rev Comm, 4. http://www.commonlii.org/ke/other/KECKRC/2002/4.html

Powell T (20 December 2017) May Faces Brexit Grilling by Powerful Committee of MPs. Evening Standard. https://www.standard.co.uk/news/politics/brexit-latest-theresa-may-to-be-grilled-on-eu-withdrawal-negotiations-by-committee-of-powerful-mps-a3723596.html

Pretty v the United Kingdom (Application No 2346/02) [2002] HUDOC (Fourth Section, European Court of Human Rights)

Rantsev v Cyprus & Russia (Application No 25965/04) HUDOC (First Section, European Court of Human Rights)

Rauch BJ et al (2017) Inhibition of CRISPR-Cas9 with bacteriophage proteins. Cell 168:150

Rawlinson MC, Donchin A (2005) The quest for universality: reflections on the universal draft declaration on bioethics and human rights. Dev World Bioeth 5:258

Reardon S (2015) Global summit reveals divergent views on human gene editing. Nature 528:173

Research and Library Division Department of Jurisconsult (2016) Bioethics and the Case-Law of the Court. Council of Europe / European Court of Human Rights, Research Report

Robinson M (2007) The value of a human rights perspective in health and foreign policy. Bull World Health Org 85:241

Rosen M (2012) Dignity: its history and meaning. Harvard University Press

Ruiz-Chiriboga OR (2013) The American Convention and the protocol of San Salvador: two inter-twined treaties-non-enforceability of economic, social and cultural rights in the Inter-American System. Neth Quart Human Rights 31:159

Schaefer GO (2 August 2016) The future of genetic enhancement is not in the west. The Conversation. http://theconversation.com/the-future-of-genetic-enhancement-is-not-in-the-west-63246

Schwartz CM et al (2016) Synthetic RNA polymerase III Promoters facilitate high-efficiency CRISPR–Cas9-mediated genome editing in *Yarrowia Lipolytica*. ACS Synthetic Biol 5:356

Sharp T (12 June 2013) Right to privacy: constitutional rights & privacy laws. Live Science. https://www.livescience.com/37398-right-to-privacy.html

Sholley JB (1951) Constitution of the United States of America. In: Cases on constitutional law. Bobbs-Merrill

Siliadin v France (Application No 73316/01) [2005] HUDOC (Second Section, European Court of Human Rights)

Stone G et al (2005) Constitutional law, 5th edn. Aspen Publishers

Tan Tek Seng v Suruhanjaya Perkhidmatan Pendidikan & Anor (1996) Malay Law J 1:261 (Kuala Lumpur Court of Appeal)

Taylor AL (2004) Governing the globalization of public health. J Law Med Ethics 32:500

Tisdall S (23 July 2017) Europe seeks a long-term answer to a refugee crisis that needs a solution now. The Observer. http://www.theguardian.com/world/2017/jul/22/divided-europe-refugee-crisis-italy-serbia-greece

Travis J (4 December 2015) Inside the Summit on human gene editing: a reporter's notebook. Science. http://www.sciencemag.org/news/2015/12/inside-summit-human-gene-editing-reporter-s-notebook

Trstenjak V, Weingerl P (eds) (2016) The influence of human rights and basic rights in private law, vol 15. Springer

Turriziani JV (2014) Designer babies: the need for regulation on the quest for perfection. Law School Student Scholarship, 595. http://scholarship.shu.edu/cgi/viewcontent.cgi?article=1595&context=student_scholarship

Tysiac v Poland (Application No 5410/03) [2007] HUDOC (Fourth Section, European Court of Human Rights)

Uchiyama M, Nagai A, Muto K (2018) Survey on the perception of germline genome editing among the general public in Japan. J Human Genet 63:745

UNESCO (11 November 1997) Universal Declaration on the Human Genome and Human Rights. http://portal.unesco.org/en/ev.php-URL_ID=13177&URL_DO=DO_TOPIC&URL_SECTION=201.html

UNESCO (16 October 2003) International Declaration on Human Genetic Data. http://unesdoc.unesco.org/images/0013/001361/136112e.pdf

UNESCO (2006) The Universal Declaration on Bioethics and Human Rights. http://unesdoc.unesco.org/images/0014/001461/146180E.pdf

UNESCO (5 October 2015) UNESCO Panel of Experts Calls for Ban on "Editing" of Human DNA to Avoid Unethical Tampering with Hereditary Traits. https://en.unesco.org/news/unesco-panel-experts-calls-ban-editing-human-dna-avoid-unethical-tampering-hereditary-traits

United Nations General Assembly, Report of the Sixth Committee, 'International Convention against the Reproductive Cloning of Human Beings (A/59/516/Add.1). https://digitallibrary.un.org/record/542699/files/A_59_516_Add.1-EN.pdf

Wade N (21 December 2017) Scientists seek ban on method of editing the human genome. The New York Times. https://www.nytimes.com/2015/03/20/science/biologists-call-for-halt-to-gene-editing-technique-in-humans.html

Wade N (19 January 2018) Scientists seek moratorium on edits to human genome that could be inherited. The New York Times. https://www.nytimes.com/2015/12/04/science/crispr-cas9-human-genome-editing-moratorium.html

Warbrick C (1989) Federal aspects of the European Convention on Human Rights. Mich J Int Law 10:698

Weisberg SM, Badgio D, Chatterjee A (2017) A CRISPR new world: attitudes in the public toward innovations in human genetic modification. Front Public Health 5

Wilkie M. 'Australian Bill of Rights Bill 2017 No., 2017' 38

World Health Organization, WHO Expert Advisory Committee on Developing Global Standards for Governance and Oversight of Human Genome Editing. https://www.who.int/ethics/topics/human-genome-editing/committee-members/en/

World Medical Association (19 October 2013) WMA Declaration of Helsinki - Ethical Principles for Medical Research Involving Human Subjects. https://www.wma.net/policies-post/wma-declaration-of-helsinki-ethical-principles-for-medical-research-involving-human-subjects/

Yong E (2 August 2017) The designer baby era is not upon us. The Atlantic. https://www.theatlantic.com/science/archive/2017/08/us-scientists-edit-human-embryos-with-crisprand-thats-okay/535668/

Chapter 6
The Dynamics of Basic Constitutional Rights in Selected Jurisdictions

Abstract This final chapter delves into the dynamics of basic constitutional human rights in the selected jurisdictions by studying the key approaches taken by each of the jurisdictions in generally dealing with human rights matters. In examining the potential regulation of future pre-implantation genetic interventions in the jurisdictions (which should take into account human rights components such as the right to life, the right to privacy (autonomy) and the right to equality and non-discrimination), the chapter introduces the concept of 'entry points of regulatory approaches'. These entry points of regulation are symbolic of concerns regarding various branches of biomedical technologies. In the chapter, I identify two spectrums of these entry points: firstly, the practical, positivistic and private law aspects; and secondly, the philosophical human rights aspects. I posit that these classifications point to the main prevailing and concerning issues that would prompt a state to regulate those technologies. Using these entry points of regulation as lenses of comparisons allows us to navigate the role of fundamental rights in the selected jurisdictions by determining how these jurisdictions prioritize the human rights components that would be relevant in pre-implantation genetic interventions.

This chapter now shifts the issues thus far into a critical evaluation of the distinct role of human rights (procured from the broad corpus of international and regional human rights instruments in Chap. 5) into the different national systems of the selected jurisdictions. The framing of these relevant particularized human rights components, which will be introduced in Sect. 6.1 below, dealing with the peculiarities relating to genetic interventions, intends to provide an infusion into the dialectical body of the constitutional framework, contemplated from the perspectives of these areas. By drawing out specific elements of comparisons, or "lenses" by which genetic interventions may be examined (I term this the 'entry points of regulation') this further lends credibility to the hypothesis made in Chap. 5, and the new 'universality' of a shared values system in the constitutional dimension.

On this basis, Sect. 6.1 begins with an evaluation of approaches within the framework of a regulatory model. The first approach taken is to use a human rights model. A model or framework of regulation is by evidentiary consideration, generally considered a paragon of virtue by upholding the Rule of Law; one of the ways in doing

© Springer Nature Switzerland AG 2019

P. L. Lau, *Comparative Legal Frameworks for Pre-Implantation Embryonic Genetic Interventions*, https://doi.org/10.1007/978-3-030-22308-3_6

so is by showing that human rights is an important component to match the significant values in the jurisdictions, despite the imbued plurality and differing social, economic and political priorities on each national agenda. Human rights as the root consideration of a regulatory model on an international and regional level thus has the propensity to influence regulatory models in different jurisdictions even across international borders. The selected jurisdictions are arguably countries in which the foothold of human rights is not a very strong, cemented one (as opposed to countries such as Germany and France, for example). It is, however, this fact that makes the exercise of comparison in this chapter immensely interesting. The underlying justification for this comparison is vital as a means of demonstrating the role of human rights as an effective tool in presenting similar values that are deemed to be meaningful in each particular jurisdiction. This would pave the way for a more open and well- rounded discourse on the approaches in forming a well-informed regulatory model for pre-implantation genetic interventions.

The second approach taken is similar to a functional analysis approach,[1] but in this instance, more pertinent to law, in that it considers the relationship between laws, human rights and the practical dimensions in the scientific or medical sphere. This is intended to provide an identification of patterns that reinforce how issues in pre-implantation genetic interventions may be solved.[2] To do this, I introduce the entry points of regulation, or regulatory approaches. I select appropriate "tools" that may be used as lenses to examine this relationship: these include reproduction as a realm that impacts on population health and control; health, safety, risk assessments and efficacy of scientific research and development; the adequacies of informed consent in medical treatments; and other regulatory dimensions that may encompass accreditation standards and the transfer and use of genomic data and genetic materials. These insights can provide a basis into how human rights and shared values that have been negotiated upon, can inform the operability of liberties within a constitutional framework. In particular, I will highlight the main human rights components that feature most prominently in this kind of bioethical discourse; namely, the right to life, the right to privacy and the right to equality/non-discrimination (collectively the Human Rights Components).

In Sect. 6.2, thus, the role and influence of fundamental rights in the different jurisdictions will be highlighted. The selected jurisdictions are grouped in accordance with the development of laws or governance in the realm of the Human Rights Components. The first group comprises the United States (US) and provides an acknowledgment that the consideration of the Human Rights Components are often fragmented and polarized in each state, largely owing to diversified interpretation of the fundamental rights contained in the Amendments to the US Constitution and pronouncements made by the US Supreme Court (where relevant). The consideration of the US is also vital due to the fluidity of its changing climate of political influence: what may traditionally have been regarded as constitutional rights worthy of protection under the Amendments in the Constitution, may, perhaps, in the near

---

[1] Michaels (2006).

[2] Lando (2017), p. 17.

future, make the transition into less fundamental rights because of political influences and consultative governmental policies that change the direction of its national agenda, possibly impacting the operability of Pre-Implantation Genetic Diagnosis (PGD) with genetic interventions.[3] In the second group, comprising the United Kingdom (UK) and Australia, I will demonstrate that both these jurisdictions have made some sophisticated progress in the discourse that bridges the Human Rights Components with biomedical technologies, informing us on potential developments of a framework in pre-implantation genetic interventions. Another interesting observation regarding UK and Australia is that each of them have advanced proposals for a national Bill of Rights. These endeavours are still being deliberated upon in UK and Australia. Finally, the third group consists of Malaysia, also a Commonwealth nation like Australia, but existing with a fascinating dualist legal system (of traditional common law and Parliamentary laws/legislation, with non-secular, religious Shari'a law); and Thailand, a constitutional monarchy with a troubled historical past of military *coup d'etat* and presently (at the time of writing), governed by a military junta. The examination of this third group will reveal that the Human Rights Components do not have a place in the prioritization of each national agenda; in addition, the development of laws or governance in biomedical technologies are either sparse, or non-existent. In fact, a startling discovery through the validation of this endeavor will reveal that laws or modes of governance in Malaysia and Thailand,[4] are often prompted by 'regulatory triggers', events that explode on such a large scale so as to justifiably warrant an immediate response and intervention by legislators.[5]

The usefulness of this grouping exercise is illustrated in Sect. 6.3, reinforcing the foundational premise of any regulatory model: the assumption that human rights discourse, and specifically, the Human Rights Components, must feature prominently in regulatory models because the potential proliferation of pre-implantation genetic interventions go beyond our present understanding about the frontiers of life, and the possible expansion of this frontier into directions that we may not previously have envisaged, the latter of which can fundamentally change the fabric of social organization as we know it. Section 6.3 thus is a detailed expansion of the constitutional (and in some cases, the political) framework of each of the jurisdictions, including its treatment of the Human Rights Components on a national level. Utilizing primarily legislative statutes, international human rights reports and court cases at the national level (including a small selection of cases from the European Court of Human Rights (ECtHR)), and other regulatory measures amounting to 'soft law', the analysis shows that the operation of these rights are additionally garrisoned upon social, ethical or religious bases.

---

[3] Farokhmanesh (12 December 2016).

[4] Australian Associated Press (20 February 2015).

[5] Umeda (6 April 2015). This was an example where Thailand passed a fast-tracked new surrogacy law after being inundated with scandal in the international community for its legislative laxities in commercial surrogacy.

The chapter concludes in Sect. 6.4 by revealing that the structural framework of each of the jurisdictions uncovers interesting, differing approaches to human rights. By demonstrating the 'entry points of regulation' here as a comparative criterion, emphasis is placed on the value of considering the features of Human Rights Components in these differing jurisdictions. This provides a more distilled idea about how these rights operate and are incorporated into the constitutional systems. The idea behind this analysis is to indicate that the approaches employed by the different jurisdictions can be used to gauge the response and reactions to the future regulation of pre-implantation genetic interventions, taking into consideration the fundamental outcomes indicated in Chap. 5 as well. Finally, as a matter of final reflection, the analysis will also reveal my interpretation of the key concerns for each jurisdiction in how they may have chosen to regulate forms of ARTs, PGD or genetic interventions thus far, consistent with the framing of the entry points of regulation.

6.1   The Key Approaches in a Rights-Based Discussion

In this Chapter, I put forward the key approaches in determining the appropriate foundational viewpoints of regulatory models for pre-implantation genetic interventions. This is premised on a rights-based approach, with the added value of functionalism in law and legal jurisprudential theories to round off on the implications of biomedical technological concerns.

The Human Rights Components within the scope of this book, as mentioned above, cover the right to life, the right to privacy, and the right to equality. The general reason why these specific principles or fundamental rights have been chosen is because they are issues that naturally flow from reproduction and reproductive elements, all of which are determinative components in considering pre-implantation genetic interventions,[6] and more so because the liberties afforded rest on individual choices over one's own bodies. As explained in Chap. 3, the recognition and acknowledgement of reproductive rights by the United Nations, bear legitimacy through the Convention for the Elimination of All Forms of Discrimination Against Women (CEDAW),[7] the only international human rights instrument (thus far) that recognizes and reinforces a woman's reproductive right.[8] The relevant Committees

---

[6] VAWnet, National Resource Center on Domestic Violence, 'Reproductive Justice, Reproductive Health and Reproductive Rights: A Framework' https://vawnet.org/sc/reproductive-justice-frameworks.

[7] Often referred to as an international bill of rights for women, the Convention was adopted by the United Nations General Assembly in 1979, and defines discrimination against women as "…any distinction, exclusion or restriction made on the basis of sex which has the effect or purpose of impairing or nullifying the recognition, enjoyment or exercise by women, irrespective of their marital status, on a basis of equality of men and women, of human rights and fundamental freedoms in the political, economic, social, cultural, civil or any other field."

[8] United Nations (1979).

in the Office of the High Commissioner for Human Rights (OHCHR) in the United Nations[9] strive to ensure the reach and international enforcement of CEDAW *vis-à-vis* the signatories to the convention. The Human Rights Components highlighted here, as the associated fundamental human rights that are connected to general women's rights in reproduction, raise very important entanglements with PGD and genetic interventions. Indeed, the OHCHR states that "women's sexual and reproductive health is related to multiple human rights, including the right to life, the right to be free from torture, the right to health, the right to privacy, the right to education, and the prohibition of discrimination."[10]

How are the Human Rights Components anchored to rights, specifically in the field of reproduction? More importantly, why are these rights important in the context of considering pre-implantation genetic interventions? In the case of the right to life, this is treated as one of the most fundamental, constitutional rights not only in many countries around the globe, but also in numerous international conventions and proclamations.[11] The right to life, however, has been at the epicenter of abortion debates, not only in the US,[12] but also within Europe,[13] simply because this challenge begs the question of women's reproductive rights juxtaposed against the status of an embryo or fetus. There is no simple answer in respect of this conflicting notion, but a variety of approaches in the interpretation of this right within the framework of constitutional rights have allowed liberal democracies to "protect the inherent, supreme and inalienable worth of human beings"[14] through the balancing of constitutional proportionality within the structure of the body politic.[15] This conundrum will undoubtedly continue in the field of debating pre-implantation genetic interventions, and how it challenges our interpretative understandings of the universal right to life.

---

[9] United Nations, Office of the High Commissioner, Human Rights 'Sexual and Reproductive Health and Rights' http://www.ohchr.org/EN/Issues/Women/WRGS/Pages/HealthRights.aspx.

[10] Ibid.

[11] For examples, see: Article 3 of the Universal Declaration of Human Rights ("Everyone has the right to life, liberty, and security of person."); Article 2 of the European Convention on Human Rights and Fundamental Freedoms ("Everyone's right to life shall be protected by law."); Article 6.1 of the International Covenant on Civil and Political Rights ("Every human being has the inherent right to life. This right shall be protected by law. No one shall be arbitrarily deprived of his life."); Article 4 of the American Convention on Human Rights ("Every person has the right to have his life respected. This right shall be protected by law and, in general, from the moment of conception. No one shall be arbitrarily deprived of his life.")

[12] Simson (2014), p. 107.

[13] For example: in Germany, Article 2 of the Basic Law for the Federal Republic of Germany states "Every person shall have the right to life and physical integrity. Freedom of the person shall be inviolable. These rights may be interfered with only pursuant to a law." In Hungary, the Fundamental Law of Hungary protects the right to life *vis-à-vis* Article II: "Human dignity shall be inviolable. Every human being shall have the right to life and human dignity; the life of the foetus shall be protected from the moment of conception."

[14] Mahlmann (2012), p. 389.

[15] Ibid 392.

Within the scope of the right to privacy (which also encompasses the principle of autonomy), both at an international level and also in many national systems, the right to privacy also constitutes a fundamental, constitutional right[16] that is intrinsic to another fundamental, constitutional right: the right to life.[17] In Chap. 1, the right to privacy and essentially, the exercise of autonomy within the scope of contemporary liberal eugenics, is illustrated *vis-à-vis* parental or individual choices made with regards to all aspects of reproduction, even if those choices have to do with discarding embryos in in-vitro fertilization (IVF) and PGD for reasons contemplated by parent/s themselves.[18] Although my view here insofar as autonomy is concerned, is hinged on the dialectic questions of the boundaries of exercising parental autonomy in PGD and genetic interventions, the choices made by "family [as] the level of implementation"[19] in liberal eugenics should not be an objectionable facet. Consistent with the principle of self-determination, privacy and autonomy, PGD and genetic interventions logically "involves no centralized decision fixing the future human type".[20] The ethics of the morality, however, are a different question entirely.

The element of consent, and more particularly, informed consent, is also another component that is integral to the evaluation of pre-implantation genetic interventions. With advancements made in medical technologies and research environments, it would not be archetypal to expect that a respect for one's right to privacy, should also make its presence felt via the doctrine of informed consent[21] which commands the stage through dimensions of clinical practice and ethics. The establishment of the doctrine of informed consent is therefore a fundamental tenet in the relationship of knowledge, communication, interaction and understanding between medical and research professionals and patients, resting on the premise that an individual may

---

[16] For example, in some of the chosen jurisdictions, such as Australia, England and the United States, the right to privacy is recognized as a constitutional right (although not explicitly stated in selected Constitutions, as the case may be) and is often understood to generally mean a right to personal autonomy, to personal and family life, and to choose whether or not to participate in the engagement of some acts or experiences.

[17] The right to life is undoubtedly one of the most universally recognized human right in the world; and where the right to life was traditionally premised on discourse relating to the protection of human life, contemporary musings are now seeking an extension of the "right" to life, to include the instrumentalities that accompany this "right". This includes the "right" to die (often overlapping with the right to privacy), and if national courts are inclined to rule on the permissibility of the extension of this right, then it will surely dramatically change the fabric of how the right to life has been traditionally interpreted in the future.

[18] Whether a parent/s choose to undergo IVF and PGD for the selection of the best possible embryo to be implanted, for reasons such as avoiding genetic mutations or disabilities, or for purposes of gender or sex selection, or to produce a savior sibling for the benefit of an older child: these reasons inform the decisions made by parent/s without intervention by states, allowing freedom of will, flexibility and voluntariness in reaching these choices.

[19] Wiesenthal and Wiener (1999), p. 383.

[20] Nozick (1974).

[21] See Article 5, Convention for the Protection of Human Rights and Dignity of the Human Being with regard to the Application of Biology and Medicine, 1997 (the Oviedo Convention).

then make full and conscious decision in consenting to any form of medical treat-ment or interventions. For example, in England, researchers at the Centre for Health, Law and Emerging Technologies (HeLEX)[22] in the University of Oxford, have formulated an improved mechanism of informed consent, referred to as the "dynamic consent" movement.[23] This emphasizes the importance of research partnerships,[24] and is presently being undertaken in the RUDY (Rare UK Diseases of bone, joints and blood vessels) study. This evolutionary metamorphosis seeks to dramatically improve the relationship between the involved stakeholders in clinical settings that involve medical research and treatments, or even within the framework of research projects heralding the beginning of developmental technologies in the eradication or treatment of diseases. The crux of the dynamic consent paradigm, with emphasis on the word "dynamic", shows that the consent process is continuously extant and changing in the equation of clinical relationships, thereby suffusing each step in the length of a particularized medical process with clarity, transparency, and, on a very crucial level, choice and autonomy.[25] This is certainly extremely desirable when we consider pre-implantation genetic interventions.

Special contemplation must also be given to the right to equality,[26] and its accom-panying principles of non-discrimination; like the other human rights considerations mentioned above also have a firm footing in many legal systems as a fundamental, constitutional right. The mounting concerns of the hypothetical simultaneous use of pre-implantation genetic interventions range mostly from the usual dystopian ideas about the 'unnatural' future of humankind, to the intensification of existing inequal-ities within the stratified layers of society.[27] These concerns on equalities have also been highlighted in Chap. 1, through the structural architecture of equality, citizenry and individual relationships within the function of society,[28] and by asking whether scientific advances do truly empower individuals. For these reasons, I advance my

---

[22] 'Centre for Health, Law and Emerging Technologies (HeLEX) — Nuffield Department of Population Health' https://www.ndph.ox.ac.uk/research/centre-for-health-law-and-emerging-technologies-helex.

[23] Teare et al. (2017), p. 816.

[24] Ibid 2.

[25] Kaye et al. (2015), p. 141.

[26] See: Article 1 ("All human beings are born free and equal in dignity and rights.") and Article 7 ("All are equal before the law and are entitled without any discrimination to equal protection of the law."), both of the UDHR; Article 26 of the International Covenant for Civil and Political Rights ("All persons are equal before the law and are entitled without any discrimination to equal protec-tion of the law."); Article 2(3) and 3 of the International Covenant for Economic, Social and Cultural Rights also contain similar equality and non-discrimination provisions as the ICCPR; Article 14 of the European Convention on Human Rights and Fundamental Freedoms ("The enjoy-ment of the rights and freedoms set forth in this Convention shall be secured without discrimina-tion on any ground such as sex, race, colour, language, religion, political or other opinion, national or social origin, association with a national minority, property, birth or other status."); Article 24 of the American Convention on Human Rights ("All persons are equal before the law. Consequently, they are entitled, without discrimination, to equal protection of the law.")

[27] Reiss and Straughan (1996).

[28] Dyson (1997), p. 46.

arguments on the basis of a balance between a functional human rights approach that takes into account each jurisdiction's entry points of regulation.

6.1.1 The Human Rights Approach in the Field of Genetic Interventions

In light of the foregoing, I provide reasons why we should employ a human rights approach, or what is more commonly known as a "rights-based approach" when considering the field of genetic interventions. In the discourse relating to human rights based approaches (HRBA) determining the purpose of this approach is likely to reveal its utility in a framework for pre-implantation genetic interventions. As I have consistently highlighted throughout the course of this work, the regulation of pre-implantation genetic interventions requires a human rights foundation for comparison because the consequences directly affect global citizens and populations. If we also look to international agencies that have been tasked in global development, the literature reveals that HRBAs, in conjunction with "genuine accountability"[29] and true cooperation between various agency actors, is an important starting point in achieving a developed status. Hence, employing a HRBA in the field of genetic interventions should be highly encouraged because it proceeds to aid in the development of values and policies in this area.[30]

From the perspective of international organizations such as the United Nations, (and clearly, an international perspective is very much needed when considering pre-implantation genetic interventions), human rights have always been considered to be integral to the development process. Approaching the necessary global development through the eyes of the HRBA is justified on the following basis: it "serves to define injustice,"[31] that "empowerment should be replaced with models that place the individual at the center of the development process",[32] or that "rights can serve as a tool that can allow advocacy movements to challenge inequity more effectively."[33] The allure of the HRBA lies in the fact that it captures the normative, pragmatic and ethical categories that relate to development practice.[34] Celestine Nyamu-Musembi and Andrea Cornwall refer to the work of J. Hausermann's discussion paper in preparation of the Government White Paper on International Development[35] that "what is distinctive about a human rights approach to development is that it works by setting out a vision of what ought to be: that is, it provides a powerful normative frame-

---

[29] Nyamu-Musembi and Cornwell (2004).
[30] Ibid.
[31] Grugel and Piper (2009), p. 80.
[32] Ibid.
[33] Ibid.
[34] Nyamu-Musembi and Cornwell (2004), p. 2.
[35] Ibid.

work to orient development cooperation."[36] In doing so, the normative frame "brings an ethical and moral dimension to development assistance."[37]

If we view this against the background of having the benefit and support of norms in international law that can be effectively agreed upon and translated into national legal systems, the HRBA has much to offer in a field of scientific development such as genetic interventions and how future populations may be affected, whether positively or negatively, by the advent of such technological interventions. It is naïve, however, to expect the implementing a HRBA will be as simple as a theoretical expatiation of its benefits, advantages and promise for future development.

The critiques on the full realization of an effective HRBA are numerous, and it would not help to delude the thinking to assume that this is the only path in which to achieve a balance in formulating a regulatory framework. For example, Varun Gauri and Siri Gloppen highlight that the HRBA is often inundated with the politics of 'compelling' "global compliance based on international and regional treaties."[38] Some of the problems they have highlighted include not only the problem with compliance on an international level, but also the saturation of the "middle class capture"[39] despite the appealing attributes of the approach. The other problem relating to the HRBA and its entrenchment in the sphere of international law is that this may border on "excessive legalism"[40] and a state of rigid formality that emphasizes formal legal requirements as opposed to practical aspects of services and access to development.[41] The reality is despite the depository of human rights conventions, treaties and legislations at an international level, the interpretation of this state of formality *vis-à-vis* judicial reasoning, state compliance, and other enforcement and accountability measures, are less clear cut. This is attributable to how these rights may be asserted by groups affected by violations, influenced by what may be known as a programming approach which depend on contextual local environments, local infrastructure elements and institutional considerations.[42] More importantly, with the burden of claiming rights often placed on these individuals as opposed to states on delivering them,[43] the formation and shift in rights consciousness,[44] from "I want", to "I am entitled to"[45] often requires awareness and re-education of how HRBA can aid accountability and transparency in a legal environment.

Applying the HRBA in the field of genetic interventions, and being able to also deal with pre-implantation genetic interventions simultaneously, is the starting

---

[36] Ibid.
[37] Ibid.
[38] Gauri and Gloppen (2012), p. 485.
[39] Ibid 485.
[40] Grugel and Piper (2009), p. 80.
[41] Ibid 81.
[42] Gauri and Gloppen (2012), p. 493.
[43] Grugel and Piper (2009), p. 85.
[44] Gauri and Gloppen (2012), p. 494.
[45] Ibid 495. Gauri and Gloppen quote Pitkin (1981), pp. 327–352.

point for "a dominant framework for contemporary social justice movements."[46] A HRBA has the ability to incorporate a variety of modalities for mobilization, as well as institutional mechanisms[47] such as to "enhance accountability oriented institutions in governments and donors",[48] to reinvigorate the educations of communities to view themselves as "rights holders",[49] or to "support legal mobilization in national courts."[50] Whether or not these approaches take on a juridical or quasi-juridical model in the international sphere, the HRBA specifically in the field of genetic interventions assist in eventually "structuring state-party negotiations regarding international regime design."[51]

6.1.2   Entry Points of Regulation: Genetic Interventions in the Selected Jurisdictions

Before we begin to examine the manner in which different constitutional systems regard the role of fundamental (and/or constitutional) rights, I first begin to frame the query by introducing the entry points for regulatory approaches or regulation (or "tools" as elements of considerations) that may be used to probe further into examining the role of these rights. This is by no means an exhaustive list but is meant to highlight the key lenses by which to frame the inquiry into pre-implantation genetic interventions. I argue that the practical operation of these key lenses then feed into how the Human Rights Components within each of the selected jurisdictions have been (or have not been) realized, and how they may apply in the field of pre-implantation genetic interventions. This influences the manner in which regulatory differences may be revealed, and the respective 'entry point' into the business of regulation. These entry points of regulation, I argue, denote the concerns of each jurisdiction regarding various branches of biomedical technologies. Because different jurisdictions will have different concerns about biomedical technologies, the entry points, or points at which they will choose to regulate, is largely dependent on the main prevailing and concerning issues. Therefore, what a state is likely to be more concerned with, whether they be ethical, social or other practical issues that relate to biomedical technologies, the state is then more likely to legislate in alignment with these concerns.

Following a detailed analysis of the treatment of human rights in each of the jurisdictions in Sect. 6.3, I will then provide an analysis of the entry points of regulation in Sect. 6.4 of this chapter. I categorizing the entry points of regulation into

---

[46] Ibid 494.

[47] Ibid 487.

[48] Ibid.

[49] Gauri and Gloppen (2012).

[50] Ibid 487.

[51] Ibid 489.

two main branches: the positivistic law aspects of regulation (the First Entry Point); or the more philosophical human rights dimensions (the Second Entry Point). There are numerous issues that may be categorized under the First Entry Point or the Second Entry Point, and I briefly provide some examples of these.

The First Entry Point by which to examine the Human Rights Components in pre-implantation genetic interventions may include the concept of informed consent, which includes not only the provision of knowledge and full information and disclosure, but also all matters relating to health, safety, risks and efficacy assessments to a potential patient. In Sects. 3.4 and 4.2.1, I highlight the fundamental tenets of the doctrine of informed consent in any form of medical treatment and medical research, which is, indeed commensurate with the Universal Declaration on Bioethics and Human Rights.[52] For example, within the United States, the crystallization of the doctrine of informed consent was given constitutional importance[53]; a leading case in this regard is *Planned Parenthood of Southeastern Pennsylvania v Casey*,[54] which stipulates the necessity for the information to be completely truthful and non-misleading.[55] It can be said that the decision in *Planned Parenthood* opened the doorway to state legislated informed consent laws, although it is also recognized in recent events that this declaration of constitutionality may come under attack at the Supreme Court level in the future.[56]

However, such is the role necessitated by the function of informed consent within any kind of medical or scientific discourse, that the 2500 year old Hippocratic Oath was recently revised for contemporary medical settings, and has now been aptly renamed the World Medical Association's Declaration of Geneva, or The Physician's Pledge.[57] One of the most interesting components of this updated pledge is the inclusion of patient autonomy within the purview of its wordings: "I will respect the autonomy and dignity of my patient." Linguistically, words such as "autonomy" and "dignity" within the sphere of biomedical technologies carry a great deal of weight. I view this inclusion as being aligned with the fundamental concept of informed consent, which necessitates full, frank and honest disclosure of all relevant information in the provision of medical treatment. However, this also suggests that despite the fiduciary duty of a medical professional to save lives and do no harm to their patient, a respect for patient's autonomy means a respect for choice, decisions, and essentially, that patient's right to privacy. As will be highlighted in the jurisdictional analysis below, the right to privacy bears strongly on the doctrine of informed consent, and much more fully within the realm of pre-implantation genetic interventions.

---

[52] UNESCO (2006). Please see Articles 6.1 and 6.2 which provides for the "prior, free informed consent" of any participants to medical treatment or medical research.

[53] Daniels et al. (2016), p. 181.

[54] 'Planned Parenthood of Southeastern Pa. v. Casey, 505 U.S. 833 (1992)' (*Justia Law*) https://supreme.justia.com/cases/federal/us/505/833/.

[55] Daniels et al. (2016), p. 181.

[56] Shear (28 June 2018).

[57] Parsa-Parsi (2017), p. 1971.

In addition to that, a full realization of how informed consent may be meaningfully achieved would also mean that assessments of safety, risks and harm are fully conceptualized to the understanding of a potential patient. Safety, risks and quality assurance standards may also be included under the First Entry Point. Where reproductive interventions are concerned, Philip Peters has written a particularly compelling case on the harms and protections that may implicate upon future children who are the product of such reproductive interventions. He states that "it is harmful to the class of future children to choose an option that is more dangerous to them, even if the more dangerous option is likely to result in the birth of a child whose life is worth living."[58] He is accurate in stating that the issue of harm and safety in reproductive interventions cannot be oversimplified and must always be a paramount consideration.[59] Although there is much truth to this statement, the issue of harms and safety must also extend, beyond the metaphysical arguments[60] on the conception of future children, but also to safety and potential harm mechanisms to the recipients of the interventions, the women (in most cases).

In the anniversary debate of the Centre for Ethics and Law in Biomedicine (CELAB) in Budapest in 2016, John Harris, a libertarian moral philosopher, provided illumination on these safety concerns[61] and as a junior doctoral researcher working in that field, I had the privilege to hear first-hand his scholarly opinions. In the debate titled "Why Gene Editing Should Not Be Stopped", Harris fielded the recognition that safety would always be a concern in any form of biomedical technological intervention into the human body. But in contemporaneous parallelism to this argument, he makes the compelling analogy of genetic interventions to the beginnings of modern medicine and advancement in early biomedical technologies, including issues relating to vaccinations, fertility treatments, gene therapy, personalized medicine and the like. For example, the vaccination issues have been subject to intense debates even before the proper medical terms were coined. In contemporary pediatric practice, vaccinations are regarded as the norm,[62] rather than the exception to the rule, although there is still open critique in the form of mostly unsubstantiated scientific evidence that form the premise of the "Anti-Vaxxers" movement.[63] Harris makes the point that vaccinations have occupied a central role in contemporary societies these days because at one point, they have been regarded

---

[58] Peters (2004), p. 4.

[59] Ibid 3.

[60] Ibid 5. The most notable metaphysical argument regarding the protection of future children who are born of reproductive interventions is the non-existence argument. The National Bioethics Advisory Council in the United States, after the birth of Dolly the sheep had this to say of a comparison between existence and non-existence: "….is problematic. Not only does it require us to compare something unknowable – non-existence- with something else, it can also lead to absurd conclusions if taken to its logical extreme. For example, it would support the argument that there is no degree of pain and suffering that cannot be inflicted on a child, provided the alternative is never to have been conceived."

[61] Center for Ethics and Law in Biomedicine (19 January 2016).

[62] Rothstein (Winter 2015, Number 44).

[63] Canal (7 March 2017).

as "safe enough" for human use, for the benefits of immunological system, and have been instrumental in eradicating diseases like polio, smallpox, pneumococcal and other diseases that once characterized the highest mortality rates for human beings. The same reasoning is equally applicable to genetic interventions, using the 'safe enough" argument as opposed to the knowledge that the procedures may be 100% foolproof. Scientific and medical research into CRISPR, for example, continues to assess the risk and safety of the technology in its human genome applications,[64] which has generally been accepted to form part of the wider debate relating to the bioethical enquiries on genome editing. On this basis, they also become instrumental in influencing the operation of the Human Rights Components because of the wide-ranging implications of safety and harm in the context of human rights discourse.

The logistical and regulatory dimensions of genetic interventions could be another element that is included under the concerns of the First Entry Point. Beyond the moral and philosophical arguments about genetic interventions, these concerns are more practical in nature, connecting scientific and medical practice and provision of services with the law and the framework of regulation.[65] These regulatory dimensions may encompass issues that relate, for example, to accreditation and licensing of genetic interventions, the range, scope and limitations in the kind of services that may be provided, the inclusion of social concerns in spheres of responsible research ethics and conduct, and the transportation and transfer of genetic materials from one location to another, amongst others. For example, in Chap. 4, I posit that the regulatory approaches that may assist in regulating biomedical technologies can also apply to pre-implantation genetic interventions. The final premise of Chap. 4 hinges on a combination of regulatory approaches to establish a pliable and adaptable regulatory framework in regulating technologies. Certainly, this is likely to be one of most important legal dimensions in regulating genetic interventions in its various spheres.

The connection of the regulatory dimensions to other aspects of logistical concerns is very practical and real. Taking the example of transporting genetic materials for research, analysis and experimentation from one location to another raises the additional component of combining scientific and biological practice with laws relating to road and transportation, or biological or hazardous substances in a particular jurisdiction. For example, within the UK, this would necessarily involve compliance with requirements by the Health and Safety Executive particularly in the handling of biological agents.[66] Although it is unlikely that genetic materials fall within the category of hazardous biological substances, possibilities may ensue that

---

[64] Rath (2018).

[65] I highlight the regulatory approaches in biomedical technologies in Chap. 4. These regulatory approaches that look to combining the force and operation of the law, with alternative means of regulation in the form of guidelines and soft laws for instance, look to a comprehensive governance system of genetic interventions that would be more flexible and adaptable.

[66] Department of Health, UK, Health and Safety Executive, 'Biological Agents: Managing the Risks in Laboratories and Healthcare Premises' 82.

extraordinary genetic intervention mechanisms involving external pathogens may also come under the regulatory framework of the Control of Substances Hazardous to Health Regulations 2002.[67] In the meantime, the European Union (EU) has established specific directives "to regulate the procurement and use of human tissues and cells for human application,"[68] but more importantly, they address issues regarding quality assurance and safety standards.

The Second Entry Point, on the other hand, may include the more philosophical aspects of reproduction. In more circumstances than can be imagined, any discussion on reproduction still continues to attract significant debate. One of the reasons why this is so, may be attributable to the fact that reproduction still has many gendered dimensions that do not always necessarily promote true development. I make this claim with the understanding that despite centuries old women's suffrage movement, the trajectory of reproductive rights legislation from as early as the seventeenth century has been deeply imbued with significant and distinct dimensions of gender, that present women's bodies in a certain functional manner.[69] These persistent views of women's bodies and reproduction conversely affect the manner in which reproductive rights legislation is framed, drafted and enforced. Despite the fact that reproductive rights legislation currently exist in many jurisdictions around the world, I observe that these legislations carry strong vestiges of patriarchy that either lead to or continue to perpetuate gender and social inequalities, even within the context of the rights in themselves. As our community of global citizens move into the height of the digital age, however, continued burning and new concerns emerge, not only for legislators, policy makers and market stakeholders, but much more so for women. The cosmopolitan shift in contemporary societies makes more acute the fact that the governance of women's bodies in reproduction (*vis-à-vis* reproductive rights legislation) has often been poorly framed and misleading in many national systems, and clearly within the framework of national legislature. This fundamental flaw therefore leads to, or continues to subdue women's positions of inequalities.

The many spectrums in which bodies are viewed as 'properties' warranting intervention by external factors, or as "fetal containers"[70] whose 'products'[71] have been subjected to a commodification process, also does little to dispel the inequalities that exist. Indeed, within the scope of reproduction, women's bodies particularly have been subjected to notions of propertization over hundreds of years and in many societies. The fact that women's bodies are being commodified have borne less concern than it does other forms of commodification in biomedical technologies.[72] Donna Dickenson instead proclaims that the acceptance of feminist theories of women's bodies as objects, rather than subjects, perpetuates implicitly the "sover-

---

[67] Joanna Marshall, 'Transport of Biological Materials', University of Edinburgh, 51.

[68] Mahalatchimy et al. (2012), p. 136.

[69] Goslinga-Roy (2000), p. 113.

[70] Maier (1989), p. 8.

[71] Goold et al. (2014), p. 1.

[72] Dickenson (2007), p. 8.

eignty of the male subject and to consign women to the role of victims."[73] On this basis, a simple supposition would suggest that the narrative in respect of the gendered dimensions of reproduction must be altered in light of burgeoning medical and scientific technologies.

The occurrence of continued and perpetuated inequalities also makes it difficult for the goals and agenda of the International Conference on Population and Development (ICPD) and the Convention on the Elimination of All Forms of Discrimination against Women (CEDAW) to be meaningfully achieved. Despite the fact that CEDAW is the only international convention that specifically deals with women's rights and has been ratified by 189 countries in the world, the discourse in national realities is very different from the theoretical expectations of its outcomes. Using the earlier mentioned HRBA in law and development,[74] which I believe is one of the foundational components for a more effective development of reproductive rights, deliverance of reproductive justice, and its legislation in developing legal systems, I believe it is possible to dismantle the biological from the persona, and to dismiss the shackles of patriarchal hegemony in reproduction and reproductive rights legislation in different legal systems, using the ICPD and CEDAW as an international comparative benchmark for the minimum realization of reproductive dimensions.

It should be borne in mind that the abovementioned considerations are but a few of the numerous aspects that need to be evaluated in pre-implantation genetic interventions. It cannot be denied that the close nexus of all the mentioned concerns, to human rights considerations (particularly the Human Rights Components) impact upon a legitimate operation and safeguarding of equality, equity and accessibility within a contemporary democratic framework.

6.2   The Role of Human Rights As Constitutional Rights in the Selected Jurisdictions

In Chap. 4, the development and growth of intellectual and scholarly discourse relating to the abortion debates and prenatal testing technologies were considered, because these are discourses that feature close resemblance to pre-implantation genetic interventions in many aspects. The chapter had put forward that the manner in which these debates had developed, and subsequently, been given a legalized, or at least, regulatory framework, had been influenced primarily by factors such as politics (due to the differing political systems in those countries), as well as socio-economic awareness *vis-à-vis* public perception and the protection of their individual rights, and the role played by medical professionals and scientists, who have now begun to realize the propensity of these technological advancements. The

---

[73] Ibid 13.
[74] Gauri and Gloppen (2012).

nature of these technologies were impactful enough within the sphere of life sciences to provide adequate grounds on which the culmination of human rights concerns that resulted should similarly be embodied within the constitutional framework. By subsequently highlighting the different approaches used in the different constitutional systems in each of its treatment of fundamental rights (which I claim includes both human rights and constitutional rights), the likes of which may be applied to the discourse on pre-implantation genetic interventions. Prior to such undertaking, however, it would first be useful to question the kind of role that would be played by fundamental rights within constitutional systems. For example, would fundamental rights (in which human rights are also included) be considered constitutional rights? Or would constitutional rights generally include a selection of fundamental rights applicable to its body of citizenry? Is there a difference between fundamental rights and constitutional rights? Where would the scope of international human rights law fit within these relationships? It would be helpful to begin an examination in this section by looking at the influential work of Robert Alexy in *A Theory of Constitutional Rights*.[75]

In this book, the position that is adopted considers that the core fundamental rights within a particular jurisdiction may incorporate some components from the overarching international human rights discourse (applicable to all persons by virtue of their status as human beings). The particularization of those fundamental rights within the countries may or may not be constitutionally embedded (enshrined within a country's national constitution), which, if so, would impart upon them a constitutional dimension, making them fundamental constitutional rights. But the question as to whether human rights are indeed constitutional rights is also worth considering. In the introduction to Robert Alexy's work, it was stated that "rights become constitutional because of their perceived substantive significance as expressions of an underlying political morality."[76] Because constitutions exist on a spectrum that straddles purely the formal, procedural or purely substantive dimensions,[77] the prioritization of rights afforded in each particular jurisdiction is very much dependent on rights that are considered the most beneficial and important in those jurisdictions. As pointed out in Alexy's work, taking the UK by way of example, the concept of parliamentary sovereignty is paramount in the operation of its constitutional system. This makes it a largely procedural constitution, which would require Parliamentary or legislative intervention in terms of the pronouncement of applicable human rights and in accordance with set procedures.[78] In Germany, however, its constitution regards human rights as supreme within the purview of its laws, making its constitution a largely substantive one.[79] These are but examples in which human rights are perceived in the constitutional context, and the level of importance attributed to its considerations.

---

[75] Alexy (2010).

[76] Ibid, xix.

[77] Ibid.

[78] Ibid.

[79] Ibid. xx

The inter-mingling relationship between constitutional systems and international human rights law is also another dimension which influences the manner in which human rights may be considered fundamental (constitutional) rights. In the triumvirate of what has become commonly known as the 'International Bill of Rights', comprising the Universal Declaration of Human Rights (UDHR), International Covenant on Civil and Political Rights (ICCPR) and International Covenant on Economic, Social and Cultural Rights (ICESCR),[80] this is likely to be the most emphatic force of the applicability of human rights globally, ratified by the largest number of countries worldwide within their individual national legal systems. Notwithstanding this international perspective, there has been debate as to how and the extent to which human rights have been properly incorporated and enforced by individual states, particularly in countries where a track record for respecting such rights is questionable.[81] Whether these human rights are, indeed, constitutional rights would very much depend on the process of "constitutionalization" put forward by Gardbaum. He identifies that in countries like Canada and the UK, the internationalized element of human rights discourse have found their way into a process of internal "constitutionalization" *vis-à-vis* the elevation of human rights to a higher level, the placement of these rights with legal status and within the given regime.[82] Another manner in which it is possible for human rights to attain a "constitutional" status is through "transformation of a particular international law regime from a purely-based treaty entity to a 'constitutional' one."[83] The European framework of the European Convention on Human Rights is a good example of this "constitutional" status, and at least provides a form of mechanism to remedy violations through the forum of the European Court of Human Rights, or other appropriate human rights committees.

In this regard, an examination of the manner in which each of the selected jurisdictions embodies the protection of fundamental or human rights within their individual constitutional frameworks reveals several telling features. For example, whether there is harmony between these concepts and how it may be adequately achieved, or the level of import of international human rights elements into a scope that may be adequately adjudged within each constitutional jurisdiction, or the regulatory approaches that may exhibit historical or cultural vestiges influencing how human rights may or may not feature in a country's constitutional framework.

The claim made here is that the affordance of human rights within a constitutional system is necessary, not only regarding pre-implantation genetic interventions, but also in moving a country's status forward in the age of development in many other aspects. A basic example may be taken in considering the rights of the

---

[80] Gardbaum (2008), p. 749.

[81] Some of these countries are primarily indicated by their developmental status, but within the purview of the selected jurisdictions, countries like Malaysia, Thailand, and China as well as other countries in Muslim Middle East and the African region, also indicate a lower threshold of prioritization for a human rights agenda within their constitutional systems.

[82] Gardbaum (2008), p. 752.

[83] Ibid 753.

child as one of the recognized components in international human rights discourse. In many countries around the world, especially under-developed, or developing countries, the intertwined relationship between child marriages and non-secular fundamentalism on a national level or the phenomenon of poverty, influences the level of protective mechanisms that may be granted to children. Would child marriages therefore not violate children's rights, particularly where parental approval or other religious authorities are the main arbiter in making decisions for children to enter into under-aged marriages? Where continued perpetuation of child marriages is afforded through poor legislative interventions in accordance with the fundamental tenets of international law, then this also means that a progressive movement towards more liberalized governance and respect for international consensus is also hampered in the process. The same may be true of pre-implantation genetic interventions, particularly so because these genetic interventions have the capacity to penetrate societal barriers and influence the operation of social orders that impact upon individual rights and liberties.

6.3   Differences in the Rights-Based Frameworks of Selected Jurisdictions (The Human Rights Components)

In this section, having considered the development of biomedical technologies in Chaps. 4 and 5, and using that as a basis for a similar justification on how the regulatory or governance framework for pre-implantation genetic interventions may also be treated, we now turn to analyse the differing approaches employed by the selected jurisdictions in how their individual national systems employ protective mechanisms of human rights generally, and the Human Rights Components specifically. The analysis here divides the selected jurisdictions into three specific groups, based on the level of development in these countries insofar as the recognition of the Human Rights Components are concerned, particularly where there is a great divide of controversial issues.

The rationale for the analysis grouping is as follows: Group One encompasses the United States, which, although enumerates the embodiment of fundamental rights within its Constitution, is often subject to saturated fragmentation of how rights and autonomy is viewed from the perspectives of its individual states. Insofar as the Human Rights Components are concerned, the United States is not a particularly strong demonstration for this aegis, although it is also acknowledged that some fundamental rights are often given more importance over others. A diversity of strong opinions, especially in the public realm, is demonstration of how the role played by the United States Supreme Court is one that is often faced with difficult constitutional questions that may not fall squarely within the purview of its jurisdiction. It is also part of my hypothesis that the role played by political agendas in the United States ranks highly on the affectation of individual rights and liberties. Group Two consists of the United Kingdom and Australia, because both national legal systems, although through different means, exhibit a relatively sophisticated degree of protective mechanisms for a variety of rights (and the accompanying Human Rights Components) that are incumbent within fertility treatments, PGD, and other similar ventures; this does not specifically mean however, that the overall human rights protection is strongly ingrained in the constitutional culture. The trajectory on how these rights and particularly, rights relating to women's bodies and the exercise of their individual autonomy, for example, has developed, has been quite exemplary *vis-à-vis* the legitimacy of the Human Embryology and Fertilization Authority (HFEA) that primarily regulates assisted reproductive technologies and its instrumentalities, and partially, the Human Rights Act 1998 (HRA) and bears recognition as a framework to be considered. The third and final group comprises Malaysia and Thailand, countries that are not known for the protection of human rights generally. Although some respite is offered in these jurisdictions concerning reproduction and its incumbents,[84,85] the more general over-arching priority appears to focus on reproductive tourism[86] and protection for the equality and emancipation of women more specifically in this context.

The hypothesis pursued through this analysis is intended to shed light on how the framework of human rights and the Human Rights Components are treated in the selected jurisdictions. What is sought as a visible outcome is to delineate a commonly negotiated shared values system, as has been illustrated in Chap. 5 from the various international and regional human rights instruments, but to also recognize the inherent plurality and cosmopolitanism of the different constitutional systems, including any 'alternative' legal systems that also exist (such as Shari'a law[87] as the Islamic legal system[88] in Malaysia).

6.3.1 Group One (1): The United States

The constitutional law framework of the United States (US) and its federated and unified single nation solution was the subject of inspiration for the Australian federation. From its long and significant history, the structure of the US Constitution's protection of individual liberties is largely found in the Amendments to the Constitution (also known as its Bill of Rights). With the exceptions of the public function and the entanglement exception, private conduct does not generally have to comply with the Constitution as the latter is only applicable to state or federal government conduct. In some circumstances, of course, Congress may be in a position to apply constitutional norms to private conduct. The Bill of Rights contained in the Amendments to the Constitution are generally applicable to state and local

---

[84] Center for Reproductive Rights (Asia) (20 February 2014).

[85] Umeda (2015).

[86] Stasi (2015), p. 17.

[87] Dahlan and Faudzi (9 March 2016).

[88] Shuaib (2012), p. 85.

governments through incorporation via the due process clause in the Fourteenth Amendment.

In the earlier parts of this book, considerable expansion was undertaken in relation to several aspects of the US legislative framework. In Chap. 3, by explaining the state of PGD laws in the US, considerable emphasis was placed on the rather haphazard approach to PGD regulation in the country. As also explained in the said chapter, the conflation of PGD as being equivalent to the abortion debates (because of the discarding of embryos) has resulted in a flawed interpretation through the variety of state legislations for PGD. Perhaps also unusually, the chapter had highlighted that to date, the US is still yet to regulate PGD on a federal law because of the polarity of debates relating to reproductive technologies and reproductive rights in the country. Further, in Chap. 4, the abortion debates in the US were given further expansion as a manner of highlighting how these debates have been interlinked to human rights and individual liberties, whether constitutionally guaranteed or not. In our contemporary environment, post *Roe v Wade*, the abortion debates still continue to be highlighted quite vigorously. In recent news regarding the nomination of Justice Brett Kavanaugh as a Supreme Court Justice by President Donald Trump,[89] concerns have been abound as to the possibility that women's reproductive rights in the US could be fundamentally altered. Whether this remains pure speculative conjecture or a real and concrete inkling of the future cannot be said with great certainty. However, since the naming of Justice Brett Kavanaugh as a Supreme Court Justice nominee, there has been a score of responses from the scholarly community regarding Justice Kavanaugh's potential influence on the bench because of his carefully reflected, diligent and solid conservative stance on a variety of issues as well as his "expansive interpretation of executive powers".[90]

Bearing in mind the applicable levels of constitutional scrutiny[91] regarding the constitutionality of a particular legislation in the US, the balance hangs between legitimate government purpose, procedural (and substantive) due process, and individual rights and liberties that therefore trigger the scrutiny tests. The kind of scrutiny tests that is triggered would very much depend on the rights concerned, and whether these rights are given protection under the Constitution. Because the narrative of US human rights discourses, whether regarding violations of constitutionality or otherwise is featured prominently in international news portals and media outreach, it is easy to mistake the US constitutional framework as a fully realized and democratic one. It is largely a substantively political constitution, but the inter-

---

[89] Bennett (12 July 2018).

[90] Ibid.

[91] In determining the constitutionality of a particular legislation, the different levels of scrutiny are applicable: the rational basis test, the intermediate scrutiny test and the strict scrutiny test, all of which necessitates a balance between the means and the ends and analysing the least restrictive ways in which redress may be had. For example, in the rational basis test, a law may be upheld if it is rationally related to a legitimate government purpose and the burden of proof lies on the entity that claims the law is not constitutional. The applicable levels of scrutiny, or which scrutiny tests that should be used, would usually depend on the kind of rights that is concerned within a particular law.

pretation of constitutional guarantees and liberties within the framework of the Human Rights Components have no doubt demonstrated some interesting perspectives. One of these is the manner in which liberties can still be founded even where the US Constitution is silent. There are two significant aspects where liberties have been found in this context; firstly, regarding the right to privacy (including the 'right' to procreate or have offspring), and secondly, regarding liberty and equal protection *vis-à-vis* sexual orientation. Besides raising the question of liberal constitutional interpretation, the following cases mostly illustrate the existence of a penumbra of rights within the substantive due process clause and the equal protection clauses of the Amendments to the Constitution.

The right to privacy is likely the most famous unwritten constitutional right in US constitutional history. Privacy, which may encompass a variety of inter-personal individual relationships, but also, more importantly, the freedom of choice and autonomy, has often been subject to a variety of interpretation as to its true meaning. Indeed, James Whitman wrote: "if privacy is a universal human need that gives rise to a fundamental human rights, why does it take such disconcertingly diverse forms?"[92] As a legitimate question that warrants answers, he continues by examining the manner in which privacy differs from the European and the American viewpoint, lending interesting perspectives to the vast interpretations of what privacy may entail.[93] In fact, it is appropriate to start this examination at the point of *Skinner v Oklahoma*[94] where the majority of the court found that the forced sterilization statute for a repeat criminal offender was a violation to the accused person's fundamental "right to procreate". The reliance by the Supreme Court in *Skinner v Oklahoma* hinged on the interpretation of the equal protection clause of the Fourteenth Amendment (instead of the due process clause of the same Amendment). In Chap. 1, I also provided reasons as to why the court perhaps employed the mechanisms of the equal protection clause instead of the due process clause, and although the judicial reasoning in *Skinner v Oklahoma* was not looked upon as favourably because it lacked "a secure constitutional foundation",[95] it nevertheless paved the way for a contemporary interpretation of a 'fundamental' right to privacy via the penumbra of the US Constitution.

This is most clearly illustrated in *Griswold v Connecticut*[96] where the majority of the Supreme Court upheld that a Connecticut statute that prohibited the use of contraceptives was unconstitutional. The reasoning employed was the due process clause in the Fourteenth Amendment. More specifically, Justice Douglas who delivered the opinion of the court stated that "the specific guarantees in the Bill of Rights have penumbras formed by emanations from those guarantees that help give them

---

[92] Whitman (2003), p. 1154.

[93] Ibid 1160–1171.

[94] 'Skinner v. Oklahoma Ex Rel. Williamson, 316 U.S. 535 (1942)' (*Justia Law*) https://supreme.justia.com/cases/federal/us/316/535/case.html.

[95] Stone et al. (2013), p. 779.

[96] 'Griswold v. Connecticut, 381 U.S. 479 (1965)' (*Justia Law*) https://supreme.justia.com/cases/federal/us/381/479/case.html.

life and substance."[97] In fact, these guarantees themselves are capable of creating the zones of privacy, what he terms the "penumbral rights of privacy and repose".[98] Justice Black, one of the dissenting justices in the case, however berated the majority for "discovering and applying a constitutional right to privacy."[99] He advocated a direct express test based on a specific violation by stipulating that the state government "has a right to invade [privacy] unless prohibited by some specific constitutional provision",[100] of which there was none. In any case, *Griswold v Connecticut* was also significant, not only because it signalled the beginning of the recognition of a right to privacy as a penumbral constitutional right, but also because Justice Goldberg raised interesting questions about the application of the Ninth Amendment as a liberty from which privacy emanates.[101,102]

However, the reasonable limit of the scope of this right to privacy has always been fraught to contention. Indeed, commentators also questioned what the right to privacy in *Griswold v Connecticut* specifically outlined, and whether it also included autonomy generally.[103] As it would later turn out, it appears to be so, and the seminal landmark decision of *Roe v Wade*[104] that later surfaced in 1973 to decriminalize abortion on the grounds of a woman's constitutional right to privacy, including personal privacy that is "fundamental" and "implicit in the concept of ordered liberty".[105] More importantly, in this case, Justice Blackmun who delivered the majority opinion of the court proclaimed:

> Where certain fundamental rights are involved, the court has held that regulation limiting these rights may be justified only by a compelling state interest and that legislative enactment must be narrowly drawn to express only the legitimate state interests at stake...[106]

In choosing not to adopt one single theory of the beginning of life, the court went on to adapt the three-stage standard test for abortions based on the "compelling" point of viability:

(a) For the stage prior to approximately the end of the first trimester, the abortion decision and its effectuation must be left to the medical judgment of the pregnant woman's attending physician.

---

[97] Stone et al. (2013), p. 843.

[98] Ibid.

[99] Beaney (1966), p. 253.

[100] Stone et al. (2013), p. 849.

[101] Ibid 850–851.

[102] The Ninth Amendment of the US Constitution reads: "The enumeration in the Constitution, of certain rights, shall not be construed to deny or disparage others retained by the people." The broad interpretation of this Ninth Amendment by Justice Goldberg was intended to protect any other additional rights that may not have been specifically protected in the Bill of Rights.

[103] Stone et al. (2013), p. 852.

[104] 'Roe v. Wade, 410 U.S. 113 (1973)' (*Justia Law*) https://supreme.justia.com/cases/federal/us/410/113/.

[105] Stone et al. (2013), p. 855.

[106] Ibid 857.

(b) For the stage subsequent to approximately the end of the first trimester, the State, in promoting its interest in the health of the mother, may, if it chooses, regulate the abortion procedure in ways that are reasonably related to maternal health.

(c) For the state subsequent to viability, the State in promoting its interest in the potentiality of human life may, if it chooses, regulate, and even proscribe, abortion except where it is necessary, in appropriate medical judgment, for the preservation of the life or health of the mother.[107]

Till date, *Roe v Wade* remains the standard for legal abortions in the US, and was further reaffirmed in *Planned Parenthood of Southeastern Pennsylvania v Casey*.[108]

In *Planned Parenthood*, the constitutionality of the Pennsylvania Abortion Control Act of 1982 came under attack, specifically elements that related to parental consent and spousal consent for abortions in the state. Although this case reiterated the importance of the doctrine of informed consent, and gave it a constitutional veracity, which opened the pathways to the legislation of informed consent state laws, the specifics of the case, which sought to control the manner in which women could obtain abortions, was questioned. Indeed, the Justices of the Supreme Court's eponymous phrase "Liberty finds no refuge in a jurisprudence of doubt"[109] set the tone for the maintaining the judgment in *Roe v Wade* in accordance with the *stare decisis* rule. Additionally, in order to maintain the position in *Roe v Wade*, the justices formulated the undue burden analysis in *Planned Parenthood*. The effect of the undue burden analysis would render a particular law invalid if "its purpose or effect is to place a substantial obstacle in the path of a women seeking an abortion before the foetus attains viability."[110] Hence, as it stands, *Roe v Wade*, and *Planned Parenthood* remain the decisive forces behind the abortion laws in the US, deriving constitutional protection (of the right to privacy) from the due process clause of the Fourteenth Amendment.[111] The crucial point to note isn't simply that the penumbra of the Constitution creates a right to have abortions, or even the reproductive rights of the woman. More specifically, the penumbral right recognizes the over-arching concept of individual autonomy and choice, the notions of control over one's bodily integrity and its instrumental dimensions.

The second penumbral 'right', which was derived from constitutional interpretation, relates not only to the right to privacy (in maintaining personal relationships) but also relies on the principles of non-discrimination and equality that is protected under the due process clause of the Fourteenth Amendment. *Lawrence v Texas*[112] would come to represent America's outspoken nightmares with homosexuality. In this case, the 'sodomy' laws in the state of Texas came under constitutional review. The word 'sodomy' bearing its original origins from the eleventh century,[113] had

---

[107] Ibid.

[108] 'Planned Parenthood of Southeastern Pa. v. Casey, 505 U.S. 833 (1992)' (n 54).

[109] Stone et al. (2013), p. 873.

[110] Ibid 880.

[111] Ibid 874.

[112] *Lawrence v Texas* [2003] Supreme Court 539 U.S. 558.

[113] Book of Genesis of the Bible, Chapter 18 & 19, Sodom & Gomorrah.

been a 'crime' tainted by misinterpretation, the fear of God and selective enforcement of the law, and became a malleable tool of anti-gay movements and legislation. With the recognition by the Supreme Court in *Lawrence v Texas* that this era of over-scrupulous discrimination is finally over (at least, legally); and with the controversial decision in *Bowers v Hardwick*[114] having been overruled, it is likely that any form of anti-gay legislation, such as the Texas "homosexual conduct" law in this case, may be a thing of the past.

The majority opinion of the Court, given by Justice Kennedy, relied on the involved parties' rights under the due process clause of the Fourteenth Amendment of the Constitution, with an emphasis of analysis in the sphere of personal "liberty"[115] which cannot be intruded upon,[116] exacerbated by the fact that the state could not demonstrate that it had any legitimate interest by which that intrusion may be justified. This critical reasoning is practical because it seeks to 'repair' the stain that was *Bowers*.

Justice O'Connor's concurring opinion, however, suggested that the court could have reached a similar conclusion by using the equal protection clause (of the same Fourteenth Amendment instead). In her opinion, she proclaims that the equal protection clause "is essentially a direction that all persons similarly situated should be treated alike".[117] Explaining her opinion by reference to past legislations subject to the rational test review, her view is that the Texas sodomy laws would also not pass this test if it were subject to scrutiny. In essence, the Texas sodomy laws positioned homosexuals in an undesirably unequal position from heterosexuals because the laws did not criminalize the latter. Although this could be interpreted as yet another avenue by which *Lawrence* could have swung in favor of the petitioners, Justice O'Connor's reliance on it was brilliantly calculated to maintain her echelon of authority. Because she had, 17 years ago, concurred with the opinion in *Bowers*, the case that *Lawrence* now explicitly overruled (and which she now maintained to stand by), she must now avoid the awkward difficulties of explaining the discordance in her stand between these two decisions. Notwithstanding her reasons, her champion of equal protection for the unconstitutional treatment by law (of homosexuals) must be lauded.

In the meantime, since the landmark decision of *Lawrence v Texas*, the case of *Obergefell v Hodges*[118] followed suit in 2015, another landmark decision that cemented the penumbral constitutional right by extending marriage to same-sex couples. The strength of these decisions from the constitutional viewpoint demonstrates that even without an express constitutional guarantee to privacy or other

---

[114] *Bowers v Hardwick* [1986] Supreme Court 478 U.S. 186.

[115] Justice Kennedy opined that all citizens are guaranteed *"life, liberty and property"* and it was this basis upon which he based his argument on the due process clause of the Fourteenth Amendment.

[116] Per Justice Kennedy: "…the right to liberty under the Due Process Clause gives them the full right to engage in their conduct without intervention of the government".

[117] Stone et al. (2013), p. 927.

[118] *Obergefell v Hodges* [2015] Supreme Court 14–556.

points of equality, positive outcomes can still be achieved through judicial reasoning and pronouncement.

6.3.2   Group Two (2): United Kingdom and Australia

6.3.2.1   The United Kingdom

The constitutional framework in the United Kingdom (UK) is an interesting one for comparative reasons, because it remains largely 'unwritten' in nature,[119] comprising a mix of both written and unwritten sources of law, such as cases, statutes, and conventions.[120] Its constitutional system is often described as "a political constitution infused with normative values"[121] and there is a tug-of-war regarding this claim. Where some scholars have argued that the UK constitution appears to be a "malleable set of political rules, which lacks any substantive normative dimension other than a basic commitment to democratic governance,"[122] others have opined that the state of UK's constitution continues to remain in flux and transition, having the flexibility to undergo constant renewal.[123] On this basis, the nature of the UK's constitution has mostly been described as a "political constitution."[124]

Notwithstanding the ingrained exercise of political power in the state of affairs in the UK, what can, however, are the key presumptions that hinge on fundamental rights on a broad level. Colm O'Cinneide identifies three core values that underlie the UK constitutional system, values that promote "political and legal accountability, and provide a normative framework which shapes how political debate and decision-making proceeds in Britain."[125] These core values comprise the presumption of liberty,[126] the imperative of representative government,[127] and the rule of

---

[119] Some scholars, however, prefer to refer to the nature of the UK constitution as being "uncodified", as opposed to being "unwritten", due to the fact that there does, in fact, exist many sources of the UK constitution that are written (such as statutes in the form of Acts of Parliament) and case law and judgments of courts.

[120] Conventions in the context of UK constitutional law refers to the special political mechanisms that govern the relationship between the different branches of the state. These are not in any written form, but fundamentally comprise rules of vital constitutional importance to the workings of the government.

[121] O'Cinneide (2012), p. 239.

[122] Ibid 2., citing Ch. 1 of Bogdanor (2009).

[123] Ibid.

[124] Ibid 1.

[125] Ibid 3.

[126] Ibid 2. O'Cinneide interprets this as the requirement for public authorities to exercise clear legal basis for any actions that may infringe upon individual human rights and liberty.

[127] Ibid 3. This value is the core underpinning of the UK's political representation system of a parliamentary democracy.

law.[128] The UK's membership into the European Union (EU) *vis-à-vis* the European Communities Act of 1972, often described as the 'Europeanization' influence, was thought to have introduced a new legal dimension into the UK's constitutional system. It is precisely this legal dimension of the incorporation of European Union law into UK law, subjecting the UK's proud constitutional parliamentary sovereignty to the prevalence of the supremacy of EU law that led to the enactment of the HRA of 1998 as a sign of UK's accession to the ECHR.[129] On this basis, the development and growth of the Human Rights Components may, in significant ways be attributed to the HRA, although the English common law prior to the entry in force of the HRA also helped to shape the foundations of early discourse relating to the Human Rights Components. Having said this, however, the UK's impending exit from the EU is likely to impact on the operability of the HRA and possibly, the manner in which human rights may be attributed in a national context.

Further, this also does not discount the significant aspects of the role of the HFEA (the HFEA is the key arbiter in dealing with all matters relating to fertility in the UK, including reproductive technologies), created under the auspices of the Human Fertilization and Embryology Act of 1990 (1990 HFEA Act). It should be noted that the 1990 HFEA Act has been significantly revamped *vis-à-vis* the Human Fertilization and Embryology Act of 2008 (2008 HFEA Act).[130] This section will elaborate on how the UK has developed its framework in terms of the Human Rights Components through an analysis of its common law jurisprudence and various legislative Parliamentary acts, the HRA, as well as the tenuous relationship with the Strasbourg courts.

At this juncture, it is incumbent to begin this analysis with one of the most impassioned human rights considerations, the right to life. The UK's common law system for the protection of human rights, prior to the enactment of the HRA, meant that although the right to life certainly existed within its constitutional confines through a universal recognition of life itself, the right was specifically protected and enforced through a medium of case laws and different Parliamentary acts. The English common law sought to give rise to protective measures of the right to life primarily through the crimes of murder and manslaughter. Within the jurisprudential context of these crimes, it can be established that the manner in which the 'right' to life was

---

[128] Ibid. The rule of law is no stranger in any contemporary constitutional framework, and similarly, in the UK, operates as the bridge between individual citizens' freedoms and the extent of authority exercised by the government.

[129] Within the ideation of constitutional theory, it is also likely that this 'Europeanization' influence is one of the factors that have led to Brexit.

[130] The amendments to the 1990 HFEA Act were made pursuant to a report issued by the House of Commons Science and Technology Committee titled "Human Reproductive Technologies and the Law." In essence, the report surmised that new challenges brought about by technological changes and "recent changes in ethical and societal attitudes" necessitated the revamping of the 1990 Act. Another significant change in the HFEA Act of 2008 is the parenthood provisions, which takes into account research on admixed embryos, removing the need for a "father", and allowing persons of same sex relationships and unmarried couples to be treated as parents of children born of surrogacy procedures.

protected hinged upon the intention behind the commission of the crimes; and the taking away of life was fundamentally explained by Edward Coke in the Institutes of the Lawes of England,[131] a seminal book of four parts that has been accepted as the foundational basis of the English common law system. This does not necessarily mean that the 'right' to a person's life was only protected if it was wilfully taken away through the commission of these crimes. The common law also had a variety of other legislative acts in which this 'right' was protected, even where the offences involved were of a civil nature, for example, in the context of a fatal car accident. The most relevant legislative acts, however, insofar as a 'right' to life is concerned, particularly in the sphere of reproductive rights, medical treatment and the natural consequences of murder or manslaughter, would be the Abortion Act 1967, the Suicide Act 1961, and to a certain extent, the Coroner and Justice Act 2009. Case law regarding withdrawal of medical treatment and/or euthanasia, such as *Airedale National Health Service Trust v Bland* (Bland)[132] would also be relevant for the analysis here.

As demonstrated in Chap. 4, the abortion debate is certainly one of the most polarizing debates, even within the scope of contemporary and evolved intellectual discourse, and the scope of the right to life is best illustrated in these arguments. The Abortion Act 1967 is relevant here because it is a reflection of a process of a multitude of years' battles for women's reproductive rights through a pro-choice movement.[133] The intensely divided and strong emotions that runs high in the abortion debates however demonstrate that even though abortion is legal in the UK (except in Northern Ireland) provided the conditions under the Abortion Act 1967 is fulfilled, scholars and commentators have pointed out that it is still not expressly recognized as a "right".[134] The regulation of abortion in the Abortion Act 1967 permits doctors to sanction and carry out abortions on medical grounds, although it has been indicated that the Department of Health in the UK in practice, allows doctors to do so without medical grounds and to ensure that abortions are provided in a safe, hygienic environment for women who require so.[135] Abortion still remains a sore topic in the UK, and involves a fundamental balancing of competing women's individual and reproductive liberties/rights, and the larger community policing of public interests. Even at this point of time, the legal status of the human embryo or fetus still remains in contention, as well as the impact of criminalization for abortion in many countries around the world, and adds to the constructively determinative elements of both pro-life and pro-choice activist groups. In the UK (excepting Northern Ireland, with its own set of laws relating to abortions), many cases are centered on particular, interpretive alternatives or understandings relating to the provisions of

---

[131] Coke (1628). See also: Coke (1642).

[132] [1993] AC 789.

[133] Humanists UK (3 February 2013).

[134] FPA, The Family Planning Association, A Sexual Health Charity, 'Abortion Law, England, Scotland and Wales', https://www.fpa.org.uk/sexual-and-reproductive-rights/abortion-rights/abortion-law.

[135] Ibid.

the Abortion Act 1967. For example, regarding any form of statutory rights for fathers in respect of the abortion decision, as indicated in cases such as *Paton v British Pregnancy Advisory Service*,[136] and *C v S*,[137] and the Scottish case of *Kelly v Kelly*.[138] At present, it has been agreed that a woman's 'right' to abortion is an uncertain certainty,[139] although within England, the situation fares much better than that in Northern Ireland.

In Northern Ireland, the 'right' to life has been singularly guarded as the sole reason why abortions are illegal. It should be borne in mind that Northern Ireland never agreed nor signed up to the Abortion Act of 1967. A recent spate of controversies regarding abortion in Northern Ireland has made the headlines of international correspondents; but the historical understanding of Northern Ireland's remarkably astute position on anti-abortion laws because of its deeply religious Catholic and Christian beliefs, is also vital. The lack of support for the Abortion Act of 1967 in the Northern Ireland Assembly also indicates why the act itself does not apply in Northern Ireland.[140] With the exception of the Infant Life (Preservation) Act of 1929, the laws relating to abortion in Northern Ireland have been half-hearted and piecemeal in nature. It was only through controversial cases such as the case of *A, B and C v Ireland*,[141] at the European Court of Human Rights, and the death of Savita Halappanavar[142] in Galway (who died because she was refused an abortion) that prompted a change of law in the Irish republic in 2013.[143] However, the circumstances under which abortions were allowed were extremely limited, and did not allow for abortions in cases of rape, incest or fetal abnormality.

Recent events of a landmark nature in Northern Ireland, however, may turn the tide of the abortion laws in the republic.[144] The Supreme Court in the UK recently heard,[145] on 24th to 26th October 2017, an appeal relating to the unconstitutionality of the Northern Ireland abortion laws (alleged to breach the European Convention on Human Rights), with the challenge brought by the Northern Ireland Human Rights Commission, with interveners Humanists UK[146] and Amnesty International. The Supreme Court is delivered its ruling in 2018,[147] and although the court did not overturn Northern Ireland's abortion laws, its ruling however signaled the urgent need to revisit said laws, and appears to put pressure on the government to deal with the issue. It is possible that this could also be influenced by a referendum held in the

---

[136] [1979] QB 276.

[137] [1988] QB 135.

[138] [1997] FLR 828.

[139] Nianias (2 November 2015).

[140] Kelly (8 April 2016).

[141] *Case of A, B and C v Ireland* [2010] Grand Chamber 25579/05.

[142] BBC News (14 November 2012).

[143] O'Connor (28 October 2017).

[144] McDonald (23 October 2017).

[145] Bulman (24 October 2017). Accessed 16 January 2018.

[146] Humanists UK (9 October 2017).

[147] Bowcott (7 June 2018).

Republic of Ireland, which demonstrated a landslide vote to repeal Ireland's near total ban on abortions and called for repeal its strict abortion laws accordingly.[148] New abortion legislation has passed in Ireland as of 1 January 2019,[149] and may assist in lifting Northern Ireland out of a highly repressive abortion scheme that oppresses, instead of empowers women.

The 'right to life' within the confines of the UK also extends to circumstances that may controversially be regarded by some factions as an interpretive and inclusive 'right' to die. This would center upon the debates whether the right to life provided for in the European Convention on Human Rights, also extends to a right to die, or whether this would perhaps fall under the right to privacy indicated in Article 8 instead. Within the UK, for example, in the Suicide Act of 1961, the most prominently recorded case at the level of the European Court of Human Rights is the case of *Pretty v United Kingdom*.[150] The Suicide Act 1961 makes it an offence for a person to assist in the suicide of another, pursuant to Section 2(1) of the Act. The applicant suffered from a terminal and debilitating motor neuron disease and wished for her husband to assist in her suicide; she sought for a declaration from the Public Prosecutor not to prosecute her husband in the intended suicide. The refusal by the Public Prosecutor led the applicant to challenge the declaration at the Strasbourg Court. One of the key reasons cited by the applicant was that Article 2[151] of the European Convention on Human Rights should also extend to a 'right' to die, in a manner determined by the applicant to be in his or her best personal interests. The court did not agree with this line of reasoning; but in all observations of fairness, the wordings of Article 2 is clear and expressly only includes a "right to life". The extension of this argument to a right to die as well is tenuous at best, and would circumvent the scope of the court's jurisdiction. Having said this, however, it is also important to bear in mind that the European Convention on Human Rights has been described as a "living" Convention; and is a convention that is capable of evolving with the necessary parameters of social change in communities.

This pertinent evolving nature of the convention becomes tested if we wish to proclaim for the extension of a person's 'right' to die in 'dignity'. The human dignity[152] argument is one that has been most commonly been touted in many aspects that relate to the human condition and circumstances in which the personhood of an individual come into question,[153] and is extended and employed in many instances regarding death and dying, and assisted suicide and euthanasia discourse. Recent events have also prompted a reinvigoration of discussions as to the interpretation of the provisions of the Suicide Act 1961. A landmark ruling by the Court of Appeal in

---

[148] McDonald et al. (26 May 2018).

[149] Byrne (3 January 2019).

[150] *Pretty v the United Kingdom* [2002] European Court of Human Rights 2346/02.

[151] Article 2 reads: "Everyone's right to life shall be protected by law. No one shall be deprived of his life intentionally save in the execution of a sentence of a court following his conviction of a crime for which this penalty is provided by law."

[152] Mahlmann (2012).

[153] Barak (2015).

the UK regarding the case of a terminally ill man named Noel Conway had made the headlines.[154] Noel Conway challenged the UK's ban on assisted suicide in the Suicide Act 1961, on the contention that it is incompatible with the European Convention on Human Rights. Stricken with a debilitating motor neuron disease that confines him to a breathing ventilator 22 h a day, Noel Conway fought for the 'right to a dignified death.' The Court of Appeal had ruled against his challenge, and Mr. Conway filed an appeal against the ruling at the country's highest court. However, at the Supreme Court, the position remain unchanged, as the court "rejected his bid to appeal against the ruling."[155] If Noel Conway's appeal had been successful, then the landscape regarding the restrictive nature of assisted suicide in the UK, which has remained for over 50 years, could have changed quite dramatically, and provoke a new flurry of debates. Until that happens, however, the status quo remains.

Another marked characteristic of the protection of the right to life within the UK, and in the future, perhaps Northern Ireland as well, has also been incumbent upon the life of the embryo/fetus and the protection of the unborn child, which historically has been granted the status of personhood. Barring the provisions of the Abortion Act of 1967 and pending the ruling to be made by the Supreme Court in 2018, the protection of the right to life is fragmented, and still does not remain a right insofar as women's reproductive rights are concerned. The illustrations of the UK's jurisprudence of cases that exemplify a protection of the right to life are, at best, piecemeal in nature.

It is also important to bear in mind that with the right to life argument presented in the abortion discourse, the right to privacy argument is conversely and simultaneously engaged. Murphy has correctly identified the condemning reality of the interpretation of 'life' by pro-life activism, stating that the perspectives of viewing this right to life are limited[156]; anti-abortion activists focused on the status of the embryo as a 'human life' look past the fact that unsafe abortions in themselves pose dangerous, injurious, and in many cases, causes death, to the women requiring such abortion. The condemnation for legalized abortions, if any, prevents women (particularly, women of lower income groups and from under-presented and marginalized sections of society) from having safe, hygienic access to proper medical abortions. In this regard, the opposing dimension of the women's 'right to life' has been ignored, in favor of the 'life' of the embryo or fetus. On another level, the conflict between the right to life, and the right to privacy in terms of the women's bodily integrity and personal autonomy, continues to remain interweaved in the circumstances.

In context of the right to privacy in UK constitutional law, and with reference particularly to the common law, this was a virtual non-existence as a concept prior to the 1990s period. English common law traditionally only recognized a varied form of the right to privacy through the doctrine of breach of confidence. The modern doctrine of breach of confidence found its jurisdictional basis in equity in early

---

[154] Dearden (18 January 2018).

[155] BBC News (27 November 2018b).

[156] Murphy (2013), p. 162.

cases such as *Coco v AN Clark (Engineers) Ltd*,[157] although a historical tracing of this doctrine can be found in as early as 1732 relating to a case concerning an abuse of confidence.[158] In *Coco v Clark*, the distinguishing points made by the judges essentially confined the 'breach of confidence' to cases of a commercial nature, which would cause detriment to the person revealing such confidential information in confidence. In further cases in English jurisprudence, such as *Saltman Engineering Co. Ltd v Campbell Engineering Co*[159] the parameters of the duty of confidence were further cemented by the judges defining what a duty of confidence amounted to. On this basis, the nature of 'privacy' in the manner of our contemporary understanding[160] did not exist.

The 'right' to privacy in the context of contemporary human rights discourse only began to gain importance after the HRA, and the House of Lords had to consider the possibilities of extending the doctrine of breach of confidence, into the realm of 'privacy' in the context of Article 8 of the ECHR.[161] The earliest cases interestingly involved celebrities; in cases such as *Campbell v MGN Ltd.*[162] and *Douglas v Hello! Ltd.*[163]; in both these cases, the celebrities concerned brought actions for a breach of their privacy against the media, the latter of which had published exclusive photographs and/or detail that were 'private' to the celebrities. However, in these cases, the courts had awarded damages to the plaintiffs on the basis that the misuse of this personal confidential information had a commercially viable interest to the 'owner' of such information. The scope for privacy in a purely personal context was not addressed. Hence, other than the existence of this particular set of circumstances, the scope of the 'right' to privacy was largely underdeveloped. The Select Committee on Culture Media and Sport issued a report stipulating as follows: ".....we firmly recommend that the government reconsider its position and bring forward legislative proposals to clarify the protection that individuals can expect from unwarranted intrusion by anyone— not the press alone— into their private lives. This is necessary to fully satisfy the obligations upon the UK

---

[157] Coco v AN Clark (Engineers) Ltd. [1969] RPC 41.

[158] Tsaknis (1993), p. 19.

[159] Saltman Engineering Co. Ltd. V Campbell Engineering Co. [1984] 65 RPC 203.

[160] By this contemporary interpretation, the purview of privacy would extend to much more than information of a confidential nature in a commercial setting. The 'privacy' shift has extended to cover a myriad of matters relating to individual autonomy and decision-making processes, bodily integrity, the healthcare and clinical setting, sexual orientation, family life and provisions, data protection, the separation of individuality in the workforce, electronic commerce and online surveillance and the internet, freedom of speech, personal data, and many other grey issues which are still being subject to interpretation.

[161] Article 8 of the ECHR provides for the "right to respect for private and family life", which also extended to a person's home and correspondence. It is often considered to be one of the convention's most open-ended provisions that may be subject to a multitude of interpretations as to what may constitute "private and family life."

[162] Campbell v MGN Ltd. [2004] UKHL 22.

[163] Douglas v Hello! Ltd. [2005] EWCA Civ 595.

under the European Convention on Human Rights."[164] The response by the government, in essence, indicated that it did not intend to enact specific legislation in the UK relating to privacy, and preferred to allow the courts to interpret any issues that may arise: "the weighing of competing rights in individual cases is the quintessential task of the court, not of government or of parliament. Parliament should only intervene if there are signs that the courts are systemically striking the wrong balance; we believe there are no such signs."[165] On this basis, the position of the 'right' to privacy in the UK thus far is limited to privacy in the context of private information bearing a commercially confidential nature. In terms of privacy in the realm of personal matters, the insurmountable task is then left to the devices of the Strasbourg court. Indeed, following this line of reasoning, the highly contested area of reproductive rights and issues in the UK, and particularly in Northern Ireland, has borne the brunt of this uncertainty.

With the exception of two cases dealing specifically with the use of Human Leukocyte Antigen (HLA) tissue typing, and PGD, (for the purposes of determining suitable embryos free from disorders to be implanted to produce 'savior siblings'[166]), and which could be interpreted to encompass the fundamental tenets of a right to privacy, other cases where the recognition of a right to privacy had prevailed (in the context of Article 8 of the ECHR) were cases brought against the UK (and one against Ireland) in the European Court of Human Rights in Strasbourg. In the HLA tissue typing and PGD cases, the domestic HFEA had to consider whether these reproductive technologies could be utilized to allow parents to select embryos for the purpose of creating savior siblings.[167] It must be noted that these cases do not, in any manner, invoke any right to privacy within English constitutional jurisprudence, but instead concerned parental decisions and the relevant authority's jurisdiction to issue licenses for the utilization of reproductive technologies. However, the underlying interpretation that is sought, is that both these cases raise legal and ethical considerations about the selection of embryos to become savior siblings; the essence of the decisions made by the parents in these circumstances, are a happenstance of the exercise of their reproductive liberties. An exercise of, and enjoyment of reproductive liberties is in concert with the aims and purpose of Article 8 of the ECHR. Indeed, the HRA in itself is a logical application of the ECHR to human reproductive issues and practices in the UK. These may include "choice of family making; adoption; infertility treatment; abortion; status of the fetus; sterilization; mental incapacity; consent; genetic testing; potential discrimination against single,

---

[164] Thomson (2003).

[165] Ibid 3.

[166] Pennings (2004), p. 311.

[167] Spriggs and Savulescu (2002), p. 289. It is believed that the term 'saviour siblings' was coined by Spriggs and Savulescu; and refer to children born of carefully selected embryos that have been created through IVF and combined with other reproductive technological method such as PGD and HLA tissue typing. These saviour siblings are often necessary and 'created' and conceived, in most circumstances, to help cure an existing sick child who requires familial DNA or blood matches for certain types of diseases.

homosexual and transsexual parents; and special environments."[168] Needless to say, the 'creation' of potential savior sibling embryos, although by now given the purview of a more regulated legal framework in the UK, still continues to raise discursive and vehement objections that center on "the first step towards designer babies."[169]

In the first notable case, the Hashmis sought the HFEA's approval to carry out not only PGD for the purpose of ensuring that they would have a child born free of hereditary beta thalassemia, but to also conduct HLA tissue typing to ensure that such embryo would be a blood match for their existing child so that they would be able to use the umbilical cord blood to save their child's life.[170] The Hashmis received the necessary approval from the HFEA via the authority's power under Section 3(1) of the 1990 HFEA Act,[171] although ultimately they were not successful in the creation of suitable embryos. The Comment on Reproductive Ethics (CORE), a pro-life activist group, challenged the decision of the HFEA in 2002, and sought for judicial review of the same.[172] In the meantime, during this period, the Hashmis continued to battle, after failure of the initial creation of embryos, to continue with the HLA tissue typing and PGD procedures. Following protracted litigation, the House of Lords in *R (on the application of Quintavalle) v Human Fertilisation and Embryology Authority*[173] ruled that the decision made by the HFEA was a lawful one and contained within the scope of its powers under the 1990 HFEA Act.

The judgment of the House of Lords focused on the background, construction, interpretation and the legislative intent of the 1990 HFEA Act. Interpretive guidance was also sought from the case of *Royal College of Nursing of the United Kingdom v Department of Health and Social Security*,[174] especially in connection with legislation that "[deals] with a controversial subject involving moral and social judgment on which opinions strongly differ."[175] There is no mention at all of Hashmis' right to privacy, nor their opportunities to found a potential healthy child or even their decision to 'save' their existing sick child; in other words, no references whatsoever were made to their reproductive rights or liberties.

---

[168] Bahadur (2001), p. 785.

[169] Spriggs and Savulescu (2002).

[170] Mulvenna (2004), p. 163.

[171] Section 3(1) of the 1990 HFEA Act prohibits the creation of an embryo, or to keep or use it, unless it is pursuant to a licence granted by the HFEA. It is upon this scope of power exercised by the HFEA that approval was granted to the Hashmis for the use of PGD and HLA tissue typing.

[172] The declaration sought by CORE was that the HFEA did not have the authority pursuant to Section 3(1) of the 1990 HFEA Act to grant a licence for the creation of embryos. The central tenet of CORE is the absolute respect for the human embryo.

[173] *R (on the application of Quintavalle) v Human Fertilisation and Embryology Authority* [2005] UKHL 28.

[174] *Royal College of Nursing of the United Kingdom v Department of Health and Social Security* [1981] AC 800.

[175] Ibid, 822.

The second notable case, with facts similar to the Hashmis' case, is the Whitakers' case.[176] In this instance, the existing child of the Whitakers, Charlie, suffered from Diamond Blackfan anemia, and his parents intended to use PGD to have another child who would be able to become a donor for his/her brother, in terms of a match for tissue and cell type. Approval was sought from the HFEA, but the HFEA rejected the approval on the basis that the Whitakers' case was relevantly very different from the Hashmis. The distinction between the Hashmis' and the Whitakers' case was the fact that in the former, the disease (beta thalassemia) was a hereditary condition, and in the latter, the disease was sporadic. What this essentially meant was that the HFEA would not be able to justify the approval based on its existing criteria in its Code of Practice[177] as well as the circumstances indicated in the 1990 HFEA Act, for the use of PGD. The HFEA also considered the concept of harm that may be imposed on the embryo or future child. On this ground of reasoning, the Whitakers were not allowed to avail of PGD procedures in the UK, and instead, travelled to the United States where they were able to do so, beyond the legislative reach of the HFEA and the HFEA Act. Like the Hashmis' case, although the Whitakers did not further litigate for judicial review on the decision of the HFEA, the HFEA, in the process of evaluating the application made by the Whitakers, did not specifically consider the Whitakers' right to exercise their reproductive liberties. Instead, the focus on the HFEA was incumbent on the specificity of Charlie's disease and the applicable procedural and substantive requirements of granting a license for PGD. Scholars have criticized the distinction made by the HFEA in Whitakers' case, from that in Hashmi and have presented concrete critiques regarding the justifications made by the HFEA as being no different from the circumstances in Hashmi.[178]

Notwithstanding these criticisms, however, the purpose of highlighting the Hashmis and Whitakers case intends to highlight the imperative applicability of Article 8 of the ECHR in the reproductive rights jurisprudence of the UK. I also emphasize that despite the stark absence of an express right to privacy, the circumstances that surround both the Hashmis and Whitakers are also a natural and progressive recognition of the doctrine of informed consent. In the UK, the doctrine of informed consent has traditionally imposed the onus on medical professionals to adequately inform patients of all consequences, risks, benefits, and the like of potential medical treatments. The quintessential marker for this onus to be met was indicated in the once quintessential *Bolam*[179] case. The UK Supreme Court in a fairly

---

[176] Sheldon (2004), p. 137.

[177] Human Fertilization and Embryology Authority, 'Code of Practice' https://ifqlive.blob.core.windows.net/umbraco-website/2062/2017-10-02-code-of-practice-8th-edition-full-version-11th-revision-final-clean.pdf.

[178] Sheldon (2004), p. 160.

[179] Bolam v Friern Hospital Management Committee [1957] 1 WLR 582. The Bolam test, as it became known in the body of negligence within tort law, was instrumental in the determinative standards of care involving skilled medical professionals.

recent landmark decision of *Montgomery v Lanarkshire Health Board*,[180] has now overruled the *Bolam* test application. The changing paradigm of informed consent *vis-à-vis* the dynamic consent movement[181] also heralds a better framework for a continuous process of consent in the medical field; in both the Hashmis and Whitakers case, their explicit informed consent to the risks and contraindications of the sought reproductive technologies should also frame the manner in which their autonomy as parents is fulfilled.

On a broader concept of privacy at the level of the Strasbourg court, cases such as *Evans v UK*,[182] and *Dickson v UK*,[183] are, so far, the only cases levied against the UK in respect of the recognition of reproductive liberties under the scope of Article 8 of the ECHR. The outcome in *Evans v UK* under which a violation of the applicant's Article 8 was rejected[184] demonstrated that the Strasbourg court recognized that the reproductive desires (or, in this case, non-desires) of the male counterpart were equally as important as the woman's, *vis-à-vis* the application of the doctrine of informed consent.[185] Indeed, the Strasbourg court reached an outcome that was practical in every sense; John Harris had this to say[186]:

> If the woman succeeds in this case then the whole basis upon which the Human Fertilization and Embryology Authority has operated thus far will be overturned.
>
> Until now, it has operated on the basis that there must be continuing consent between a man and a woman in every stage of the reproductive process.
>
> If she succeeds in this case, then she will have established that the man's role ends once the egg is fertilized.

Notwithstanding the recognition of Ms. Evans' reproductive rights under Article 8 of the ECHR, arguably a more difficult balancing of the liberties of Ms. Evans as well as her former partner, the emerging tell of the court's judgment indicates that the UK HFEA had, to the very best of its abilities and pursuant to the scope of exercise of its powers under the HFEA Act, taken into account principles of shared responsibility between a man and a woman within the field of reproduction. This 'respect' for the shared roles between man and woman, however, interestingly does not apply to decisions made regarding abortions.

---

[180] Montgomery v Lanarkshire Health Board [2015] UKSC 11.

[181] Minari et al. (2014), p. 118.

[182] Evans v United Kingdom, Application No. 6339/05.

[183] Dickson v United Kingdom, Application No. 44362/04.

[184] Amongst other claimed violations of the ECHR, the applicant claimed that her Article 8 right had been violated because she could not implant embryos that she had 'created' with her former partner as he had withdrawn his consent to the continued storage and use of the embryos by the applicant after their relationship had ended.

[185] The first instance decision of the High Court, and thereafter, the House of Lords in the Evans case, both upheld the existing IVF law in the UK *vis-à-vis* the HFEA Act, which required the consent of both parties for the IVF process to continue, including the implantation of embryos created under the IVF procedures.

[186] BBC News (11 September 2002).

In *Dickson v UK*, the Strasbourg court arrived at a decision, which concurred, on the violation of the applicant's Article 8 right. The applicant, who was serving a mandatory life sentence for murder, applied for the access to artificial insemination facilities to have a child with a woman he had met through the prison pen-friend scheme. The application in Dickson's case turned on Section 47 of the Prisons Act 1952 in the UK; the Secretary of State was imbued with the authority to "make rules for the regulation and management of prisons.......... and for the classification, treatment, employment, discipline and control of persons required to be detained therein."[187] When the applicant's claim to the Secretary of State was refused on several occasions, he filed a claim with the European Court of Human Rights. The applicant's claim succeeded and the court proclaimed that there had indeed been a violation of both Articles 8 and 12 of the ECHR. In arriving at their judgment, the Strasbourg court stated that the UK did not perform an adequate balance of considering competing public and individual interests, and the refusal by the Secretary of State (on several occasions) were unjustified simply on the ground of offending the public opinion. In this manner, what would have been viewed by the UK as a run-of-the-mill administrative responsibility and prison management, insofar as the applicant was concerned, was viewed by the Strasbourg courts to be move beyond the boundaries of 'simple' domestic control and regulation.

The flowing consequences from the determination of these cases at the European Court of Human Rights signals that the realm of privacy in the UK must be expanded with more flexibility. It is right to raise concerns about the evolution of human rights protection as medical and scientific technologies advances. However, it is also incumbent on all relevant stakeholders to first, understand more holistically the implication that privacy does not simply impact information or personal data, but also encompasses the realm of personal individual liberty and decision-making processes, and the exercise of personhood in a respected manner. Secondly, stakeholders must put into measures safeguards that promote the safest, best and most egalitarian way possible for the use of the technologies, whilst maintaining the necessary impetus and commercially viable incentivization to continue carrying out research and development in accordance with ethical guidelines of practice.

Finally, in respect of the right to equality and the determination of principles of non-discrimination, it should be mentioned that this is likely the human rights component with the smallest library of jurisprudential cases by which to examine. Colm O'Cinneide[188] starts by first questioning if equality is, indeed, a constitutional principle to be reckoned with. He correctly identifies that the position in the UK with regards to the status of equality is unclear, although it has been recognized that equality is a core value that is prevalent in the UK constitutional order based on the rule of law. He quotes Baroness Hale in *Ghaidan v Godin-Mendoza*[189]: "democracy is founded on the principle that each individual has equal value."[190] Notwithstanding

---

[187] Section 47(1) Prisons Act 1952 (c.52).

[188] O'Cinneide (14 September 2011).

[189] Ghaidan v Godin-Mendoza [2004] UKHL 30.

[190] Ibid, 132.

our ability to trace the historical common law background into UK's respect for the right to equality, a more pressing concern arises because common law adjudication indicates that it is, at best, functional on a case-by-case basis. With the ratification *vis-à-vis* the HRA, and specifically concerning Article 14[191] of the European Convention on Human Rights, however, questions also arise as to whether the principle, and the 'right' of equality, has been incorporated into efficacious and practical means. Indeed, the semi-controversial view of John Stanton-Ife reminds us that we should not mistake the parameters of Article 14 as enshrining a formal or even constitutional right to equality that is applicable to all member states of the European Union.[192] In fact, in the precursor to his article, he says:

> A constitutional principle of equality should not be recognized......For it to be the case that a right should be protected as a constitutional right, the putative right should protect an intrinsic value, or should be an instrumental means to the achievement of another value, or values clearly identified or should serve some rhetorical function.[193]

This may legitimately be accurate, at least from the constitutional aspect in the manner described by Stanton-Ife. He says that "Article 14 is a non-discrimination rule that has no essential reference to a principle of equality properly so-called: it demands equality in the application of independent granted rights and cannot be violated in isolation from other convention rights."[194] If we agree that this is accurate, and that the right to equality is distinct from the anti-discrimination right, then it must also follow that there is no deliberate and wilful expression of "equality" in the scope of Article 14. In this manner, the reconciliation between the European measures of equality, with the UK's 'constitutional' recognition to equality, is only minutely bridged on a case-to-case basis. Bearing in mind the fairly limited scope of equality application in the UK at this juncture, the struggle within the UK has been largely focused on gender equality.

A series of earlier cases from the 1980s period tested out the UK judiciary's approach to equality and/or non-discrimination principles. Although the facts of the case are not particularly relevant in this context, *Abdulaziz, Cabales and Balkandali v the United Kingdom*[195] was one of these cases that dealt with equality between men and women in terms of immigration and settlement in the UK. But the contemporary trajectory of the recognition of a dynamic diversity in the UK can perhaps be attributed to the case of *X, Y and Z v the United Kingdom*,[196] a case that touched upon how the fundamental recognitions of transsexual individuals was treated in the sphere of the margin of appreciation attributed to individual member states. In this

---

[191] Article 14 reads: "the enjoyment of the rights and freedoms set forth in this Convention shall be secured without discrimination on any ground such as sex, race, colour, language, religion, political or other opinion, national or social origin, association with a national minority, property, birth or other status."

[192] Stanton-Ife (2000), p. 133.

[193] Ibid 134.

[194] Ibid 135.

[195] Abdulaziz, Cabales and Balkandali v The United Kingdom (1985).

[196] X, Y and Z v The United Kingdom (1997).

case, a post-operative transsexual individual (female to male transsexual) wished to be recognized as the biological father of a child born out of artificial insemination procedures (under the HFEA Act). Although the court took into account that the notions of "family life" within the scope of Article 8 of the convention was indeed subject to contemporary and evolutionary interpretation in changing times, the court acquiesced to the wider margin of appreciation for the UK to determine the applicability of the applicant's status as the "father" in coherence with family law in the UK. Finding adequate national remedies whereby the child would carry the applicant's name, the applicant would be able to apply for joint custody of the child, the corresponding Article 14 was also dismissed accordingly, and the court found that the applicant's convention rights had not been violated. This case is significant, because it took place in the pre-HRA 1998 era and laid the foundational groundwork for a significant number of judicial cases in the UK that have been forced to revisit interpretations of family, privacy and equality and non-discrimination as well. *X, Y and Z v the United Kingdom* is unusual because it deals with a form of reproductive liberty that may not have been considered before: the right of a transsexual individual to claim a 'biological' relationship with a child born through artificial insemination procedures. I believe that this interestingly begs the question whether the applicant's liberty in the course of the reproduction process would be recognized in an equivalent manner to a woman's reproductive right and liberty. In this manner, questions of equality within the fluid changes of 'gender' needs to be revisited and a series of cases brought against the UK after *X, Y and Z v the United Kingdom*[197] signals that the tides of social change have contributed to a better understanding on the protection of equality in this context. A significant and notable legislative pronouncement *vis-à-vis* the *Gender Recognition Act* 2004 in the UK is proof of this transcendence. The Council of Europe's Sexual Orientation and Gender Identity (SOGI) Unit is no doubt a contributory watchdog to the UK's liberalized recognition of gender equality.

From a non-discrimination perspective, the Supreme Court decision in *Bull and another v Hall and another*[198] is worthy of examination; Baroness Hale makes the distinction between direct and indirect discrimination in the case,[199] in balancing the conflicting European convention provisions of Article 9, and Articles 8 and 14. The essential highlight in the case was that the provisions of Equality Act 2010 stipulates that "a provision criterion or practice is indirectly discriminatory if the person who applies it "cannot show it to be a proportionate means of achieving a legitimate aim.""[200] Certainly it was found in this case that save for the exercise of the Bulls' religious beliefs, their discriminatory behavior and business policies did not set out

---

[197] See: Sheffield and Horsham v the United Kingdom (Application No: 31-32/1997/815-816/1018-1019); Christine Goodwin v the United Kingdom (Application No. 28957/95); Parry v the United Kingdom (Application No. 42971/05); R and F v the United Kingdom (Application No. 35748/05); and Grant v the United Kingdom (32570/03).

[198] Bull and another v Hall and another (2013) UKSC 73.

[199] Ibid, para. 16–40.

[200] Ibid, para. 41.

to achieve a legitimate aim. It would certainly be interesting to measure future cases in the UK by reference to Baroness Hale's seminar judgment on the purview of discrimination and equality.

In the meantime, UK equality and non-discrimination laws also extend into the realm of work and employment. The enactment of the Equality Act of 2010 has been an instrumental force in the contemporary development of equality and diversity in the UK, bringing together a large number of legislations in the UK and consolidating them into a single unified act. The merger of the legislation comprised in the Equality Act 2010 is: the Equal Pay Act 1970, the Sex Discrimination Act 1975, the Race Relations Act 1976, the Disability Discrimination Act 1995, the Employment Equality (Religion or Belief) Regulations 2003, the Employment Equality (Sexual Orientation) Regulations 2003, the Employment Equality (Age) Regulations 2006, the Equality Act 2006 Part 2, and the Equality Act (Sexual Orientation) Regulations 2007. A number of significant cases that deal with inequalities, particular gender inequality in employment issues include cases like *Lewis v HSBC Bank plc*[201] and *Grant v HM Land Registry and Another.*[202] From the outset, however, what can be distinctly observable in relation to the Equality Act 2010 is its emphasis within the sphere of employment. In terms of identifying how reproductive rights may be protected at the constitutional level, the Equality Act 2010 is of little assistance. However, it is my hypothesis that a fundamental recognition of equality at the very beginning, is, however, the path to a realization for better protective mechanisms of women's reproductive rights. In a similar manner that the recognition of equality in gender has evolved through the last decade, it is a foreseeable contribution that this would also likely extend into the sphere of reproductive rights equality and non-discrimination.

With the UK's impending exit from the European Union, however, and with many scholars believing that the EU has been instrumental in keeping the national system in check to comply with convention rights,[203] there is a degree of uncertainty in how human rights mechanisms will continue to be protected in the UK. Sandra Fredman raise the foreseeable possibilities that the UK government's proposed repeal bill following Brexit may contain some undesirable clauses that may grant government ministers "the power to amend or repeal primary legislation in order to disentangle national law from EU law."[204] The fear expressed is that all such equality and non-discrimination legislation, the Acts of Parliament in the UK, may be subject to repeal and leave a gaping vacuum that cannot be immediately sealed. Further to this, the chair of the Equality and Human Rights Commission (EHRC) in the UK, David Isaacs, had this to say:

> The Government has said that it will retain our domestic equality protections after Brexit, but as things stand, we will lose the safety net European law currently provides to ensure

---

[201] Lewis v HSBC Bank plc UKEAT/0364/06/RN and UKEAT/0412/06/RN.

[202] Grant v HM Land Registry and Another [2011] EWCA Civ 769.

[203] Fredman et al. (2015).

[204] Ibid 8.

our protections aren't eroded. That's why we want to see a new constitutional right to equality enshrined in the Withdrawal Bill.[205]

As it presently stands, the future of the protection of the Human Rights Components within the UK may be on shaky ground, but the retention of safeguards for protections to continue unabated continue to be pushed for by the EHRC and many other human rights organizations in the country.

6.3.2.2   Australia

The operation of the legal system in Australia is an engaging one, combining elements from its Commonwealth heritage *vis-à-vis* the British system when it used to be a collection of colonies, as well as strong elements of federalism akin to the United States (US) political system of government. The evolution of Australia's constitutional framework as a system of colonies from the early nineteenth century begun from its British rule, with Great Britain having established six (6) colonies (now known as "Territories" or "States") in Australia: New South Wales, Tasmania, Western Australia, South Australia, Victoria and Queensland.[206] Having drawn inspiration from its own Westminster system of government, the British projected this onto the colonies then, where each of the Territories developed their own Parliamentary systems.[207]

In the late 1870s, a movement for a federated Australia emerged,[208] and the British drew heavily from the political system of the US, which, at that time, combined a single federated nation from thirty eight (38) states.[209] On this basis, the Constitution of the Commonwealth of Australia[210] came into effect in 1901 (which was, in fact, a British Act of Parliament passed in Australia). This Constitution remains Australia's fundamental document that establishes its federalist system of government, and is supplemented by conventions entered into by elected politicians, as well as rulings of Australia's apex court, the High Court. The remnants of Australia's British colonization and rule were completely eradicated with the passing of the Australia Act 1986, with the aim to "bring constitutional arrangements affecting the Commonwealth and the States into conformity with the status of the Commonwealth of Australia as a sovereign, independent and federal nation."[211]

Because of the relationship between Australia's federal Commonwealth government and the States, and particularly in the interpretative exercise of the Australian

---

[205] Watts (15 October 2017).

[206] Clark (1905).

[207] Australia and Australian Government Solicitor (2016).

[208] Aroney (2009).

[209] Australia and Australian Government Solicitor (2016).

[210] Commonwealth of Australia Constitution Act (The Constitution) https://www.legislation.gov.au/Details/C2013Q00005/Html/Text.

[211] Australia Act 1986 (No. 142, 1985) https://www.legislation.gov.au/Details/C2004A03181/Html/Text.

Constitution, the country has witnessed its fair share of internal constitutional conflicts over the years.[212] (This is similar to the position of the US federation as well). Scholars have attempted to espouse on the interpretation of Section 118 of the Australian Constitution, which reads: "Full faith and credit shall be given, through the Commonwealth to the laws, the public Acts and records, and the judicial proceedings of every State."[213] Kirk points out that from the outset, the applicable choice of law within the Australian national context is, in itself, a constitutional issue,[214] (although it had not been considered in that nature prior to the Breavington case) making their developments through cases such as *Breavington v Godleman*,[215] *Lipohar v The Queen*,[216] and *John Pfeiffer Pty Ltd v Rogerson*.[217] Kirk had presented his solutions on possible interpretations of Section 118 in the determination of how the "full faith and credit" could be attributed to individual State legislations, and puts forward what is called "the Full Approach".[218] Nevertheless, Section 109 of the Australian Constitution, is however, clear that Commonwealth statutes generally prevail over State laws.[219] On this basis, the study of Australian constitutional law has often been described as challenging not only in relation to conflicting laws between Commonwealth and States, but in some instances, although rarely so, inconsistency in laws between the different States. I posit that this stems from its unique positioning as a former British colony that incorporates significant constituents of the British legal system whilst simultaneously embodying a federated political system similar to the US.

The determination of human rights protection in Australia, therefore, is also challenging because of these reasons. Just like in the UK, Australia does not embody a specific and separate Bill of Rights in relation to the protection of fundamental rights and liberties, but attempts to hold itself up to international standards of protection. Instead, the protection of human rights in the country is culminated through the Australian Constitution, the common law (the vestiges of the British Commonwealth system), and a variety of specific legislations or statutes relating to specific human rights issues, such as non-discrimination. These include, amongst others, the Racial Discrimination Act 1975, Sex Discrimination Act 1984, the Australian Human Rights Commission Act 1986 (AHRC Act), Age Discrimination Act 1992, and the Disability Discrimination Act 1992. Pursuant to the AHRC Act, the Australian Human Rights Commission was established in 1986 to champion the education, awareness, promotion and advancement of human rights in the country. Although Australia is party to several international human rights treaties, such as the ICCPR, ICESCR, CEDAW and others, the recent 2018 Human Rights Watch

---

[212] Kirk (2003), p. 247.

[213] Commonwealth of Australia Constitution Act (n 210).

[214] Kirk (2003), p. 247.

[215] Breavington v Godleman (1988) 169 CLR 41.

[216] Lipohar v The Queen (1999) 200 CLR 485.

[217] John Pfeiffer Pty Ltd v Rogerson (2000) 203 CLR 503.

[218] Kirk (2003), pp. 261–273.

[219] Hill (2005), p. 39.

report on Australia[220] indicates that Australia's main challenges in human rights protection currently relates to asylum and refugee rights, and the rights of the indigenous community (the Aboriginal and Torres Straits Islanders community), which are subject to on-going and heated dialogue in the public sphere. Indeed, former Social Justice Commissioner, Michael Dodson, has been a lifelong activist in his work for the rights of the Australian indigenous communities, and in his work, has highlighted the fundamental necessity for Australia's increased protection of this minority group.[221]

Similarly to the position in the UK, the National Human Rights Committee (NHRC) in Australia has been attempting to propose a Federal human rights act.[222] Professor Lee identifies, however, that not all members of the NHRC is in agreement that this sort of legislation for the protection of human rights is needed.[223] In the sphere of national dialogue relating to this separate Australian Bill of Rights, Lee notes that suggestions have rallied around either an entrenched Bill of Rights into the Australian Constitution (like the Amendments to the Constitution in the US), or a specific statute relating to human rights, (like the Human Rights Act 1998 in the UK).[224] In his article, he charts a projected idea of how Australian constitutional law may markedly be changed on the assumption that a federal human rights act is established and adopted. Although the development of this idea has been temporarily halted due to political agendas in the Australian government, Lee believes that the advocates for this legislation will continue to rally on.[225] Despite these testy issues, Australia's track record in championing and developing protections in respect of the Human Rights Components, (and particularly in the development of reproductive technologies (PGD) and the advancement of research and developmental frameworks for genetic interventions), has been laudable.

As a nation that has been making advances in life sciences and biotechnologies, statistics indicate that despite it being the smallest continent in the world, Australia is also one of the fastest growing countries to lead in biotechnological advancements, ranking fifth in the world, and has the largest listed biotech sector in proportion to GDP in the world.[226] The Australian Trade and Investment Commission is a proud purveyor of Australian biotechnology, and with its "flexible regulatory regimes",[227] Australia is ranked competitively on a global level in this industry. It is

---

[220] Human Rights Watch, 'World Report 2018: Rights Trends in Australia' https://www.hrw.org/world-report/2018/country-chapters/australia.

[221] Dodson (1993).

[222] Lee (2010), p. 88.

[223] Ibid 89.

[224] Ibid.

[225] Ibid 109.

[226] Hive Legal, 'Australia's Life Science Sector – Snapshot, Trends and Our Work' http://hivelegal.com.au/australias-life-science-sector-snapshot-trends-and-our-work/.

[227] Australian Trade and Investment Commission, 'Biotechnology A Powerhouse for Science and Innovation'   https://www.austrade.gov.au/International/Buy/Australian-industry-capabilities/Biotechnology.

therefore not surprising that its legal and regulated regimes have also grown in tandem with this economic development. However, the protection of the Human Rights Components have not always had a smooth journey, despite the flexibility afforded by its regulatory framework in the field of biotechnologies.

In Australia, the consideration of the right to life has mostly been focused on euthanasia and/or end-of-life issues; but the abortion debates have also been equally antithetical, as societal and religious perceptions about the beginnings of the conception of life (in utero) has permeated the discourse. In this regard, the discussion of the treatment of the right to life and the right to privacy in Australia will be dealt with simultaneously. At this juncture of time, the abortion debates in Australia are also critical in the determination of whether and how the Human Rights Components, particularly in connection to reproductive rights and liberties, have helped to shape the aforesaid debates and perhaps, even vice versa. The provision of the Australian framework on its safeguards and measures in the protection of the right to life is based on a universal recognition *vis-à-vis* the UDHR and the ICCPR.[228] Its Commonwealth heritage means that, like in the UK, Australia also initially used the Offences Against the Persons Act 1861 in terms of beginning to frame the abortion debates by its criminalization. In recent years, however, the abortion debates have been subjected to tumultuous fracas, with the situation being described as a "battle of life and death"[229] for anti-abortion activists, juxtaposed against the pro-choice movements that accuse the Australian Turnbull government of undermining reproductive rights.[230] Although a recent briefing paper by the New South Wales Parliamentary Research Service indicates that a majority of Australians are generally in support of reproductive choice *vis-à-vis* abortions,[231] the transition for the legalization of abortions across Australia has not been an easy one. In New South Wales and Queensland, abortions are still illegal unless they come within the exceptions stipulated in those states.[232,233] The depth of importance placed on the right to life features prominently in these anti-abortion movements, which makes this a continuous human rights issue[234] that pits "life" against "choice" (privacy).[235] Of

---

[228] Australian Human Rights Commission, 'Right to Life' https://www.humanrights.gov.au/right-life.

[229] Watson (29 December 2013).

[230] Barlow (6 July 2017).

[231] Tom Gotsis and Laura Ismay, 'Abortion Law: A National Perspective, Briefing Paper No. 2/2017' *NSW Parliamentary Research Service* https://www.parliament.nsw.gov.au/researchpapers/Documents/Abortion%20Law.pdf.

[232] Gotsis and Ismay (n 231).

[233] In other Australian states, abortion is legal up until a period of time, or such other expressly stipulated condition in the relevant State statutes. For example, in the Australian Capital Territory and South Australia, abortion is legal if medically agreed upon by two doctors that it would be necessary for the benefit of the woman's physical or mental health. In Victoria, Tasmania, Western Australia and the Northern Territory, abortion is legal up until a specified period of time, after which special considerations may apply.

[234] Sifris and Belton (2 June 2017).

[235] International Women's Development Agency (IWDA) (26 April 2015).

particular note in this context is the highly significant Zoe's Law Bill in the Parliament of New South Wales,[236] which indeed demonstrates most succinctly how the right to life is often inter-connected with the right to privacy, bodily integrity and personal autonomy in reproductive matters.

In Chap. 4, the tremendous impact of Zoe's Law was highlighted, because it would dramatically change the landscape of not only abortions in Australia, but also how 'personhood' was viewed for an in-utero foetus. What cannot, however, be denied, is that the scope of the recognition of women's reproductive rights within Australia's human rights context, is likely to be marginalized in the event the bill is successful; although the reintroduced version of Zoe's Law Bill does appear to be more palatable, speculations remain rife about the unintended future consequences that it may result in.

Even without the possible gloom that overhangs Zoe's Law Bill, the legal cases in Australia thus far demonstrates that the issue of abortion is a controversial, and in some instances, arbitrary one. On a larger scale, it may even be said to be a policy-political issue,[237] as evidenced by the nifty maneuver of the Zoe's Law Bill. Two legal cases are worth highlighting at this juncture, illustrating the severity of the situation in Australia with regards to the procurement of abortions. In what has been described a landmark case where a medical professional was convicted of providing an unlawful abortion (and therefore, subjected to prosecution because of the arbitrary nature of abortion laws) after a period of almost three decades, the case of *R v Sood*[238] heralded to many that a fundamental change was much needed. Briefly, in *R v Sood*, abortion doctor Dr. Suman Sood was the first doctor since 1971 to have been convicted of providing an unlawful abortion.[239] Although Dr. Sood was cleared of the charge of manslaughter of the foetus, she was charged for two counts of procuring an unlawful abortion. Dr. Suman Sood had been disqualified from practising medicine in New South Wales for a period of 10 years. The outcome of the *R v Sood* case provoked calls for abortion law reform from the medical profession, highlighting that "doctors who want to practice with good will and in good faith to provide women with a medical service they seek shouldn't feel they could be found guilty of a criminal offence when they do that."[240] In *R v Brennan and Leach*,[241] a woman was charged for procuring her own abortion, and her husband treated as an accessory after the fact.[242] The fact that this was so, shows clearly that the importance of personal choices, and particularly choices made by women about their own reproduction, is still treated as a matter of regulation by the state. I posit that this, in itself, raises seriously misguided notions about reproductive liberties, and indeed, is a

---

[236] Crimes Amendment (Zoe's Law) Bill 2017.

[237] Graham et al. (2016), p. 335.

[238] R v Sood [2006] NSWSC 1141.

[239] Dean (2007), p. 138.

[240] Ibid 159.

[241] R v Brennan and Leach [2010] QDC 329.

[242] Petersen (2014), p. 238.

"systemic coercion of women to continue an unwanted pregnancy by threatening criminal punishment".[243]

The outcome of the reintroduced Zoe's Law Bill, and in cases like *R v Sood*, and *R v Brennan and Leach*, for example, would, in a deeply fundamental way, also further impact and alter how the right to privacy operates in Australia. At the heart of the matter, the element of choice afforded to women over their own bodily integrity is called into question, and subject to policing by the state in an all-encompassing manner. Australia's situation is a far cry from the restrictive drudgeries of third world, less developed nations, but the depth of its ingrained politicization of women's bodies,[244] such as that of Zoe's Law Bill on its reintroduction by Rev. Fred Nile, serves only to illustrate the standards of societal patriarchy imposed upon matters that should hinge upon personal decision-making and individual autonomy. Indeed, in Chap. 1, I stress upon the importance of autonomy, and question how we may unveil thinly veiled disguises of liberalism to reveal the true concept of autonomy in reproductive matters. As a manner of further recap, I recount that the fundamental concept of autonomy, and how it is tied to a woman's reproductive choice and liberty, is the key underlying reason for the right to privacy.

In the context of Australia, scholars have similarly echoed the foundational path of individual autonomy to choice[245]; Australia's constitutional "right" to privacy does not exist at all (in the same manner that a "right" to privacy also does not exist in the UK), but has, instead, evolved through the doctrine of breach of confidence and privacy protection for proprietary information, trade secrets and the like. It was only through the High Court case of *Australian Broadcasting Corp. v Lenah Game Meats Pty. Ltd.*[246] (*Lenah Game Meats*) in 2001 that the Australian legal jurisprudence began to question if a right to privacy did indeed exist pursuant to torts law. Despite the outcome of the *Lenah Game Meats* case, there was no clear positioning on the treatment of the "right" to privacy in Australia. In *Lenah Game Meats*, the judiciary was divided as to the necessity to recognize the right to privacy in the Australian jurisprudential context, divided between the liberal, utilitarian Mills approach[247] to privacy, versus the Kantian approach of privacy being of personal integrity enjoyed *qua* human existence,[248] questioning the adequacy of the existing doctrine of breach of confidence. Nevertheless, although it appears that Australia still doesn't embody a constitutional right to privacy till date, the devolvement of legal cases that take precedence from its Anglo-Saxon heritage,[249] coupled with the continued rallies for reform of abortion laws in states like New South Wales to

---

[243] Kerr (2014), p. 24.

[244] Graham et al. (2016).

[245] Kerr (2014).

[246] Australian Broadcasting Corporation v Lenah Game Meats Pty Ltd [2001] HCA 63.

[247] Richardson (2002), p. 392.

[248] Ibid 390.

[249] Ibid 393.

emphasize the importance of choice and bodily integrity, should be viewed as positive indicators of the treatment of privacy as a "right".[250]

With respect to the right to equality and/or principles of non-discrimination, particularly where women's reproductive choices and decisions are concerned, the literature indicates that Australia still has some way to go before making its mark more widely felt.[251] Because reproductive choices in Australia are often built around the gender equality discourse, the feminist interpretation of the right to equality and principles of non-discrimination are relevant here.[252] Taking into account that a large number of legislative statutes in Australia, similarly to the UK, are targeted at the protection of equality in the employment sphere,[253,254] we must question if the trajectory of equality is also given an equal measure of weight in other spheres of women's lives. The recent Human Rights Watch World Report indicates that the rights of women particularly are at risk in Australia, but this appears to be focused on areas of indigenous communities.[255] An earlier report from November 2017 raised the disturbing allegations of involuntary sterilizations on women with cognitive and mental disabilities.[256] On the assumption that the veracity of these reports are accurate, and with Australia being a signatory of CEDAW and having ratified the same, the allegations of the reports bring to the surface disturbing trends of inequality in the country. Indeed, a veritable point highlighted by Katherine Kerr indicates that the criminalization of abortions in Australia is a form of discrimination in accordance with the provisions of CEDAW.[257]

Issues relating to involuntary sterilizations, and women's reproductive rights as illustrated in the preceding paragraphs, necessitate the drawing of a conclusion that the measure of equality from the purely biological aspect, is considerably lacking. In the course of action for equality that seeks equal footing with male counterparts, the juxtaposition is also that specific equality aspects are hinged solely on the biological aspect (reproductive rights and liberties). It is pursuant to these biological aspects that the feminist interpretation of abortions is put forward. Kerr states, "Equal citizenship necessitates that a woman's choice for self-determination be legally respected, not criminalized. A woman cannot access other fundamental rights if her health, including her reproductive life, is beyond her control."[258] Further, the fact that abortion is criminalized "relegates women to a secondary class of citizenship by exerting criminal responsibility to her for a factor of biology."[259] In this

---

[250] Women's Agenda (16 August 2017).

[251] Women's Electoral Lobby, 'Reproductive Rights' (*Women's Electoral Lobby*) http://www.wel.org.au/reproductive_rights.

[252] International Women's Development Agency (IWDA) (26 April 2015).

[253] Sex Discrimination Act 1984.

[254] Workplace Gender Equality Act 2012.

[255] 'World Report 2018: Rights Trends in Australia' (n 220).

[256] Cody and Nawaz (9 November 2017).

[257] Kerr et al. (2014), p. 31. See Articles 12.1 and 16(1)(e) of CEDAW accordingly.

[258] Ibid 32. Kerr quotes from: Cook and Dickens (2003), pp. 2–3; Banyard (2010), pp. 181–182.

[259] Ibid 31.

regard, the relegation to citizenship on an unequal level can be interpreted, from a constitutional viewpoint, to be a relegation based on discrimination. If this appears to sound feminist in nature, it surely is. Kerr skilfully charts a feminist analysis of abortion laws in Queensland[260] by referring to the views of Robin West,[261] and there is more than an ounce of accuracy in some of these propounded theoretical ides.

West's article is a highly relevant and eye-opening instrument in highlighting the quasi-legal treatment of women from a patriarchal standpoint. Using emotive appeals and carefully impassioned words that evoke imaginative visuals, touching on sexual violence, mental health, pregnancy and maternal well-being (both physical and spiritual), West's work proposes a reframing of abortion discourse by capitulating it onto pregnant women and any contextual decision-making that emanates from her.[262] The conceptualization of equality from the feminist perspective is framed in terms of problems that arise from patriarchy and male dominance, traditionally within the spheres of government and the continual acknowledgement of the woman's role into motherhood. The maintenance of abortion as a crime within the purview of State legislation, West proclaims, continues to restrict women's reproductive choices and ignores the roles they intend for themselves in the future; suggesting therefore that the conservative, patriarchal values imposed on women's fertility and sexuality is a manner of relegating women to the role of mothers if they become pregnant.[263]

The highly disturbing allegations from as recent as 2015 must also be discussed at this juncture: the involuntary sterilizations of women, despite Australia's obligations under CEDAW and other international human rights instruments.[264] In an examination by the United Nations before the Universal Periodic Review in 2015, concerns were raised about the involuntary sterilization practices in Australia,[265] of women with mental disabilities. A further United Nations report from November 2017 indicated that involuntary sterilizations were also rife amongst women from Australia's indigenous community, and particularly more so where these women also suffered from mental disabilities.[266] It is also worth noting that Australia does not have any laws in place regarding the prohibition against forced sterilizations of this nature, and the United Nations' definition of forced sterilizations are likely to

---

[260] Ibid 32.

[261] West (1988), p. 1.

[262] Kerr et al. (2014), p. 32.

[263] Ibid 33.

[264] Cody and Nawaz (9 November 2017).

[265] Jabour (10 November 2015).

[266] Australia has had a long and troubled past with its people from the indigenous communities, such as the aboriginal and Torres Straits Islanders. Studies conducted by the Australian Institute of Aboriginal and Torres Strait Islander Studies have revealed a comprehensive history of discrimination, ill treatment, unemployment, marginalization and also genocide, not only throughout the period of British colonization of Australia, but also continuing in present day on a smaller, although no less, discriminatory scale.

be treated as torture.[267] On a collective examination of all these instances, the brute force of discrimination rages forth on several fronts: on the basis of sex (being women), disabilities, and minority or racial groupings. This is clearly an area of human rights that is lacking efficacy and force in the country, although there is a glimmer of hope to the commitment made by Australia to increase awareness in these areas.[268]

6.3.3 Group Three (3): Malaysia and Thailand

The consideration of the constitutional framework of the Human Rights Components in Malaysia and Thailand will reveal the "modernization struggle"[269] that is characteristic of most contemporary Southeast Asian societies. Malaysia and Thailand are both part of the Association of Southeast Asian Nations (ASEAN), often commonly referred to in Southeast Asia as the 'European Union of the Southeast'. Similarly to the European Union, the aims and purposes of ASEAN began as a form of economic co-operation between countries in the region, and aimed to improve and further socio-economic and financial growth in the nations concerned. Concerns relating to issues of human rights have ranked low in priority, but until recent years (in 2012), and particularly through international pressure, have resulted in the birth of the ASEAN Human Rights Declaration (AHRD).[270] Despite this establishment, however, the human rights track record in the region remain at peril: recent examples of continued human rights violation have prompted concerns that the ASEAN is essentially a "toothless tiger",[271] and human rights scholars in Asia has called for its non-interventionist approach into the affairs of its member states to be revisited.[272]

In recent times, some of these human rights violations have included the mass exodus of Rohingya refugees from Myanmar into countries like Thailand, Malaysia and Indonesia, with the Myanmar government being criticized on the international front for allowing the perpetuation of the violations against these people.[273] In the Philippines, the reign of President Duterte's 'war' against drugs has been described as resembling a criminal enterprise, because of the extreme measures of violence deployed to pursue this campaign.[274] Malaysia struggles with allegations of high-

---

[267] Jabour (2015).

[268] Cunningham (16 February 2017).

[269] Baharuddin (2008), p. 1.

[270] Association of Southeast Asian Nations, 'ASEAN Human Rights Declaration and the Phnom Penh Statement On The Adoption of the ASEAN Human Rights Declaration (AHRD)' http://www.asean.org/storage/images/ASEAN_RTK_2014/6_AHRD_Booklet.pdf.

[271] Levett (13 January 2007).

[272] Dasgupta (15 February 2018).

[273] Newland (July 2015, Issue No. 13).

[274] Gomez (3 July 2017).

level political corruption[275] and a violation of constitutional rights and liberties, for example, relating to religious freedom[276] and freedom of speech.[277] Thailand's strict *lèse-majesté* laws has seen recent operation after the death of their much-loved and revered monarch, King Bhumibol Adulyadej, when speculation and concerns were rife about the ascension of his son, Maha Vajiralongkorn.[278] These examples, and more that emanate from the region, show that the regional human rights discourse is lacking depth and enforcement by state parties, and needs to be rigorously developed in the near future. Particularly for Malaysia and Thailand, the World Reports of 2018 from the Human Rights Watch indicate that the common human rights concerns in Malaysia[279] and Thailand[280] are mostly centered on freedom of expression, attacks against human rights defenders, police and military violence and the criminal justice system, and the treatment of refugees and asylum seekers. Therefore, it is not surprising that other facets of human rights considerations, such as reproductive rights, are low on the rung of Malaysia's and Thailand's political and socioeconomic agenda.

A further commonality between Malaysia and Thailand is the existence of a dualist 'legal' system in the countries. In the case of Malaysia, its adoption of the British Westminster system of Parliamentary government and democracy was the impetus for its constitutional framework. Like Australia, and the US, Malaysia's historical and political structures began as a group of states known as the Federation of Malaya.[281] In a system of government that recalls that of the UK, but with a written Federal Constitution,[282] Malaysia's *Rukunegara* (or national principles)[283] is additionally beleaguered to uphold the rule of law by enumerating "a five point national philosophy"[284] as a means of promoting national unity in the country. An additional feature of its legal system, however, is the employment of Shari'a law,[285] (or sometimes also commonly known as Islamic law), as a fundamental tenet of religious adjudication pertaining to Islam and Muslims in Malaysia. As a

---

[275] Wright (12 September 2016).

[276] Morningstar News (23 February 2018). See also: AsiaNews.it (21 September 2005); Davidson (27 February 2018); and The Economist (22 September 2016).

[277] See also the case of *Public Prosecutor v Azmi Bin Sharom (Criminal Ref. No. 06-5-12/2014(W))* (6 October 2015) http://www.kehakiman.gov.my/judgment/file/PRESS_SUMMARY_PP_V_AZMI_SHAROM.pdf.

[278] Agence France-Presse, Bangkok (9 June 2017). See also: Holmes (29 October 2016).

[279] Human Rights Watch, 'World Report 2018: Rights Trends in Malaysia' https://www.hrw.org/world-report/2018/country-chapters/malaysia.

[280] Human Rights Watch, 'World Report 2018: Rights Trends in Thailand' https://www.hrw.org/world-report/2018/country-chapters/thailand.

[281] Lee (2004), p. 226.

[282] Bari and Shuaib (2009).

[283] The five-point philosophy of the Rukunegara is as follows: Belief in God, Loyalty to King and Country, Upholding the Constitution, Rule of Law and Good Behaviour and Morality. See further: http://www.perdana.org.my/spotlight2/item/rukun-negara-the-national-principle-of-malaysia.

[284] Yatim (1995), p. 27.

[285] Saliba (June 2011).

Muslim-majority country, the relationship between Islam and human rights in Malaysia is often subject to animated discourse; and it is not surprising that in many instances of contemporary Malaysian society, the tussle between Malaysia's civil law[286] and Shari'a law has been subject to "fragmented 'domains of control' within social life in Malaysia".[287] This will be elaborated on in the sections below.

Thailand's legal system is also fascinating, in that it exists through installation by a military junta. The 'design' of the military junta's legal system in Thailand (pursuant to a coup d'etat in May 2014 following months of political crisis involving Thaksin Shinawatra)[288] has become so deeply embedded in Thai life[289] that international concerns have been raised about its deepening grip on Thai political agenda and its broken promises to restore law and order in the country within a year from the coup.[290] Its recently drafted 2016 Constitution[291] following the epoch of the 2007 Thailand Constitution[292] has been subject to much discourse because of its apparently controversial contents, primary of which included the appointment of an "outsider prime minister" within the context of Thailand's internal political rife between the National Council for Peace and Order (NCPO) and the Democrat Party, with the military-led National Legislative Assembly (NLA),[293] and the inclusion of a "long-term blueprint for national reform"[294] and the formation of a crisis panel. Concerns relating to the manner in which the draft Constitution had been prepared and the referendum held on voting for the Constitution was also provoked by allegations of bias and one-sided communication.[295]

With the prevalence of political instability and crises of 'constitutionalism' in these countries, the displacement of the rule of law (temporary or otherwise) and with it, the respect and concern for human rights considerations, is not surprising.

---

[286] 'Civil law' in this context does not denote the civil law codified system of governance that is prevalent in most European states. Instead, 'civil law' in the context of Malaysia refers to laws applicable to all citizens of the country. Like the UK, Malaysia's legal system is a common law jurisdiction having reference to legislative pronouncements and decisions of the courts *vis-à-vis* the doctrine of judicial precedent.

[287] Baharuddin (2008), p. 4.

[288] BBC News (2011), 'Profile: Thaksin Shinawatra' http://www.bbc.com/news/world-asia-pacific-13891650.

[289] Themgumpanat and Tanakasempipat (21 May 2017).

[290] Human Rights Watch, 'World Report 2016: 'Thailand' https://www.hrw.org/world-report/2016/country-chapters/thailand.

[291] Draft Constitution of Thailand 2016 (Unofficial English translation) http://constitutionnet.org/sites/default/files/thailand-draft-constitution_englishtranslation_june_2016.pdf.

[292] Constitution of the Kingdom of Thailand (2007) https://www.samuiforsale.com/law-texts/thai-law-translations-constitution-of-thailand.html.

[293] Panaspornprasit (2017), p. 356.

[294] Ibid 357.

[295] Ibid.

6.3.3.1   Malaysia

As a former British colony in the eighteenth to twentieth centuries up until 1957, Malaysia's legal system is similar to the UK's. The role of human rights within the sphere of Malaysian legal and political discourse has mostly centered on debates that relate to its applicability, simply because of "power relations and the diversity of belief systems and interests"[296] within Malaysia. The plurality of Malaysia's peoples, cultures and beliefs system make up a complex demography in the country; but despite Malaysia's positioning in Asia, particularly, as a strong and growing developmental and economic nation that has transformed itself following the Asian financial crisis in 1997,[297] the glitter of its development has masked the underlying local responses to fundamental rights and liberties. Malaysia is signatory to a variety of international human rights instruments, but the perplexity of its relativist position on human rights has been summed up as follows by former Finance Minister, Tun Daim Zainuddin:-

> For Malaysia, the protection of human rights and fundamental freedoms, consonant with the principles enshrined in the Universal Declaration of Human Rights (UDHR), is guaranteed in the Malaysian Constitution. Malaysia, however, believes that human rights and fundamental freedoms would be meaningless if the country is destabilized by social, political and economic choice……. there is a need to review the various human right instruments and also the standards of human rights…. Such a review should also take into account the peculiarities of national values, religions, customs, tradition, social and economic systems in a particular country, and attempts should be made to harmonize human rights in a balanced manner, moving away from the present unhealthy dominance of Western values and concepts.[298]

Saliha Hassan and Caroline Lopez identifies Malaysia's position on the global human rights narrative, which continues to take inspiration in the relativist manner, placing immense emphasis on "Asian values" and prioritizing community as opposed to individual liberties.[299] Indeed, the literature of global human rights narratives in Malaysia has revealed three main areas of debate: the indivisibility of human rights,[300] the universalist and relativist approach in human rights,[301] and the "mainstream interpretation of human rights and its connection with globalization."[302] The purported "Asian values" of Malaysian society is less than surprising because of its colonial past; these values "emphasize deference to authority, acceptance of relatively strict government control and non-interference by one nation in the internal affairs of others as the basis for their concomitant understanding of human

---

[296] Hassan and Lopez (2005), p. 111.
[297] World Bank Global Knowledge and Research Hub (December 2017).
[298] Hassan and Lopez (2005).
[299] Ibid 120.
[300] Ibid 111.
[301] Ibid.
[302] Ibid 112.

rights."[303] In an era where its independence was newly achieved, these Asian values undoubtedly lent some sense in the navigation of Malaysia's position on the world map; but I posit that the "Asian values" cannot continue to dominate the discourse in purported cultural, religious and historical relativism in contemporary democracy.

In considerations of fairness, the reliance on "Asian values" and 'bashing' of the West has fallen into decline,[304] and the enactment of Malaysia's Human Rights Commission of Malaysia Act 1999 (HRCM Act)[305] and the establishment of the Human Rights Commission of Malaysia (SUHAKAM)[306] under the act, was meant to address the protection of human rights in Malaysia. (The HRCM Act has since undergone amendments, and the latest version of this act is known as the Human Rights Commission of Malaysia (Amendment) Act of 2009.)[307] Read together in conjunction with the Federal Constitution of Malaysia (Federal Constitution),[308] which provides for the basic fundamental rights and liberties of Malaysian citizens, the intention was for SUHAKAM to not only oversee the protection of human rights in Malaysia, but to also "be given as broad a mandate as possible.... to inquire into violations of not *some*, but *any* human rights."[309] It is this continued relativist approach to human rights that has led Malaysia to its present state; where it remains one of the very few countries in the world where the majority of the population needs affirmative action and "protection" from the minority.[310]

In the years during British colonization, and subsequently, following its independence, the unusual affirmative action movement in Malaysia, characterized in the Federal Constitution under Article 153,[311] became a fundamental tool in the socio-economic development of the Malay community in the country. Historically, the Malays were the under-privileged members of the Malaysian community despite comprising the majority of the population in Malaysia, and part of the reasons for this enactment of Article 153 was a culmination of special economic privileges accorded to the Malays under British rule. The addition of Amendment 8A[312] to Article 153 continues to be in contention, as it created what is known as the "quota

---

[303] Ibid 120.

[304] Ibid 122.

[305] Human Rights Commission of Malaysia Act 1999 (Act A597).

[306] SUHAKAM is the acronym for Suruhanjaya Hak Asasi Manusia Malaysia; or in English, the Human Rights Commission of Malaysia. The Commission was established in 1999, with its first inaugural meeting held on 24 April 2000. The formation of the Commission was in response to Malaysia's active involvement in the United Nations Commission on Human Rights in 1995 to 1995, and the changing political climate in the country, and government concerns raised necessitated the establishment of an independent national human rights institution.

[307] Human Rights Commission of Malaysia (Amendment) Act 2009 (Act A1353).

[308] Federal Constitution of Malaysia 1957.

[309] Hassan and Lopez (2005), p. 129.

[310] Torii (1997), p. 209.

[311] Constitution of Malaysia 145.

[312] Ibid 148.

system" in Malaysia, whereby educational institutions in Malaysia were "required to ensure the reservation of such proportion of such places for Malays and natives of any of the States of Sabah and Sarawak."[313] Under the National Economic Policy in Malaysia in 1970, this "special position" was given effect to, but the raging discourse of late has focused on the indefinite application of the special position in contravention of the "sunset clause" contained in the National Economic Policy.[314,315] Until today, this remains a highly sensitized and controversial issue for non-Malays in Malaysia, and the treatment of the "special position" has, in practice, converted from an accorded 'privilege' to a "right",[316] spanning sectors such as education, housing, socio-economic welfare, private industries, politics and the public sector. In this regard, it is not surprising that recent political democracy in Malaysia has focused on the "Malay supremacy" and nationalistic tendencies, creating a highly polarized rift in different sections of society. Clearly, this state of affairs demonstrates that the right to equality, one of the Human Rights Components, is very much flawed.

In this context, Malaysia does not have a very good human rights enforcement track record. Whether it is attributable to other political concerns, which are certainly more pressing at this juncture, this is not to excuse its approach towards a more deep-seated respect for human rights. In the past 8 years, the consideration of human rights in the country deteriorated tremendously amidst claims of high-level governmental abuse of process, corruption and criminal breach of trust. Persons who were regarded as 'dissidents' of the government rhetoric were persecuted, freedom of speech was stifled,[317] and an increasing 'Islamization' narrative by the then ruling government was used to divide communities and propagate over-sensitivity of Islam and the Malay supremacy.[318] With all these events, issues that relate to the Human Rights Components were rarely highlighted simply because they were not priorities within the national context. The rhetoric previously emphasized by the former government regime was "Cash is King",[319] unabashedly reinforcing that anything was possible at the right price (monetarily).

For example, the right to privacy and equality does not strictly exist within the Malaysian context, and it is also evident from the manner in which individuals' liberty to exercise personal autonomy in many dimensions of life is not permitted. Despite the express recognition of a person's right to equality under Article 8 of the Malaysian Federal Constitution,[320] which states that "all persons are equal before the law and entitled to the equal protection of the law", the realities are very contrary

---

[313] Ibid.

[314] Jomo (2004).

[315] Zubedy (21 June 2012) http://www.freemalaysiatoday.com/category/opinion/2012/06/21/nep-the-good-and-the-bad/.

[316] Lee (2016), p. 511.

[317] Nazlina (14 January 2016). See also: Agence France-Presse (20 February 2018).

[318] Malm (19 February 2018).

[319] The Straits Times (18 May 2018).

[320] Constitution of Malaysia.

to the express pronouncement. As an example, an illustration of this point regarding the lack of recognition of equality and privacy can be observed from the treatment of the LGBTQ[321] community. LGBTQ persons continue to be persecuted under the archaic Sections 377A, 377B and 377C of the Penal Code, a legacy of the British colonial rule, which criminalizes "carnal intercourse against the order of nature".[322] In India, another former British colony, the Supreme Court, in a historic, landmark judgment, recently ruled on the unconstitutionality of Section 377 of the Indian Penal Code (similar to Malaysia's), heralding a new era of rights recognition in the country.[323] Because of Malaysia's position as an Islamic nation, it is unlikely that a similar event will happen in the country. Meanwhile, in recent news that highlighted the travesty of child marriages in Malaysia,[324] Islam fundamentalists have proclaimed that the LGBTQ communities pose more of a problem than the "common"[325] child marriages. The constitutionally recognized "special" status of the Malays, although subject to differing debates on whether it is a right or a privilege, further contributes to inequalities in the country. Interestingly, despite these restrictions that culminated from the outdated Penal Code, abortions were legalized in 1989 in the Penal Code (Amendment) Act 1989, which provided the criteria under which abortions could be obtained and carried out. However, the social stigma that is attached to abortions is such that it is still considered a taboo topic and it would in practicality be difficult to obtain an abortion in a public hospital. The cultural and religious permeations into Malaysian society, especially because Malaysia is a multi-cultural nation, make it such that issues such as bodily integrity and personal autonomy, or equality considerations, often have very specific, contextual, patriarchal backgrounds. This demonstrates that human rights necessitate a comprehensive reform in Malaysia. With the recent political regime change in the country,[326] however, particularly where the Human Rights Components are concerned, there is hope that things may change in the future.

6.3.3.2 Thailand

One of the most prodigious facts to be recognized regarding Thailand is that it has never, in its history, been the subject of colonization. Contemporary scholarly work attribute this unique feature to Thailand's remarkable ability to negotiate diplomatic efforts strategically, which allowed it to remain as a buffer state, remarkable at a time when almost all other Southeast Asian countries fell to a multitude of colonization efforts by the West. Because of this, it is likely that Thailand has retained its

---

[321] LGBTQ is the acronym for lesbian, gay, bisexual, transgender and queer used to denote primarily non-heterosexual or non-cisgender community members.

[322] Penal Code 1936 (Act 574) 314.

[323] BBC News (6 September 2018a).

[324] Busby (2 July 2018).

[325] Zurairi (4 July 2018).

[326] Heydarian (11 May 2018).

distinctive and unique culture. However, a further study of its laws, legislative enforcement and political structures also reveal that it is likely this highly nationalistic characteristic that has not only contributed to internal and domestic political strife, but also to the fact that it has had 18 redrafted constitutions since 1932. Unsurprisingly, very much like Malaysia, Thailand's national priorities are a constant tug-of-war between economic and political in nature, leaving other issues largely neglected. Its issues of political dynasties,[327] its histories navigating through numerous transitional periods,[328] the strength of its monarchised military,[329] and its continuing problems of inequalities due to political structures and at one point of time, client-elitist policies[330] have all contributed to "the full blossom of democratization"[331] being far from achieved in the country. The full culmination of its legal history is, indeed, a most fascinating one.[332]

In a recent report by the Human Rights Watch, issues regarding human rights violations in the country were raised and Thailand's NCPO (the military junta) has yet again failed to restore democratic rule to the country.[333] In the foregoing paragraphs of this Sect. 5.3.3 it was highlighted that Thailand's newly drafted 2016 constitution[334] is still an interim constitution and contains many questionable provisions that reveal improprieties. Although considerable positivity was indicated with both the 1997 and 2007 Thai Constitutions,[335] because it managed to chronicle Thailand's recognition of international human rights, and subsequently led to the establishment of the National Human Rights Commission, the issue of Thai constitutionalism still remains in flux because its constitutional redrafts often reflect "the power of very different political movements."[336]

Andrew Harding and Peter Leyland, in their work on Thailand's constitutional system have likely produced the most comprehensive resource on Thai constitutionalism in the English language.[337] Before hastening a discussion on the Human Rights Components in Thailand, an introduction to the work of Harding and Leyland is helpful because Thailand's constitutional history is a unique one that has come to influence the "contemporary functioning of Thai constitutionalism."[338] In their work, Harding and Leyland identify the peculiarities and turmoil of Thailand's political history and whether through a recognition of the unusual elements of Thai constitutionalism, despite the shortcomings of the Thai Constitution of 1997 and

---

[327] Thananithichot and Satidporn (2016), p. 340.

[328] Panaspornprasit (2017).

[329] Chambers and Waitoolkiat (2016), p. 425.

[330] Hewison (2014), p. 846.

[331] Panaspornprasit (2017), p. 351.

[332] Hickling (1972), p. 8.

[333] Human Rights Watch 'World Report 2018' (n 280).

[334] Draft Constitution of Thailand 2016 (n 291).

[335] Constitution of the Kingdom of Thailand (2007).

[336] Nolan (2012), p. 2.

[337] Harding and Leyland (2011).

[338] Nolan (2012), p. 1.

2007, these were often described as the most 'democratic' constitutions because of the extensive redrafts the Constitution Drafting Assembly in Thailand had undertaken. In essence, Harding and Leyland provide a comprehensive account of comparisons between the 1997 and 2007 Thai Constitutions, but also contend that "Thai electoral politics can qualify as democracy in form only."[339] Many other pieces of scholarly work also recognize that democratization is a constant struggle in Thailand; some scholars have attributed the overthrow of Thaksin Shinawatra's government in 2014 to the inability of Thai nobility, aristocracy and what is commonly known as "royalists"[340] to accept what is deemed as contemporary challenges to the long-held Thai status quo. In fact, the contemporary challenges of inequalities in Thai society have been consistently connected to its political structures. Despite Thailand's increasing rates of economic growth, the level of poverty is not equally reduced, particularly in rural areas that "continue to suffer the highest levels of absolute poverty."[341] This is attributed to the "structuring of a political system that has been, by and large, exclusionary and dominated by political and economic elites who are suspicious of electoral democracy."[342] On this basis, and a range of literature of economic inequalities that prevail in the country, it is not surprising that there is "serious failure in terms of equity"[343]; the presence of the stronghold military junta continues to perpetuate these inequalities, and like Malaysia, the Human Rights Components are simply not considered to be pressing issues to be dealt with.

Although it cannot be denied that Thailand has indeed gone through monumental changes in its political dimensions, and have entered into and ratified various international human rights treaties, the practical implementation of human rights is very similar to the situation in Malaysia. Despite an impassioned defense of Thailand's human rights recognition in its substantive laws and policies enumerated within the constitution,[344] (which include the guarantees of non-discrimination, right to life and the right to privacy), the practicalities often do not reflect the constitutional guarantees. Not only is law enforcement weak, but the mechanisms of institutional enforcement are also questionable.

In terms of stifling freedom of speech and expression, similar to the situation in Malaysia, Thailand's military junta has been known to charge individuals for any form of speech or expression that criticizes the military government, even if such discussions were academic in nature, and much more so, their *lèse-majesté* laws.[345] Vitit Muntarbhorn highlighted the disparities of several judgments made in the Thai Constitutional Courts, including controversial issues regarding individuals with

---

[339] Harding and Leyland (2011), p. 49.

[340] Hewison (2014), p. 857. Hewison states: "Royalists imagined a diminution of the monarchy's role and its political centrality, and this resulted in a decade-long and still unfinished political struggle over the nature of Thailand's democratization."

[341] Ibid 849.

[342] Ibid.

[343] Ibid 853.

[344] Muntarbhorn (2005), p. 358.

[345] Human Rights Watch, 'World Report 2018' (n 280).

disabilities intending to seek judicial positions, regarding gender discrimination of women who are obliged to change their maiden names to their husbands' names upon marriage, as well as declaration of assets by politicians.[346] Despite the constitutional guarantees of non-discrimination and equality, the reflection not only by the Constitutional Courts but also within the practice of the laws therefore indicates to us otherwise.

The role of women within Thai society is also particularly disturbing. Although women generally enjoy many rights as men in the country, the societal view of women indicate that more often than not, they are not treated equally because of cultural barriers that exist. At the very outset, the predominant Thai practice of Buddhism (which is central to most of the lives of Thai people, and can be very different from other forms of Buddhism practiced in other parts of the world) already regards a woman as incapable of attaining Nirvana unless they are reborn as men. The Thai reverence to Buddhism and ordaining young men as monks view the woman's inability to do so as an inferior positioning within the cultures of communities and societies.[347] In addition to this, despite the constitutional guarantees of equality, Liza Romanow highlights that the role of women and girls in education, politics and the work force are still demonstrably low.[348] In fact, within Thailand's burgeoning and internationally infamous sex trade and sex trafficking, women, especially women of the lower income bracket or from poor or smaller rural villages, have often been forced into prostitution at some point of time.[349]

Bearing this in mind, it therefore comes as no surprise that one of the primary Human Rights Components discussed in this chapter, the right to privacy (based upon a woman's ability to make choices regarding her body) *vis-a-vis* abortion in Thailand is, at the outset, illegal. The limited provisions that govern abortions are contained in Chapter III, Sections 301 to 305 of the Thailand Penal Code,[350] and the wordings used at the beginning of the chapter makes it clear that abortion is an "offence." The pertinent sections read as follows:-

Section 301 Any woman, causing herself to be aborted or allowing the other person to procure the abortion for herself, shall be imprisoned not out of three years or fined not out of six thousand Baht, or both.

Section 302 Whoever, procures abortion for a woman with her consent, shall be punished with imprisonment not exceeding five years or fined not exceeding ten thousand Baht, or

---

[346] Muntarbhorn (2005), p. 363.

[347] Kittiwut Jod Taywaditep, Eli Coleman and Pacharin Dumronggitigule, 'The International Encyclopedia of Sexuality: Thailand' http://www.sexarchive.info/IES/thailand.html. Women still continue to be regarded as the weaker sex, and within the practice of Thai Buddhism, women's body parts and particularly menstruation are regarded as elements that are sacrilegious and harmful to monks, and other learned men. This mentality therefore continues to perpetuate gender inequalities, and to some extent, also gender segregation, of women from men.

[348] Romanow (2012), p. 44.

[349] Ibid 59.

[350] Thailand Penal Code, Thailand Criminal Law (Unofficial English Translation) https://www.samuiforsale.com/law-texts/thailand-penal-code.html.

both. If such act causes other grievous bodily harm to the woman also, the offender shall be punished with imprisonment not exceeding seven years or fined not exceeding fourteen thousand Baht, or both. If such act causes death to the woman, the offender shall be punished with imprisonment not exceeding ten years and fined not exceeding twenty thousand Baht.

Section 303 Whoever, procures abortion for a woman without her consent, shall be punished with imprisonment not exceeding seven years or fined not exceeding fourteen thousand Baht, or both. If such act causes other grievous bodily harm to the woman also, the offender shall be punished with imprisonment of one to ten years and fined of two thousand to twenty thousand Baht.

If such act causes death to the woman, the offender shall be punished with imprisonment of five to twenty years and fined of ten thousand to forty thousand Baht.

Section 304 Whoever, attempts to commit the offence according to Section 301 or Section 302, first paragraph, shall not be punished.

Section 305 If the offence mentioned in Section 301 and Section 302, be committed by a medical practitioner, and:

• It is necessary for the sake of the health of such woman; or
• The woman is pregnant on account of the commission of the offence as provided in Section 276, Section 277, Section 282, Section 283 or Section 284 the offender is not guilty.

Medical anthropologist, Andrea Whittaker, has conducted extensive studies in Thailand regarding abortions and women's reproductive health and decision-making processes, and the major themes that have emerged on decision-making processes have connections to the material conditions and environment, the considerations of gender, and notions of motherhood and children in Thai communities.[351] Although it is undeniable that there is a deep-rooted belief in Thai culture that abortion constitutes a grievous sin in accordance with Buddhist beliefs,[352] and prevalently much more so among rural women, a combination of other problems also contribute to Thailand's restrictive abortion laws.

In the preceding paragraphs that outline Thailand's rocky political history, abortion has similarly been politicized largely through the history of its penal codes from 1973 onwards, presenting varying views on how to prevent abortions in the first place by providing proper access to contraception,[353] the debates regarding Buddhist proscription against the destruction of sentient life,[354] as well as concerns about "corrupt Western influence" regarding free sex and the destruction of Thai culture.[355] As such, activist groups continue to lobby for abortion law reform in Thailand, calling for an enlargement of the exceptions provided in Section 305 of

---

[351] Whittaker (2002b), p. 1.
[352] Whittaker (2002a), p. 46.
[353] Ibid 47.
[354] Ibid 48.
[355] Ibid.

the Thailand Penal Code, for example, covering situations where there might be serious genetic disorders, where there may be socio-economic difficulties, or mental health problems on the part of prospective mothers, or where women themselves may have AIDS or HIV infection, or failures of contraceptive devices.[356] Whittaker correctly identifies this misconception: "Abortion remains associated in the popular imagination as un-Buddhist, a sinful act of prostitutes and promiscuous students, not an issue that affects the lives of all Thai women."[357] The additional portrayal of abortions by women in the media also further lends aggravation to why women in Thailand are often afraid to seek out abortions and bear the deeply judgmental social stigma.

It cannot be denied that Thailand's internal political and economic affairs are in a state of turmoil at present. Despite the glitter of tourism and wonderment from the Land of Smiles, its state of respecting human rights leaves very much to be desired, even more so with the issues that have been raised by Human Rights Watch,[358] and which have been highlighted on an international level. Clearly, in a similar manner as its southern neighbor, Malaysia, its constitutionally recognized rights and guarantees are badly enforced and adjudicated, and its approach to human rights recognition and enforcement is equally flawed.

6.4   The Entry Points of Regulation in Pre-implantation Genetic Interventions

The diversified constitutional interpretations of human rights protections in Sect. 6.3 has revealed that it may be possible to consider regulating pre-implantation genetic interventions on similar standards. These standards would clearly need to take into account the Human Rights Components, which I have established to be concerns that are pertinent in pre-implantation genetic interventions. However, bearing this in mind, it is likely that the main regulatory difference and constitutional concerns between these selected jurisdictions and the potential applicability to pre-implantation genetic interventions, would be the appropriate entry point of regulation: either its inclusion *vis-à-vis* the practical and positivistic private law aspects (perhaps an express form in the Constitution), or from a more philosophical, theoretical human rights aspect. I also identify, besides these points of regulation, that two other future considerations are necessary to frame the discourse of regulation in pre-implantation genetic interventions. These are the status of the embryo, and the woman's body embodiment in undergoing the simultaneous marvel and barriers of artificial pregnancies.

---

[356] Ibid 51.

[357] Ibid.

[358] Human Rights Watch, 'World Report 2018' (n 280).

Based on an analysis of the framework of fundamental (constitutional) rights in Sect. 6.3 above mentioned, I make the following preliminary observations. Firstly, that although culture and/or religion do make their appearance into the ethical and philosophical discourse of the Human Rights Components, these cultural or religious dimensions appear more on the forefront of very specific issues (such as abortion or same sex marriages as these are very much rooted in religious dogma), as opposed to a wholesale and broad sweeping objection to human rights per se (example: the US, and partially, Australia, Malaysia and Thailand). Secondly, the tendency to politicize some human rights is apparent, particularly where federal agendas by ruling governments are concerned. Hence, political structures are capable of influencing the manner in which fundamental constitutional rights may be regarded (example: the US, Malaysia and Thailand). Thirdly, it would be wise to determine whether it would be preferable to have either expressly enumerated fundamental rights within the constitution (but with poor legislative laxity) (example: Malaysia and Thailand), or a more liberal, implied, judicial interpretation to extend a right from the constitution into a particular sphere (example: the US), or specific legislative responses to specific rights issues (example: the UK and Australia). All these observations are pertinent in how a selected jurisdiction may choose to regulate pre-implantation genetic interventions, a relatively emerging field of biomedical technologies that warrants critical thought for future deployment. The ways in which fundamental rights are treated in these selected jurisdictions are also likely to similarly apply to the regulation of pre-implantation genetic interventions. Clearly, in the case of some of the selected jurisdictions, the fundamental constitutional framework for the protection of the key Human Rights Components appear to be less than desirable.

In the meantime, Fig. 6.1 illustrates the main dual entry points that regulations for pre-implantation genetic interventions are likely to be concerned with in the selected jurisdictions. This is based on the analysis of each of their respective constitutional profiles in Sect. 6.3 of this chapter. The spectrum indicated at the bottom of Fig. 6.1 indicates the jurisdictions that are either more concerned with the positivistic law aspects of regulation (the First Entry Point); or the more philosophical human rights dimensions (the Second Entry Point), with some jurisdictions falling in between these spectrums.

From my analysis, the UK demonstrates an equal measure of interest in both the First and Second Entry Points. For example, the establishment of the HFEA under the purview of the HFEA Act acts as the appropriate regulatory authority to oversee all matters relating to fertility treatments and technologies, demonstrating its stand in respect of the First Entry Point. However, the ethical and philosophical concerns that have shaped the operation of the HFEA Act and the infrastructural framework of regulation in the UK also appears to be duly considered in the Second Entry Point. On an overall basis, the European approach, which emphasizes personhood, dignity and fundamental liberties, is also likely to fall under the Second Entry Point of regulations.

In the meantime, the US and Australia are likely to fall in between the First and Second Entry Point of regulation, both demonstrating strong federalist governmental

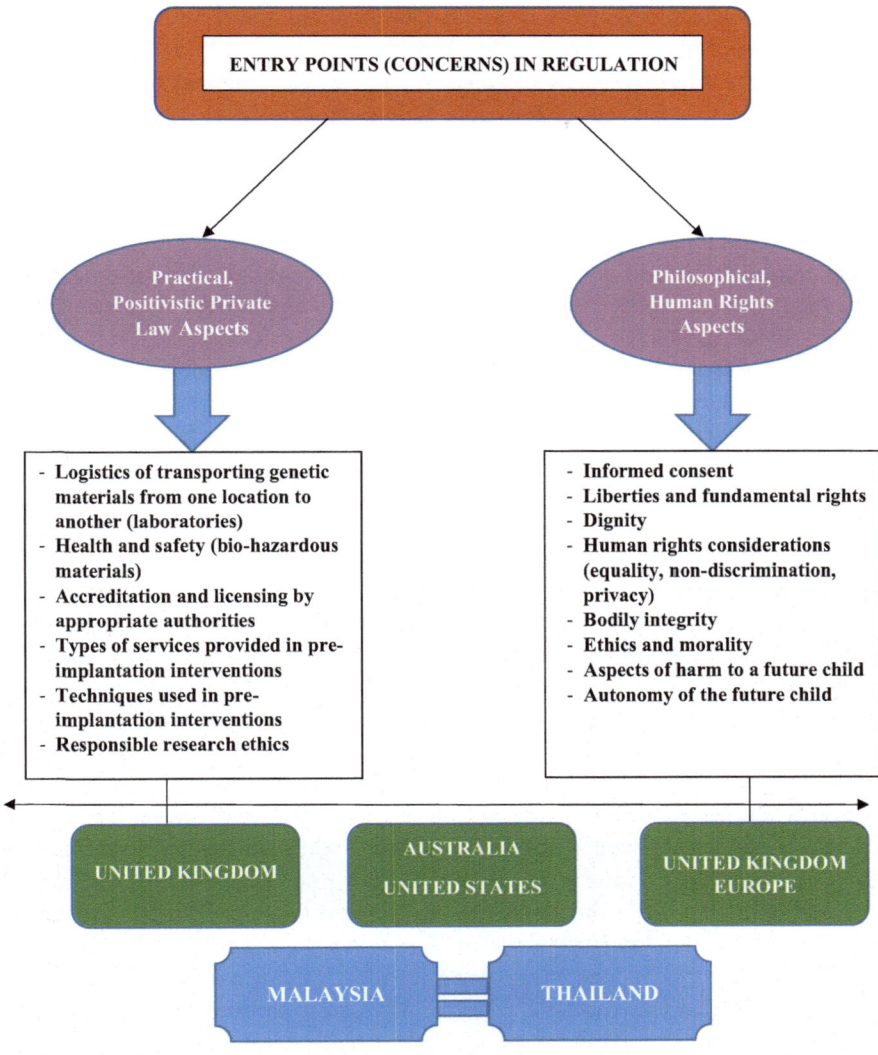

Fig. 6.1 Prospective entry points in regulation for pre-implantation genetic interventions

approaches in legislation. The existence of the federalist framework also makes it such that a constant struggle between state and federal legislation is a reality. This therefore necessitates a continuous and constant negotiation between state and federal governments in coming to a concrete understanding regarding some vital issues, which should be governed on a federal level. The representation of Malaysia and Thailand are meant to show that there is a material state of fluidity and changeability of concerns in those countries. These states are more than likely to regulate as a response to an external phenomenon. A good example of this is the manner in which

Thailand fast-tracked its surrogacy laws to ban commercial surrogacy for foreigners after the furor created by an international incident.

The downside to this speedy response to international admonishment, however, does not mean that the new Thai surrogacy laws are adequate. In fact, as the ban extends to commercial surrogacy in respect of foreigners, this does not mean that the same criteria extend to Thai couples that wish to engage surrogate mothers. In Malaysia, like Thailand, fertility tourism has in recent years proven to be financially hardy in contributing to part of its economic development. In this regard, it is certainly within the interest of the country not to strictly legislate on some biomedical technologies because the influx of foreigners that may wish to take advantage of the less strict laws in Malaysia is a financial boon to the industry. Indeed, it may even be accurate to state that the commercialized aspects of biomedical technologies, where the products of such technologies are treated as commodities and services, very much like a regular business transaction, have led to this lack of legislation, or legislative laxity.[359]

In addition to the discussion on the regulatory points of entry, there are two other issues that necessitate some greater reflection too because these have often been swept aside due to difficulties in finding consensus. This includes the status of the human embryo, and the woman's bodily experience in artificial pregnancies, both of which appear to be inadequately framed. The reasoning for this inadequacy, however, is likely to find justification in the fact that the diversified interpretations of the human embryo's status, and the experience of women in pregnancies, particularly artificial pregnancies in the case of IVF, have drawn such polarized views that it may continue to face challenges in coming to a consensus. Regarding the status of the human embryo, for example, the understanding on the conception of the embryo's life makes a very significant contribution to how abortion debates and reproductive treatments and legislations are framed. The medical or scientific acceptance of the viability of the human embryo when it attains a form of humanized personhood may not be compatible with the religious dimensions that interpret life to begin from conception. Such is the challenge presented in determining the status of the human embryo that there is no one right way to define its legal status. Indeed, even within the more liberalized approach of the European Court of Human Rights, Professor Judit Sándor highlights the court's opinion[360]: "it is 'neither desirable, nor even possible as matters stand, to answer in the abstract, the question whether the unborn child is a person for the purposes of the right to life provision of the Convention'."[361]

In respect of the woman's bodily experience in artificial pregnancies *vis-à-vis* IVF, the salient perceptions of this embodiment is blurred between bodily integrity and physical enduring, and the achievements or final product of the artificial technologies. This aspect is very often overlooked and may result in inadequate

---

[359] Whittaker and Speier (2010), p. 363.

[360] Sándor (2015), p. 355.

[361] Ibid. Also quoting the case of *Vo v France* (Application No. 53924/00) from the European Court of Human Rights, judgment dated 8 July 2004.

protections in the law that also nurtures and supports the psychological and emotional well-being of women in pregnancies. With the exception of the doctrine of informed consent, but with the fixation that occurs on the end product, that is, a fully functioning human person resulting from the implantation of embryos, the woman's experience is not given cynosure. The physical aspects of the bodily experience: from tricking the human body into believe that it is pregnant (with hormone injections and drugs), to implanting the embryo into the womb, to ensuring the continuity of embryo growth within a body that has been artificially prepared and treated to receive the embryo,[362] does not always equate success. Notwithstanding the miracle of artificial technologies that would now enable women to have children when they previously could not, the bodily embodiment does not adequately take into account the emotions and fears of failure, the differing responses of the body to treatment, and the underlying premise that continue to frame a patriarchal view of bodies in reproduction.

In hoping to achieve a more globalized biomedical approach in the manner indicated in Chap. 5, this analysis hopes to inform contemporary legal and ethical discourse in how pre-implantation genetic interventions could be considered for the future implementation. Indeed, it is not too much to be confidently idealistic that this is achievable. Judit Sándor believes that it is necessary to avoid the key traps of advancing the human gametes discourse and trying to fit them within an existing legal framework.[363] She identifies these key traps that must be avoided when considering the legislative elements: firstly, by avoiding the personalizing body parts or human gametes and simply viewing these from the perspective of human rights,[364] and secondly, by rejecting gametes, embryos and other by-products of reproduction as commodities.[365] These comparative elements indicated above are but analogies that serve to illustrate the manner in which pre-implantation genetic interventions may be regulated in the different jurisdictions. It is not enough however to simply suggest that these technologies be regulated, but also that they be regulated properly and effectively.

References

Abdulaziz, Cabales and Balkandali v The United Kingdom (1985) European Court of Human Rights Application no. 9214/80; 9473/81; 9474/81, HUDOC
Age Discrimination Act 1992
Agence France-Presse (20 February 2018) Malaysian Artist Jailed for Clown Face Caricature of PM Najib Razak That Went Viral. South China Morning Post. https://www.scmp.com/news/asia/southeast-asia/article/2133990/malaysian-artist-jailed-clown-face-caricature-pm-najib

---

[362] Ibid 359.
[363] Ibid 353.
[364] Ibid 354.
[365] Ibid.

Agence France-Presse, Bangkok (9 June 2017) Man Jailed for 35 Years in Thailand for Insulting Monarchy on Facebook. The Guardian. http://www.theguardian.com/world/2017/jun/09/man-jailed-for-35-years-in-thailand-for-insulting-monarchy-on-facebook

Alexy R (2010) A theory of constitutional rights. Oxford University Press, Oxford

Aroney N (2009) The constitution of a federal commonwealth: the making and meaning of the Australian Constitution. Cambridge University Press, Cambridge

AsiaNews.it (21 September 2005) Malaysia Malay converts to Christianity. http://www.asianews.it/news-en/Malay-converts-to-Christianity-cannot-renounce-Islam-4157.html

Association CG Press (24 October 2017) Northern Ireland abortion law "Inhuman and Degrading", Supreme Court Told, The Irish News. http://www.irishnews.com/news/northernireland-news/2017/10/24/news/northern-ireland-abortion-law-inhuman-and-degrading-supreme-court-told-1170249/

Association of Southeast Asian Nations, ASEAN Human Rights Declaration and the Phnom Penh Statement On The Adoption of the ASEAN Human Rights Declaration (AHRD). http://www.asean.org/storage/images/ASEAN_RTK_2014/6_AHRD_Booklet.pdf

Australia Act 1986 (No. 142, 1985). https://www.legislation.gov.au/Details/C2004A03181/Html/Text

Australia and Australian Government Solicitor (2016) *Australia's constitution: with overview and notes by the Australian Government Solicitor*

Australian Associated Press (20 February 2015) Thailand Bans commercial surrogacy, The Guardian. https://www.theguardian.com/world/2015/feb/20/thailand-bans-commercial-surrogacy

Australian Broadcasting Corporation v Lenah Game Meats Pty Ltd [2001] HCA 63

Australian Human Rights Commission 'Right to Life'. https://www.humanrights.gov.au/right-life

Australian Human Rights Commission Act 1986

Australian Trade and Investment Commission, Biotechnology A Powerhouse for Science and Innovation. https://www.austrade.gov.au/International/Buy/Australian-industry-capabilities/Biotechnology

Bahadur G (2001) The Human Rights Act (1998) and its impact on reproductive issues. Hum Reprod 16:785

Baharuddin SAB (2008) 7 Competing domains of control: Islam and Human Rights in Malaysia. Islam and Human Rights in Practice: Perspectives Across the Ummah 8:108

Banyard K (2010) The equality illusion: the truth about men and women today. Faber and Faber, London, pp 181–182

Barak A (2015) Human dignity: the constitutional value and the constitutional right. Cambridge University Press, Cambridge

Bari AA, Shuaib FS (2009) Constitution of Malaysia: text and commentary. Pearson Prentice Hall, Upper Saddle River

Barlow K (6 July 2017) Federal government accused of undermining reproductive rights. Huffington Post. http://www.huffingtonpost.com.au/2017/07/06/federal-government-accused-of-undermining-reproductive-rights_a_23018464/

BBC News (11 September 2002) IVF Wrangle Cases Go to Court. http://news.bbc.co.uk/2/hi/health/2249442.stm

BBC News (24 June 2011) Profile: Thaksin Shinawatra. http://www.bbc.com/news/world-asia-pacific-13891650

BBC News (14 November 2012) Abortion "Would Have Saved Wife. http://www.bbc.com/news/uk-northern-ireland-20321741

BBC News (6 September 2018a) Historic India Ruling Legalises Gay Sex. https://www.bbc.com/news/world-asia-india-45429664

BBC News (27 November 2018b) Terminally ill Noel Conway loses Supreme Court Appeal. https://www.bbc.com/news/uk-england-shropshire-46359845

Beaney WM (1966) The right to privacy and American Law. Law Contemp Probl 31:253

Bennett B (12 July 2018) How Brett Kavanaugh Could Change the Supreme Court—and America. Time. http://time.com/5336621/brett-kavanaugh-supreme-court/

Bogdanor V (2009) The new British Constitution. Hart, Oxford

Bolam v Friern Hospital Management Committee [1957] 1 WLR 582

Book of Genesis of the Bible, Chapter 18 & 19, Sodom & Gomorrah

Bowcott O (7 June 2018) Northern Ireland Abortion Law Clashes with Human Rights, Judges Say. The Guardian. https://www.theguardian.com/world/2018/jun/07/supreme-court-dismisses-bid-to-overturn-northern-ireland-abortion-laws

Bowers v Hardwick [1986] Supreme Court 478 U.S. 186

Breavington v Godleman (1988) 169 CLR 41

Bull and another v Hall and another (2013) UKSC 73

Bulman M (24 October 2017) Supreme Court to Scrutinise Northern Ireland's "degrading and Humiliating" Abortion Laws. The Independent. http://www.independent.co.uk/news/uk/home-news/northern-ireland-abortion-law-uk-supreme-court-scrutiny-women-terminations-degrading-humiliating-a8016921.html

Busby M (2 July 2018) Marriage of 11-Year-Old Girl to a 41-Year-Old Man Provokes Backlash in Malaysia. The Independent. https://www.independent.co.uk/news/world/asia/child-marriage-malaysia-11-woman-man-41-wedding-underage-unicef-provokes-backlash-a8426311.html

Byrne L (3 January 2019) Concerns as Irish Abortion Services Start. BBC News. https://www.bbc.com/news/world-europe-46737593

Campbell v MGN Ltd. [2004] UKHL 22

Canal G (7 March 2017) Everything You Need to Know About the Anti-Vaxxer Movement. Global Citizen. https://www.globalcitizen.org/en/content/everything-you-need-to-know-about-the-anti-vaxxer/

Case of A, B and C v Ireland [2010] Grand Chamber 25579/05

Center for Ethics and Law in Biomedicine (19 January 2016) Anniversary Bioethics Debate on Gene Editing. https://celab.ceu.edu/events/2016-01-19/anniversary-bioethics-debate-gene-editing

Center for Reproductive Rights (Asia) (20 February 2014) Malaysia. https://www.reproductiverights.org/our-regions/asia/malaysia

Centre for Health, Law and Emerging Technologies (HeLEX) — Nuffield Department of Population Health. https://www.ndph.ox.ac.uk/research/centre-for-health-law-and-emerging-technologies-helex

Chambers P, Waitoolkiat N (2016) The resilience of Monarchised Military in Thailand. J Contemp Asia 46:425

Christine Goodwin v the United Kingdom (Application No. 28957/95)

Clark AI (1905) Studies in Australian constitutional law. CF Maxwell, Melbourne

Coco v AN Clark (Engineers) Ltd. [1969] RPC 41

Cody A, Nawaz M (9 November 2017) UN Slams Australia's Human Rights Record. The Conversation. http://theconversation.com/un-slams-australias-human-rights-record-87169

Coke E (1628) The first part of the institutes of the Lawes of England. Or, a commentarie upon Littleton, not the name of a lawyer onely, but of the law it selfe. Societe of Stationers, London

Coke E (1642) The second part of the institutes of the lawes of England. Containing the exposition of many ancient, and other statutes; Whereof you may see the particulars in a table following. Miles Fletcher and Robert Young for Ephraim Dawson

Commonwealth of Australia Constitution Act (The Constitution). https://www.legislation.gov.au/Details/C2013Q00005/Html/Text

Constitution of the Kingdom of Thailand (2007). https://www.samuiforsale.com/law-texts/thai-law-translations-constitution-of-thailand.html

Cook R, Dickens BM (2003) Human rights dynamics of abortion law reform. Hum Rights Q 25(1):2–3

Crimes Amendment (Zoe's Law) Bill 2017

Cunningham M (16 February 2017) Well Done! Australia Adds More Funds for Sexual and Reproductive Health Program. Global Citizen. https://www.globalcitizen.org/en/content/australian-government-announces-further-funding-fo/

Dahlan R, Faudzi FS (9 March 2016) The Syariah Court: Its Position Under The Malaysian Legal System. Mondaq. http://www.mondaq.com/x/472794/trials+appeals+compensation/The+Syariah+Court+Its+Position+Under+The+Malaysian+Legal+System

Daniels CR et al (2016) Informed or misinformed consent? Abortion policy in the United States. J Health Politics Policy Law 41:181

Dasgupta S (15 February 2018) A New "ASEAN Way": finding a regional solution for human rights violations in Rakhine State. The Kenan Institute for Ethics at Duke University. https://kenan.ethics.duke.edu/a-new-asean-way-finding-a-regional-solution-for-human-rights-violations-in-rakhine-state/

Davidson D (27 February 2018) Muslims Guard Bishop Heckled by Zealous Youths at Kuching Court Complex. The Malaysian Insight. https://www.themalaysianinsight.com/s/40086/

Dean R (2007) Erosion of access to abortion in the United States: lessons for Australia. Deakin Law Rev 12:123

Dearden L (18 January 2018) Noel Conway: terminally ill man wins right to challenge court ruling preventing "dignified Death". The Independent. http://www.independent.co.uk/news/uk/home-news/noel-conway-latest-terminally-ill-right-to-die-court-challenge-ruling-motor-neurone-suicide-a8165606.html

Department of Health UK, Health and Safety Executive, 'Biological Agents: Managing the Risks in Laboratories and Healthcare Premises' 82

Dickenson D (2007) Property in the body: feminist perspectives. Cambridge University Press, Cambridge

Dickson v United Kingdom, Application No. 44362/04

Disability Discrimination Act 1992

Dodson M (1993) Social justice for indigenous peoples, 3rd edn. Aboriginal Research Institute Publications, Underdale

Douglas v Hello! Ltd. [2005] EWCA Civ 595

Draft Constitution of Thailand 2016 (Unofficial English Translation). http://constitutionnet.org/sites/default/files/thailand-draft-constitution_englishtranslation_june_2016.pdf

Dyson F (1997) Can science be ethical? N Y Rev Books 44:46

Equality Act 2010 (merging the Equal Pay Act 1970, the Sex Discrimination Act 1975, the Race Relations Act 1976, the Disability Discrimination Act 1995, the Employment Equality (Religion or Belief) Regulations 2003, the Employment Equality (Sexual Orientation) Regulations 2003, the Employment Equality (Age) Regulations 2006, the Equality Act 2006 Part 2, and the Equality Act (Sexual Orientation) Regulations 2007)

Evans v United Kingdom, Application No. 6339/05

Farokhmanesh M (12 December 2016) How a Trump administration threatens women's health. The Verge. https://www.theverge.com/2016/12/12/13904032/trump-womens-reproductive-health-affordable-care-planned-parenthood

Federal Constitution of Malaysia (Federal Constitution) 1957

FPA, The Family Planning Association, A Sexual Health Charity, Abortion Law, England, Scotland and Wales. https://www.fpa.org.uk/sexual-and-reproductive-rights/abortion-rights/abortion-law

Fredman S et al (2015) The potential challenges to equality law in the UK. Oxford Human Rights Hub. https://ohrh.law.ox.ac.uk/wordpress/wp-content/uploads/2015/08/here.pdf

Gardbaum S (2008) Human rights as international constitutional rights. Eur J Int Law 19:749

Gauri V, Gloppen S (2012) Human rights-based approaches to development: concepts, evidence, and policy. Polity 44:485

Gender Recognition Act 2004

Ghaidan v Godin-Mendoza [2004] UKHL 30

Gomez J (3 July 2017) One Year Later, Duterte Remains a Human Rights Nightmare. The Diplomat. https://thediplomat.com/2017/07/one-year-later-duterte-remains-a-human-rights-nightmare/

Goold I et al (2014) The human body as property? Possession, control and commodification. J Med Ethics 40:1

Goslinga-Roy GM (2000) Body boundaries, fiction of the female self: an ethnographic perspective on power, feminism, and the reproductive technologies. Fem Stud 26:113

Gotsis T, Ismay L. Abortion Law: A National Perspective, Briefing Paper No. 2/2017. NSW Parliamentary Research Service. https://www.parliament.nsw.gov.au/researchpapers/Documents/Abortion%20Law.pdf

Graham M et al (2016) Women's reproductive choices in Australia: mapping federal and state/territory policy instruments governing choice. Gend Issues 33:335

Grant v HM Land Registry and Another [2011] EWCA Civ 769

Grant v the United Kingdom (32570/03)

Griswold v. Connecticut, 381 U.S. 479 (1965) (*Justia Law*) https://supreme.justia.com/cases/federal/us/381/479/case.html

Grugel J, Piper N (2009) Do rights promote development? Global Soc Policy Interdisciplinary J Public Policy Soc Dev 9:79

Harding A, Leyland P (2011) The constitutional system of Thailand: a contextual analysis. Hart Publishing, Oxford

Hassan S, Lopez C (2005) Human rights in Malaysia: globalization, national governance and local responses. In: Wah FLK, Ojendal J (eds) Southeast Asian responses to globalization: restructuring governance and deepening democracy. Institute of Southeast Asian Studies, Singapore

Hewison K (2014) Considerations on inequality and politics in Thailand. Democratization 21:846

Heydarian RJ (11 May 2018) A peaceful revolution in Malaysia. Al Jazeera. https://www.aljazeera.com/indepth/opinion/peaceful-revolution-malaysia-180511140532987.html

Hickling RH (1972) The legal system of Thailand. Hong Kong Law J 2:8

Hill G (2005) Resolving a true conflict between state laws: a minimalist approach. Melb Univ Law Rev 29:39

Hive Legal, Australia's Life Science Sector – Snapshot, Trends and Our Work. http://hivelegal.com.au/australias-life-science-sector-snapshot-trends-and-our-work/

Holmes O (29 October 2016) Thailand's Crackdown on "Insults" to the Monarchy Spreads Abroad. The Guardian. http://www.theguardian.com/world/2016/oct/29/thailand-bhumibol-monarchy-insults-law

Human Fertilization and Embryology Authority, Code of Practice. https://ifqlive.blob.core.windows.net/umbraco-website/2062/2017-10-02-code-of-practice-8th-edition-full-version-11th-revision-final-clean.pdf

Human Rights Commission of Malaysia (Amendment) Act 2009 (Act A1353)

Human Rights Commission of Malaysia Act 1999 (Act A597)

Human Rights Watch, World Report 2016: 'Thailand'. https://www.hrw.org/world-report/2016/country-chapters/thailand

Human Rights Watch, World Report 2018: Rights Trends in Australia. https://www.hrw.org/world-report/2018/country-chapters/australia

Human Rights Watch, World Report 2018: Rights Trends in Malaysia. https://www.hrw.org/world-report/2018/country-chapters/malaysia

Human Rights Watch, World Report 2018: Rights Trends in Thailand. https://www.hrw.org/world-report/2018/country-chapters/thailand

Humanists UK (3 February 2013) Abortion and Sexual and Reproductive Rights. https://humanism.org.uk/campaigns/public-ethical-issues/sexual-and-reproductive-rights/

Humanists UK (9 October 2017) Humanists UK to Intervene in Northern Ireland Abortion Supreme Court Case. https://humanism.org.uk/2017/10/09/supreme-court-grants-humanists-uk-permission-to-intervene-in-northern-ireland-abortion-case/

International Women's Development Agency (IWDA) (26 April 2015) Reproductive Rights, Abortion & Zoe's Law: why freedom of choice is still feminism's biggest fight. https://iwda.

org.au/reproductive-rights-abortion-zoes-law-why-freedom-of-choice-is-still-feminisms-big-gest-fight/

Jabour B (10 November 2015) UN examines Australia's forced sterilisation of women with disabilities. The Guardian. http://www.theguardian.com/australia-news/2015/nov/10/un-examines-australias-forced-sterilisation-of-women-with-disabilities

John Pfeiffer Pty Ltd v Rogerson (2000) 203 CLR 503

Jomo KS (2004) The new economic policy and interethnic relations in Malaysia. UNRISD, Geneva

Kaye J et al (2015) Dynamic consent: a patient interface for twenty-first century research networks. Eur J Hum Genet 23:141

Kelly J (8 April 2016) Why are Northern Ireland's abortion laws different? BBC News. http://www.bbc.com/news/magazine-35980195

Kerr K (2014) Queensland abortion laws: criminalising one in three women. QUT Law Rev 14:24

Kirk J (2003) Conflicts and choice of law within the Australian Constitutional Context. Federal Law Rev 31:247

Lando H (2017) Alf Ross and the functional analysis of law. SSRN Electronical J 17

Lawrence v Texas [2003] Supreme Court 539 U.S. 558

Lee HP (2004) Competing conceptions of rule of law in Malaysia. In: Asian discourses of rule of law: theories and implementation of rule of law in twelve Asian countries, France and the U.S. Routledge Taylor & Francis Group, London

Lee HP (2010) A Federal Human Rights Act and the reshaping of Australian Constitutional Law. UNSW Law J 33(1):88

Lee H-A (2016) Affirmative action regime formation in Malaysia and South Africa. J Asian Afr Stud 51:511

Levett C (13 January 2007) Toothless Tiger ASEAN hopes to replace polite silence with a roar. Sydney Morning Herald. https://www.smh.com.au/news/world/toothless-tiger-asean-hopes-to-replace-polite-silence-with-a-roar/2007/01/12/1168105177847.html

Lewis v HSBC Bank plc UKEAT/0364/06/RN and UKEAT/0412/06/RN

Lipohar v The Queen (1999) 200 CLR 485

Mahalatchimy A et al (2012) The legal landscape for advanced therapies: material and institutional implementation of European Union Rules in France and the United Kingdom. J Law Soc 39:131

Mahlmann M (2012) Human dignity and autonomy in modern constitutional orders. In: Rosenfeld M, Sajo A (eds) The Oxford handbook of comparative constitutional law. Oxford University Press, Oxford

Maier KE (1989) Pregnant women: fetal containers or people with rights? Affilia 4:8

Malm S (19 February 2018) Malaysian Rapper Probed over Lunar New Year Dog Video. Mail Online. http://www.dailymail.co.uk/news/article-5407651/Malaysian-rapper-probed-Lunar-New-Year-dog-video.html

Marshall J, Transport of biological materials. University of Edinburgh, p 51

McDonald H (23 October 2017) Supreme Court to hear challenge to Northern Ireland Abortion Law. The Guardian, http://www.theguardian.com/world/2017/oct/23/supreme-court-to-hear-challenge-to-northern-ireland-abortion-law

McDonald H, Graham-Harrison E, Baker S (26 May 2018) Ireland votes by landslide to legalise abortion. The Guardian. https://www.theguardian.com/world/2018/may/26/ireland-votes-by-landslide-to-legalise-abortion

Michaels R (2006) The functional method of comparative law. In: Reimann M, Zimmermann R (eds) The Oxford handbook of comparative law. Oxford University Press, Oxford

Minari J et al (2014) The emerging need for family-centric initiatives for obtaining consent in personal genome research. Genome Med 6:118

Montgomery v Lanarkshire Health Board [2015] UKSC 11

Morningstar News (23 February 2018) Highest Court in Malaysia to hear appeal of Christian Converts from Islam. https://morningstarnews.org/2018/02/highest-court-malaysia-hear-appeal-christian-converts-islam/

Mulvenna B (2004) Pre-implantation genetic diagnosis, tissue typing and beyond: the legal implications of the Hashmi Case. Med Law Int 6:163

Muntarbhorn V (2005) Rule of law and aspects of human rights in Thailand- from conceptualization to implementation? In: Peerenboom R (ed) Asian discourses of rule of law. Routledge, Abingdon

Murphy T (2013) Health and human rights. Hart Publishing, Oxford

Nazlina M (14 January 2016) Prosecution wraps up case in Azmi Sharom Sedition Trial. The Star. https://www.thestar.com.my/news/nation/2016/01/14/azmi-sharom-trial-prosecution/

Newland K (July 2015, Issue No. 13) Irregular Maritime migration in the Bay of Bengal: the challenges of protection, management and cooperation. International Organization for Migration, Migration Policy Institute. http://publications.iom.int/system/files/pdf/mpi-iom_brief_no_13.pdf

Nianias H (2 November 2015) It's 2015 and British women's right to safe abortions is still uncertain. The Debrief. https://thedebrief.co.uk/news/real-life/2015-british-womens-right-safe-abortions-still-uncertain/

Nolan MA (2012) The constitutional system of Thailand: a contextual analysis. Aust J Asian Law 13:1

Nozick R (1974) Anarchy, State, and Utopia. Basic Books, New York

Nyamu-Musembi C, Cornwell A (2004) What is the 'right-based approach' all about?: Perspectives from international development agencies. Institute of Development Studies, Brighton

O'Cinneide C (14 September 2011) Equality: a constitutional principle? (UK Constitutional Law Association) https://ukconstitutionallaw.org/2011/09/14/colm-ocinneide-equality-a-constitutional-principle/

O'Cinneide C (2012) The human rights act and the slow transformation of the UK's political constitution. Annales U. Sci. Budapestinensis Rolando Eotvos Nominatae 53:239

O'Connor A (28 October 2017) How the death of Savita Halappanavar changed the abortion debate. The Irish Examiner. http://www.irishexaminer.com/analysis/how-the-death-of-savita-halappanavar-changed-the-abortion-debate-461787.html

Obergefell v Hodges [2015] Supreme Court 14–556

Panaspornprasit C (2017) Thailand: the historical and indefinite transitions. Southeast Asian Aff:351

Parry v the United Kingdom (Application No. 42971/05)

Parsa-Parsi RW (2017) The revised declaration of Geneva: a modern-day physician's pledge. J Am Med Assoc 318:1971

Penal Code 1936 (Act 574) 314

Pennings G (2004) Saviour siblings: using preimplantation genetic diagnosis for tissue typing. Int Congr Ser 1266:311

Peters PG (2004) How safe is safe enough? Obligations to the children of reproductive technology. Oxford University Press, Oxford

Petersen K (2014) Decriminalizing abortion- the Australian experience. In: Rowlands S (ed) Abortion care. Cambridge University Press, Cambridge

Pitkin H (1981) Justice: on relating public and private. Polit Theory 9:327–352

Planned Parenthood of Southeastern Pa. v. Casey, 505 U.S. 833 (1992) (*Justia Law*) https://supreme.justia.com/cases/federal/us/505/833/

Pretty v the United Kingdom [2002] European Court of Human Rights 2346/02

Public Prosecutor v Azmi Bin Sharom (Criminal Ref. No. 06-5-12/2014(W)) (6 October 2015). http://www.kehakiman.gov.my/judgment/file/PRESS_SUMMARY_PP_V_AZMI_SHAROM.pdf

R (on the application of Quintavalle) v Human Fertilisation and Embryology Authority [2005] UKHL 28

R and F v the United Kingdom (Application No. 35748/05)

R v Brennan and Leach [2010] QDC 329

R v Sood [2006] NSWSC 1141

Racial Discrimination Act 1975

Rath J (2018) Safety and security risks of CRISPR/Cas9. In: Schroeder D et al (eds) Ethics dumping. Springer International Publishing, New York

Reiss MJ, Straughan R (1996) Improving nature? The science and ethics of genetic engineering. Cambridge University Press, Cambridge

Richardson M (2002) Whither breach of confidence: a right of privacy for Australia. Melb Univ Law Rev 26:381

Roe v. Wade, 410 U.S. 113 (1973) (*Justia Law*) https://supreme.justia.com/cases/federal/us/410/113/

Romanow L (2012) The women of Thailand. Glob Majority E-J 3:44

Rothstein A (Winter 2015, Number 44) Vaccines and their critics, then and now. The New Atlantis. https://www.thenewatlantis.com/publications/vaccines-and-their-critics-then-and-now

Royal College of Nursing of the United Kingdom v Department of Health and Social Security [1981] AC 800

Saliba I (June 2011) What Is Sharia Law? | Law Library of Congress. https://www.loc.gov/law/help/sharia-law.php

Saltman Engineering Co. Ltd. V Campbell Engineering Co. [1984] 65 RPC 203

Sándor J (2015) The ethical and legal analysis of embryo preimplantation testing policies in Europe. In: Sills ES (ed) Screening the single euploid embryo. Springer International Publishing, New York

Sex Discrimination Act 1984

Shear MD (28 June 2018) Supreme Court Justice Anthony Kennedy Will Retire. The New York Times. https://www.nytimes.com/2018/06/27/us/politics/anthony-kennedy-retire-supreme-court.html

Sheffield and Horsham v the United Kingdom (Application No: 31-32/1997/815-816/1018-1019) and

Sheldon S (2004) Hashmi and Whitaker: an unjustifiable and misguided distinction? Med Law Rev 12:137

Shuaib FS (2012) The Islamic legal system in Malaysia. Pacific Rim Law Policy J 21:85

Sifris R, Belton S (2 June 2017) Australia: abortion and human rights. Health and Human Rights Journal. https://www.hhrjournal.org/2017/06/australia-abortion-and-human-rights/

Simson RS (2014) What does the right to life really entail-a framework for depolarizing the abortion debate. Connecticut Public Int Law J 14:107

Skinner v. Oklahoma Ex Rel. Williamson, 316 U.S. 535 (1942) (*Justia Law*) https://supreme.justia.com/cases/federal/us/316/535/case.html

Spriggs M, Savulescu J (2002) Saviour siblings. J Med Ethics 28:289

Stanton-Ife J (2000) Should equality be a constitutional principle? King's Law J 11:133

Stasi A (2015) Maternal surrogacy and reproductive tourism in Thailand: a call for legal enforcement. Ubon Ratchathani Law 8:17–36

Stone GR et al (2013) Constitutional law, 7th edn. Wolters & Kluwer, Alphen aan den Rijn

Taywaditep KJ, Coleman E, Dumronggitigule P. The International Encyclopedia of Sexuality: Thailand. http://www.sexarchive.info/IES/thailand.html

Teare HJA et al (2017) The RUDY study: using digital technologies to enable a research partnership. Eur J Hum Genet 25:816. http://www.nature.com/doifinder/10.1038/ejhg.2017.57

Thailand Penal Code, Thailand Criminal Law (Unofficial English Translation) https://www.samui-forsale.com/law-texts/thailand-penal-code.html

Thananithichot S, Satidporn W (2016) Political dynasties in Thailand: the recent picture after the 2011 general election. Asian Stud Rev 40:340

The Economist (22 September 2016) Taking the rap - religious freedom in Malaysia. https://www.economist.com/news/asia/21707565-malaysias-culture-tolerance-under-threat-taking-rap

The Straits Times (18 May 2018) "Cash Is King": The Fall of Malaysia's First Couple. https://www.straitstimes.com/asia/se-asia/cash-is-king-the-fall-of-malaysias-first-couple

Thepgumpanat P, Tanakasempipat P (21 May 2017) Three years after Coup, Junta is deeply embedded in Thai Life. Reuters. https://www.reuters.com/article/us-thailand-military/three-years-after-coup-junta-is-deeply-embedded-in-thai-life-idUSKCN18G0ZJ

Thomson M (2003) Confidence, privacy and damages: Hello! To Clarity. New Law Journal

Torii T (1997) The new economic policy and the United Malays National Organization- with special reference to the restructuring of Malaysian Society. Dev Econ XXXV(3):209–239

Tsaknis L (1993) The jurisdictional basis, elements and remedies in the action for breach of confidence- uncertainty abounds. Bond Law Rev 5:18

Umeda S (6 April 2015) Thailand: New Surrogacy Law | Global Legal Monitor. http://www.loc.gov/law/foreign-news/article/thailand-new-surrogacy-law/

UNESCO (2006) The universal declaration on bioethics and human rights. http://unesdoc.unesco.org/images/0014/001461/146180E.pdf

United Nations (1979) Convention on the elimination of all forms of discrimination against women. http://www.un.org/womenwatch/daw/cedaw/cedaw.htm

United Nations, Office of the High Commissioner, Human Rights, Sexual and Reproductive Health and Rights. http://www.ohchr.org/EN/Issues/Women/WRGS/Pages/HealthRights.aspx

VAWnet, National Resource Centre on Domestic Violence, Reproductive justice, reproductive health and reproductive rights: a framework. https://vawnet.org/sc/reproductive-justice-frameworks

Vo v France (Application No. 53924/00)

Watson M (29 December 2013) Battle of life and death never ends for anti-abortion campaigners. The Age. http://www.theage.com.au/victoria/battle-of-life-and-death-never-ends-for-antiabortion-campaigners-20131228-300tv.html

Watts J (15 October 2017) UK Government Watchdog Pushes for New British "right to Equality" to Stop Brexit Leading to More Discrimination. The Independent. http://www.independent.co.uk/news/uk/politics/brexit-discrimination-laws-right-to-equality-uk-equalities-watchdog-eu-a7999461.html

West R (1988) Jurisprudence and gender. Univ Chic Law Rev 55:1

Whitman JQ (2003) The two western cultures of privacy: dignity versus liberty. Yale Law J 113:1151. 73

Whittaker A (2002a) The struggle for abortion law reform in Thailand. Reprod Health Matters 10:45

Whittaker A (2002b) "The Truth of Our Day by Day Lives": abortion decision making in rural Thailand. Cult Health Sex 4:1

Whittaker A, Speier A (2010) "Cycling Overseas": care, commodification, and stratification in cross-border reproductive travel. Med Anthropol 29:363

Wiesenthal DL, Wiener NI (1999) Ethical questions in the age of the new eugenics. Sci Eng Ethics 5:383

Women's Agenda (16 August 2017) Reproductive freedom is unfinished business in Australia: Tanya Plibersek. https://womensagenda.com.au/politics/reproductive-freedom-unfinished-business-australia-tanya-plibersek/

Women's Electoral Lobby, Reproductive Rights (*Women's Electoral Lobby*) http://www.wel.org.au/reproductive_rights

Workplace Gender Equality Act 2012

World Bank Global Knowledge and Research Hub (December 2017) Malaysia economic monitor: turmoil to transformation 20 years after the Asian Financial Crisis. https://openknowledge.worldbank.org/bitstream/handle/10986/28990/MalaysiaEconomicMonitor2017.pdf?sequence=6&isAllowed=y

Wright T (12 September 2016) 1MDB Scandal Around Malaysian Prime Minister Najib Puts Spotlight on Wife. Wall Street Journal. http://www.wsj.com/articles/1mdb-scandal-around-malaysian-prime-minister-najib-puts-spotlight-on-wife-1473606895

X, Y and Z v The United Kingdom (1997) Grand Chamber, European Court of Human Rights Application No. 21830/93, HUDOC

Yatim R (1995) Freedom under executive power in Malaysia: a study of executive supremacy. Endowment Publications, Washington, D.C.

Zubedy A (21 June 2012) NEP: the good and the bad. Malaysia Today. https://www.malaysia-today.net/2012/06/21/the-nep-the-good-and-the-bad/

Zurairi AR (4 July 2018) Kelantan Deputy MB: child marriage "Common", LGBT a Bigger Issue. Malay Mail. https://www.malaymail.com/s/1648815/kelantan-deputy-mb-child-marriage-common-lgbt-a-bigger-issue

Chapter 7
Conclusion

7.1 Lessons: Unlearned and Learned

I believe that there are lessons that can be learnt from the points made in my work. In doing so, however, there is a need to first 'unlearn' some misconceptions and judgments. I argue that the regulation of prospective pre-implantation genetic interventions is a near-distant one that will prove to be extremely challenging and divisive. Such is the length and breadth of ethical, legal and social implications (ELSIs) provoked by the possibilities of future developments, that it is hard to see the forest from the trees. Since biotechnologies have made their mark on humanity generally, these challenges will continue to mount. It is, however, possible to temper these fires by unlearning and learning notions of what we may be used to. Ultimately, the outcomes that are hoped for include the following: *firstly*, a reinterpretation of the meaning of eugenics in contemporary settings; *secondly*, strong governance, stable legal and/or regulatory frameworks and concise international outlooks in formulating future governance of pre-implantation genetic interventions; and *thirdly*, the achievement of a contemporary 'universal' legal application for pre-implantation genetic interventions, that may be applicable and transmutable in different constitutional settings, leading at multi-dimensional points that factor in all aspects of human development.

7.1.1 Lessons Unlearned

We should begin with the lessons that we should unlearn, to present a clean slate in which to absorb lessons to be learned. In this instance, I present three perceptions that have been misconceived and/or borne negative judgments from past histories. These should be altered in order to allow a reconfiguration of reflective legislative efforts. The first perception is in connection with the term 'eugenics'. Historically,

© Springer Nature Switzerland AG 2019
P. L. Lau, *Comparative Legal Frameworks for Pre-Implantation Embryonic Genetic Interventions*, https://doi.org/10.1007/978-3-030-22308-3_7

eugenics earned its pejorative nature through one of the largest scale genocides in human history, fueled only by a madman's desire to populate the world with a homogeneity regarded as superior. In contemporary settings, therefore, it is easy to cry out "eugenics" at the faintest glimpse of characteristic or trait selection, which suggests a desire for genetic or social improvements. However, I also argue that we do, in fact, make 'eugenic' choices on a daily basis, which we do not condemn. The question is whether this is because we have rationalized these choices to be integral to part of the daily living of human experience; or whether we have simply attributed to 'eugenics' a normatively accepted negative perception that is characterized by genocide.

My research suggests the necessity of rationalizing the purpose for which certain selection practices are sought, instead of simply fixating on outcomes or effects in isolation. Translating this into the sphere of PGD, or pre-implantation genetic interventions, means that distinctions must be clearly drawn between medical (therapeutic) and non-medical (non-therapeutic) treatments, before a particular practice is characterized within the scope of eugenics. Judit Sándor states most succinctly that we assume the infallibility of medical criteria in determining what is eugenics, and what is not.[1] This must no longer be the case, and contemporary debates on eugenics need to employ a more rationalized narrative on the multi-layered interpretation and meaning of eugenics. This could be achieved through a deeper engagement with a variety of stakeholders, members of the public, philosophers and scientists, and no longer confine them to the medical profession.

The second perception that should be addressed is the narrative of fear and disconcert about biomedical technologies generally, and genetic interventions specifically. The presentation of the future of humankind through a dystopian lens, whether through the works of science fiction, or through scholarly acclamations, has done little to assuage the culture of fear that has descended on our humanity. Although largely characterized as a fear of the unknown, I surmise instead that this fear of the unknown is further fueled by a fear of misconceptions that we think we know or project. Nicholas Carleton describes the fear of the unknown as follows: "an individual's propensity to experience fear caused by the perceived absence of information at any level of consciousness or point of processing."[2] Despite what we believe current or future technologies may be capable of, it is precisely this uncertainty, the notion of a glimpse of possibilities, that attracts the largess of fear. Instead, I put forward the claim that as a global humanity, we cannot now dissuade the waves of technological advancements that come upon us. The mistaken fear of technological advancements is that it may erode how we understand ourselves as human beings, to lose what it means to be human. But I agree with Elizabeth Falck, who argues:

> People fear technology because they fear that it threatens what it means to be human, but they've forgotten something: technology is human. Technology is the human extended phenotype; it is any tool that enables people to extend themselves by expressing and/or synthe-

---

[1] Sándor (2015), p. 355.
[2] Carleton (2016), p. 5.

sizing a meme.[3] What people truly feat in the wake of technological progress are not tools but the exponentially growing offensive of competing memes those tools propagate.[4]

Hence, my response to the narrative of fear of genetic interventions is simple. Fear may simply be biological in nature, chemical responses to specific external or internal stimuli, or fear may be projected as part of our behavioral conditionings.[5] But the narrative of fear is useless in the face of changing global paradigms. We should not fear technologies and what they are capable of, because technologies are a creative output of the human mind. We should, instead, fear what we, as humanity, will do with technologies.[6]

Finally, the third perception that may be inscribed is the conception of universal values and universality in human rights. I assert that the meaning of universality ascribed to human rights from the conception of the Universal Declaration of Human Rights (UDHR) has shifted dramatically over the last twenty years. In the UDHR, the universal element of human rights was predicated on its applicability to "all human beings" and "everyone".[7] As an international human rights document that has repeatedly asserted its ideals, it has consistently striven to meet the ideals of global human rights recognition for everyone. But the UDHR was drafted seventy years ago, and notwithstanding its central repository as a global human rights standard, the realities of contemporary societies reflect otherwise. Hence, I have posed that the conception of universality in the UDHR, and other international human rights instruments be revisited, by employing alternative mechanisms of reinterpretation and refocusing on newly negotiated shared values.

7.1.2   Lessons Learned

With the unlearning or reinterpretation of the concepts mentioned above, the navigational journey in this book reveals connections between law, science and technology and human rights, all important narratives as part of our global development as world citizens. Despite the claims made by Dr. He Jian Kui (which, as yet, have to be fully verified), pre-implantation genetic interventions, therefore, may not have feasibly and successfully occurred. Practical experimentation of gene editing on

---

[3] The idea of the meme was first introduced by Richard Dawkins in his book *The Selfish Gene*. In essence, the meme is a replicator of information, as crucial to our being human as is our DNA. More particularly, as Elizabeth Falck states in her work 'Technology and the Memetic Self', memes are "ideas that can be expressed and replicated. Memes are instruction, behaviors, inventions, cultural traditions and stories." Please see (n 4).

[4] Falck (2014), p. 232.

[5] Adolphs (2013), p. R79.

[6] Greenfield (2012).

[7] The Eleanor Roosevelt Papers Project, 'The Universal Declaration of Human Rights' https://www2.gwu.edu/~erpapers/humanrights/lessonplans/.

non-viable human embryos have given us a glimpse into possibilities,[8] but science in itself, continues to evolve on a daily basis, and some scientists remind us that we do not sufficiently understand the implications of gene editing on human beings thus far.[9] However, I reiterate that our aim is to be prepared to address the important conversations and challenges in the road ahead for pre-implantation genetic interventions.

I advance three major points to summarize the fundamental premises of this work. *Firstly*, to reinterpret the meaning of eugenics in contemporary settings, and to make distinctions that would differentiate practices that may appear to have eugenic outcomes, but are based on rationalized purposes for avoidance of harm and disease. This would have a marked implication on the predominant debates that surround embryo selection within PGD. In addition, the separation of eugenics from its strong, defining, negative past, allows for a more informed realization of well-reasoned individual autonomy. This is consistent with the fundamental tenets of human rights discourse.

Secondly, to focus on the need for strong governance, stable legal and/or regulatory frameworks and concise international outlooks in formulating how pre-implantation genetic interventions may be governed in the future. To achieve this with maximum dexterity, my findings in the triumvirate of Chaps. 3, 4 and 5 are inter-connected in this endeavor, and lead into each other, to lend support to suggestions for effective global governance of biomedical technologies. My findings impart political, social and cultural dimensions that have led to why regulation for PGD has been shaped in particularized ways in the selected jurisdictions. I further claim that laws alone, although highly desirable, may not be adequate to deal with the rapid faces of changing technological developments. In the field of biomedical technologies, under which genetic interventions are included, legal responses cannot be the responsibility of the legislature alone. I propose parallel regulatory approaches that may be used for biomedical technologies. These regulatory approaches, together with the existence of laws, will not only involve multiple stakeholder interests, which is a desirable facet in regulation, but also can be effective as temporal plugs to inform and influence the scope of future legislation. It is also possible for these approaches to counter the challenges in regulating biomedical technologies. I also reiterate the importance of values that should be guaranteed in a constitutional space, and their necessity in considering a legal or regulatory framework. This may be extracted from a variety of international instruments that deal with international biomedical laws by determining its main principles; its shortcomings; and how these principles and shortcomings may be overcome in considering the formulating of future regulation for genetic interventions.

Thirdly, to strive towards the achievement of a contemporary 'universal' legal application for pre-implantation genetic interventions, that may be applicable and transmutable in different constitutional settings. Whilst examining the individual constitutional systems of the selected jurisdictions, I introduce the novel aspects of

---

[8] Cyranoski and Reardon (2015).

[9] Belluck (2017).

what I term "entry points of regulation". I use these entry points as an illustrative mechanism to demonstrate the concerns of the selected jurisdictions that would prompt them to regulate pre-implantation genetic interventions. How these concerns (or entry points) would come about, however, is largely dependent on the kind of role that is played by human rights in the different constitutional systems. A glimpse into each selected jurisdiction's constitutional framework, how they each deal with issues such as abortions and reproductive autonomy, and non-discrimination literature and discourse, are all indicators that similar standards could be applicable to the regulation of pre-implantation genetic interventions.

The findings of my research presents the reader with a picture on how a new form of universality of newly negotiated shared values may be achieved. Essentially, the outcome that is sought is to converge constitutional values and the protection of fundamental rights, with international outlooks of negotiated shared values. The reinterpretation of universal legal standards for protective mechanisms, because they will be renegotiated on a large scale level (taking into account the peculiarities of constitutional frameworks) is likely to be a more useful model of applicable legal standards that is modern, fresh and in alignment with contemporary advancements of medical and scientific technologies.

It is also necessary to emphasize two specific points here in connection with universality. First, I do not accept that human rights are not universal. Rather, I am of the firm belief that human rights are so fundamental that they <u>must</u> be universal; because of this, therefore this reevaluation is timely and highly necessary. As a global community of citizens, not only has our mindsets and environments evolved, but also because the points of this evolution have culminated from our creative desires. We no longer have the limitations of the past with current science and technology, but we must continue to retain the fundamental attributes of human rights. I claim that just as science and technology has evolved, so too should the international human rights regime, and some of its earlier formulated ideals. And this can be achieved, as I have presented, through a large scale international renegotiation and revisiting of a common ground for newly shared values. Secondly, I do not accept the argument of "Asian values"[10] in contemporary democratic Asian societies as a valid vindication to detract from <u>all</u> human rights values. In light of issues of severe human rights violations, accountability and transparency that continue to plague many Asian societies, the "Asian values" argument is no longer tenable. In fact, historical trajectories have feasibly demonstrated that Asian values may sometimes have been invoked to justify a vilification against "Western" ideas and propaganda, in order to continue to maintain political regimes in a certain manner.[11] What remains tenable, however, is understanding how these "Asian values" may have come about through a historical and cultural process of societal development. This enables the values to find a place in the normative framework of newly negotiated human rights ideals, distinct from the notion of cultural relativism.[12]

---

[10] Bello (1998).

[11] Sen (1997), p. 33.

[12] Donnelly (1984), p. 400.

In the meantime, Fig. 7.1 illustrates the bird's eye view of this book's roadmap, and how each chapter is connected to each other, and the basic points of summary for the main concerns that have been raised.

7.2   Other Regulatory Implications

7.2.1   Implications on Relevant Stakeholders

The considered ramifications on stakeholders insofar as regulatory frameworks are concerned must also be mentioned. Stakeholder analysis[13] and impact in a complete manner is beyond the purview of this book, although it has been mentioned throughout. It is, however, possible to briefly outline the implications of the contours of a regulatory model on a particular group of stakeholders. R. Edward Freeman first came up with the idea of the stakeholder theory as part of strategic management for corporations.[14] Briefly, he defines a stakeholder as "any group or individual who can affect or is affected by the achievement of the organization's objectives."[15] Over time, this stakeholder approach has evolved and found resonance not only within corporations and businesses, but also in many other aspects of social endeavours, including charities. I also claim that having an integral stakeholder approach is equally applicable to pre-implantation genetic interventions, because of the manner in which the latter can affect or is affected by groups or individuals in society.

Therefore, when we consider framing a regulatory system for pre-implantation genetic interventions, identifying who the potential stakeholders are, and how to effectively engage them in the regulatory process, is not only a meaningful enterprise that will have altruistic social outcomes, but to also demonstrate transparency and accountability. Within the field of medical and scientific research alone, the emphasis on stakeholder engagement cannot be stressed enough. Regulatory theories[16] that differ in the jurisdictions[17] remind us that motivations and incentives that may be peculiar in one domain, operate in a manner that is indigenous to those localized environments in the contextual framework of particular industries and the impact of globalization on targeted regulation in these environments. The role that human rights plays in connection to these stakeholders may reveal telling regulatory theories on the shaping of laws or governance; whether initiated from the perspectives of scientists or legislators, or whether they act as a limitation to the operability of private law. In any event, the reactionary effects by these stakeholders, as well as projected consequences impacting these stakeholders as a result of the exercise of

---

[13] Schmeer (1999), p. 1.

[14] Freeman (1984).

[15] Ibid 275.

[16] Sunstein (1993).

[17] Drahos (2017).

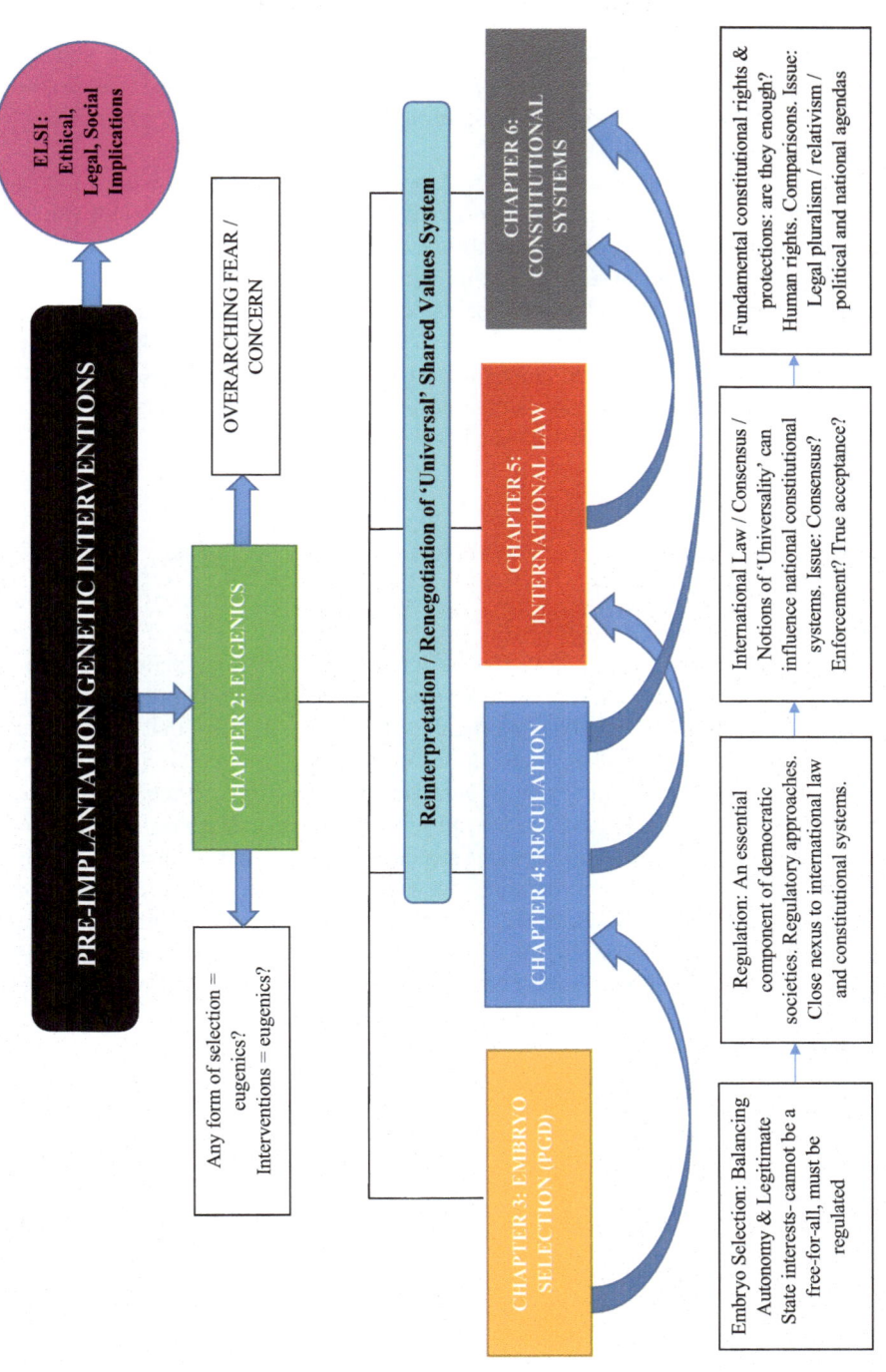

Fig. 7.1 The roadmap

technologies (especially where the concept of harm and safety is concerned) also encumbers the procedural aspects that these scientific and medical developments must markedly consider. The reaction and lapse of time through science and the remedies provided by technologies can similarly be contextualized within the framework of human rights components, through an examination of legal comparative and regulatory theory.

In the conclusions to Chaps. 5 and 6 respectively, I emphasize the need for an inclusionary and more internationalized newly negotiated system of shared values that are capable of transcending international boundaries. This, of course, is predicated upon the necessity to regulate pre-implantation genetic interventions. Besides the medical professionals and research scientists that may be involved in the clinical practice and determination, a range of other stakeholder interests are also desirable to project and realize better outcomes for the use of technologies. These stakeholders may comprise, narrowly, legislators and policy makers, researchers and scientists, and the general public viewed as a body individual.

7.2.2 Implications on Reproduction and Reproductive/ Maternal Health

Another facet worth contemplating for the future deals with the gendered dimensions in reproduction that may, in some aspects, be seen as liberating women, but in other aspects, simultaneously characteristic of the patriarchal tone that is imparted in most reproductive discourse. Implications on reproduction can be, indeed, very real, because pre-implantation genetic interventions involve a variety of clinical invasions into the woman's body in pursuit of pregnancy and birthing. The aspiration in advancing appropriate governance structures in this field is to provoke a critical paradigm shift in how reproduction is viewed, especially the patriarchal view of women's bodies in reproduction, that impact on socio-legal inequalities within a community of citizens. How reproductive health, practices, and rights legislation is framed and drafted (and by whom), over the governance of female bodies within the social and political sphere, should be closely examined.

International organizations devoted to women's rights, such as the Centre for Reproductive Rights,[18] the Global Fund for Women[19] and the United Nations Population Fund,[20] amongst many others, champion effortlessly for the continued realization of women's reproductive autonomy over the last few decades, which include access to reproductive and maternal healthcare. The World Health Organization (WHO) information as at 2015[21] indicates that maternal mortality is

---

[18] Center for Reproductive Rights (2014).

[19] Global Fund for Women, 'Global Fund for Women' (*Global Fund for Women*) https://www.globalfundforwomen.org/.

[20] United Nations Population Fund (2014).

[21] World Health Organization, 'WHO|Maternal and Reproductive Health' (*WHO*) http://www.who.

still one of the main concerns in reproductive and maternal healthcare resulting from pregnancy complications in low-resource settings. The data indicates that with the advancement of medical technologies in our day and age, most of these fatalities could have easily been prevented but for the appropriate access to reproductive and healthcare in these countries.[22] Problems of infertility (which could possibly be treated *vis-à-vis* assisted reproductive technologies) are also prevalent in low-resource countries, particularly secondary infertility in parts of Africa and Latin America, and South Asia.[23] In semi-conservative Islamic states like Malaysia, access to contraceptives is more restricted, particularly for Muslim women, although the access to reproductive and maternal healthcare appear to be more positive than in its South American or African counterparts. But access is not simply the only issue here; the information provided, costs, and availability of a range of reproductive health tools and technologies, especially in villages or rural areas, is significantly on the low end of the range. These shortcomings have resulted in high inequality discrepancies between rich and poor, the urban and rural, and certainly even within the same national context. Part of the WHO's efforts in its Global Strategy for Women's, Children's and Adolescents' Health (2016–2030)[24] (also known as the Every Woman Every Child (EWEC) Global Strategy) is an attempt to collate data and information regarding sexual and reproductive health and rights in line with the ICPD. This strategy is still presently ongoing.

Questions of appropriate access and standards of reproductive and maternal healthcare would also be prevalent from the reproductive justice perspective if we examine women who have been incarcerated in the penitentiary systems. In Brazil, for example, studies have indicated that women who have been incarcerated "have called attention to diverse problems tied to gender inequality and to the need to reduce the different forms of violence that multiply in prisons and lead to serious health depredations for this population."[25] In circumstances of incarceration, the conceptual framework already begins with a deprivation of individual liberty; and although it appears to be logical that despite such deprivation of liberty, the continuation of other individual rights would subsist, the realities of violence and violation of women's human rights within the prison system would indicate otherwise.[26]

The consideration of the role and position of women in the legislature, judiciary, political sphere and other interest or industry groups is another demonstrable illustration why dimensions of gender within the context of justice is a prevailing theme. It is not only in the discourse on gender equality; but also from the social perspective, because the continued signification of women's (under)representation in these

---

int/gho/maternal_health/en/.

[22] World Health Organization, 'WHO|World Health Organization' http://gamapserver.who.int/gho/interactive_charts/mdg5_mm/atlas.html.

[23] Inhorn (2009), p. 172.

[24] World Health Organization, 'GHO|Global Strategy for Women's, Children's and Adolescents' Health (2016–2030)' (*WHO*) http://apps.who.int/gho/data/node.gswcah.

[25] Diuana et al. (2016), p. 2041.

[26] Roth (2017).

areas has the unfortunate effect of continuing a narrative that women are 'not enough'. Although I do not proclaim that an inclusion of women into the sphere of legislative considerations will completely eradicate the disparities in gender considerations; what may be observable instead is an offering of a more immersive experience of the manner in which legislations are made. With women as the main beneficiaries under such legislation, at the very least, factors that pertain to reproduction should also be considered by women who have the capacity to influence the experiential dimensions of such legislation.

Leading modern democracies in the United States and the United Kingdom also demonstrate a less than stellar record of gender equality on the bench. The UK Supreme Court's first female President was Baroness Hale, appointed in 2017[27] and she has highlighted the issues of gender imbalance in the courts from as early as the mid-1990s period. In the United States, the current members of the Supreme Court comprise 3 women: Justice Ruth Bader Ginsburg, Justice Sonia Sotomayor, and Justice Elena Kagan.[28] In interesting data compiled by the American Constitution Society for Law and Policy, what is now being referred to as "the Gavel Gap"[29] has been revealed in respect of race and gender in state courts in the United States, demonstrating that courts are not representative of the citizens they represent; gender being clearly one of these factors.[30] Within the framework of the nomination of judges by member states to the European Court of Human Rights, a study reveals that "while there is a strong proportion of candidates that support the notion that states do not differentiate according to gender… There is a comparable proposition that contrarily indicates that the world of international judicial appointments is far from gender neutral."[31]

In other spheres of public and political office, for example, relating to the composition of the legislature, studies have revealed similarities that indicate a constant theme. If these examples of the representation of women in various sectors of public and social life are any indication, there is no reason to believe that legislation that seeks to govern reproductive autonomy will also be gender neutral. Far from being a subject matter that intends to dismantle the patriarchy,[32] the underlying aims would instead enhance focus on reproductive autonomy by painstakingly unearthing its current problems at the roots.[33]

---

[27] BBC (2017).

[28] Supreme Court of the United States 'Current Members' https://www.supremecourt.gov/about/biographies.aspx.

[29] American Constitution Society for Law and Policy, 'The Gavel Gap' (*The Gavel Gap*) http://gavelgap.org.

[30] George and Yoon (2014) http://gavelgap.org/pdf/gavel-gap-report.pdf.

[31] Hennette-Vauchez (2015), p. 195.

[32] Haraway (2000).

[33] Dickenson (2007).

7.2.3   Implications on Scientific/Medical/Healthcare Research and Development

One of the general conundrums of healthcare provision, and by that extension, medical and scientific innovations that have the aim of improving health, is its necessary relationship with the law. Crossing into territories of legislative capacity, public policy, and in some cases, political lobbying,[34] means that the provision of healthcare (impacting on people's lives) will always be subject to a framework of governance. The same should be true of pre-implantation genetic interventions. In recent times, global scientific communities have been faced with the challenging task of how to govern CRISPR,[35] whilst simultaneously engaging and encouraging further research and development (R&D) of the technology.[36] There is arguably reasonable trepidation that a broad and overly restrictive approach to gene editing, (and all extensions of its applications) could stifle further scientific and medical innovation and R&D.

There are three main areas of concern in terms of regulatory implications on further innovation and R&D. First, relating to a policy framework for health-related research; secondly, in connection with clinical trials that involve human subjects; and thirdly, the access to, and promotion of medical and scientific innovations. The United Kingdom (UK) has issued an extensive policy framework for health and social care research.[37] Some of the key matters addressed in this policy framework include patient-centric approaches such as informed consent, safety and efficiency of treatments, ethically and scientifically sound research processes and outcomes, and the registration of research data and materials with adequate privacy safeguards.[38] The Nuffield Council on Bioethics also issued an ethics-based report on genome editing[39]; and a further report on genome editing and human reproduction.[40] In the United States, the President's Council on Bioethics, and other organizations like the Committee on Human Gene Editing, and the National Academies of Sciences, Engineering, and Medicine, also presented their findings through the publication of a book that considered the scientific, medical and ethical considerations of human gene editing.[41] All these, and more, demonstrate the awareness that policy frameworks are necessary to 'contain' the unlawful proliferation of these technolo-

---

[34] RT Question More (2017).

[35] Committee on Science, Technology, and Law, Policy and Global Affairs and National Academies of Sciences, Engineering, and Medicine (2016).

[36] Harris (2015).

[37] Department of Health (England) and others, 'UK Policy Framework for Health and Social Care Research'.

[38] Ibid 4.

[39] Nuffield Council on Bioethics (2016).

[40] Nuffield Council on Bioethics (2018).

[41] Committee on Human Gene Editing: Scientific, Medical, and Ethical Considerations and others (2017).

gies, ideally within the purview of the law, or other appropriate forms of regulatory mechanisms.

Secondly, in terms of clinical trials that involve the use of human subjects, urgent consideration must be given to specific regions where financial, or other incentivization methods are peddled by unscrupulous research companies, in exchange for the seemingly voluntary participation of human subjects. Mark Barnes and Nick Wallace, in their study, outline comprehensive legal and ethical guidelines when dealing with clinical trials in developing countries.[42] They correctly identify that in many developing countries, the laws, or lack thereof, that relate to proper guidelines in the use of human subjects in clinical trials, "may offer reduced protections to human subjects, and local regulatory bodies may lack the capacity to monitor clinical trials adequately."[43] This is one of the factors that explains the reproductive or fertility tourism phenomena that takes place in Malaysia and Thailand. Although it may be arguable that the subject matter of clinical trials in pre-implantation genome editing would be the human embryo, and thus, exclude the applicability of guidelines or laws that govern clinical trials involving human subjects, this does not make the circumstances less urgent.

Thirdly, and finally, the access to, and promotion of medical and scientific innovations are often subject to a balancing exercise, contrasted against the protection of human rights, distributive justice, and the gains of commercialization through intellectual property protections. In a 2012 report issued by the World Intellectual Property Organization (WIPO),[44] the latter specifically recognized the challenges that ensue from an intersection between medical innovations, trade and intellectual property protection. Indeed, the greatest challenge is "to establish an environment that stimulates health innovation while ensuring widespread access to new, more effective products to address unmet global health needs."[45] Part of this endeavor makes it incumbent for legislators and policy makers to focus on a two-pronged outcome: the first being access to innovations—and in the case of pre-implantation genetic interventions, where such access is extended to segments of the population that would otherwise be unable to afford such treatments. The second prong would be the allowance of profits or gains from medical innovations that would continuously encourage avant-garde approaches to eradicating genetic diseases, whilst ensuring that commercial exploitation is reasonable and well-regulated.

[42] Barnes and Wallace (2017), p. 25.

[43] Ibid 248.

[44] World Trade Organization, World Health Organization and World Intellectual Property Organization(2013)https://www.wto-ilibrary.org/intellectual-property/promoting-access-to-medical-technologies-and-innovation_63a4aa65-en.

[45] Krattiger (2013).

7.3 Rebooting the Future

In making the final remarks to this book, I am inspired by Sheila Jasanoff's use of the phrase "reboot the future", because this is what my research calls for: to reboot our interpretations of eugenics, to reboot our notion of universality of human rights, to reboot conversations about women's bodies in reproduction and reproductive autonomy, to reboot the manner of engagement with concerned stakeholders in pre-implantation genetic interventions. I refer to Sheila Jasanoff's "ethics of invention",[46] calling for a solid institutional framework that has been built and sustained by states as part of a legal order. Understanding that it is often easier said than done, Jasanoff states that we may begin by inducing a fundamental reboot into our views of ethics within the context of science and technologies[47]; particularly because there is a highly "complex relationship between our technologies, our societies, and our institutions, and the implications of those relationships for ethics, rights and human dignity".[48] We must also take into account that the variables of each regulatory space are often inundated with issues such as plurality and different ideals of morality and public opinion. Technology, in its evolutionary capabilities and design, as a product of human creativity, are rarely politically neutral or detached from democratic institutions; often embedding itself within human life.[49] This, however, is the task for bioethicists, policy makers, lobbyists, and legislators, in establishing and extending a workable, sustainable regulatory or governance framework that monitors the use and limits of these specific technologies, with the input from relevant stakeholders.

In the meantime, a cursory look at major international news dailies is indicative of the leaps and bounds technological advancements have made in the last decade alone. Scientific journals continue to churn out experimental studies and new perspectives on the curative elements and applications of genome editing technologies. By all accounts, these should represent positive tidings for the future of humankind. Jasanoff is careful, however, to question the "meaning and value of human nature",[50] and whether there may be a failure to "connect law to life"[51] because laws should ideally serve as a comprehensive normative framework for technological innovation and regulation. Certainly, there is also a darker side to technologies when we do not adequately frame the right questions and boundaries on how they should, or may be regulated.

---

[46] Jasanoff (2016).

[47] Ibid 86.

[48] Jasanoff (2016).

[49] Ibid 19.

[50] Ibid 5.

[51] Agamben (2005), p. 86.

Radical opponents to technological advancements, such as Theodore Kaczynski[52] (also known as the Unabomber)[53] from the late 1970s to early 1990s period, and the Neo-Luddites movement established from 1990,[54] have so far indicated a realistic prescience about the current state of humanity and technology. The shift in current technological patterns and behaviors, particularly concerning technological giants such as Google and Facebook are already indicative of people's awareness and knowledge of the foothold of technologies.[55] Rather than focus on the doom and gloom of the inevitability of automation in modern lives,[56] legal responses that have been carefully drafted and reflected upon can be the salve to problems. For example, the European Union (EU) adopted a *lex specialis* in the form of Regulation (EC) No. 1394/2007 to deal with human tissues, cells, and genes that are used for the development of therapeutic medicinal products in regenerative medicine.[57] The EU also found it necessary to responding to issues of unethical data collection, misuse of personal information, and the right to privacy, *vis-à-vis* the General Data Protection Regulation 2016/679 (GDPR)[58] (which recently came into force on 25 May 2018). These are examples that presented the riposte as a further need for protection for, and from, technologies. In the same vein, pre-implantation genetic interventions also have the capacity to breach barriers of the previously known, and an equal legal response is necessitated to commensurate the medical technological advancements that change the nature of human lives.

Anticipating the future,[59] or at least a substantial part of it, through legal regulation and foresighting (laws and a combination of suitable regulatory approaches), coupled with careful reasoning, adequate protection of human rights, whilst responding with full contemplation of the ELSIs of pre-implantation genetic interventions, is a challenging task ahead. When we reach the cusp where all theoretical possibilities become tangible, this foresightedness bears its full weight in gold.

References

Adolphs R (2013) The biology of fear. Curr Biol 23:R79
Agamben G (2005) State of exception (Kevin Attell tr). The University of Chicago Press
American Constitution Society for Law and Policy, 'The Gavel Gap' (*The Gavel Gap*). http://gavelgap.org

---

[52] History, 'Unabomber: Ted Kaczynski: Facts and Summary' (*history.com*) http://www.history.com/topics/unabomber-ted-kaczynski.

[53] Kaczynski (1991), p. 34.

[54] Glendinning (1991), p. 6.

[55] Bartlett (2018).

[56] Badham (2017).

[57] Mahalatchimy et al. (2012), pp. 131, 134.

[58] Regulation (EU) 2016/679 of the European Parliament and Of the Council 2016.

[59] Laurie et al. (2012), p. 1.

Badham V (2017) We can beat the robots - with Democracy|Opinion|The Guardian. https://www.theguardian.com/commentisfree/2017/nov/03/we-can-beat-the-robots-with-democracy

Barnes M, Wallace N (2017) Laws and ethics affecting clinical trials in Africa. J Health Care Law Policy 19:25

Bartlett J (4 March 2018) Will 2018 be the year of the Neo-Luddite? The Guardian. http://www.theguardian.com/technology/2018/mar/04/will-2018-be-the-year-of-the-neo-luddite

BBC News (21 July 2017) First woman appointed as UK's top judge. https://www.bbc.com/news/uk-40679293

Bello W (21 July 1998) "Asian Values" Democracy. Focus on the Global South. https://focusweb.org/node/365

Belluck P (4 August 2017) Gene editing for "Designer Babies"? Highly Unlikely, Scientists Say. The New York Times. https://www.nytimes.com/2017/08/04/science/gene-editing-embryos-designer-babies.html

Carleton N (2016) Fear of the unknown: one fear to rule them all? J Anxiety Disord 41:5

Center for Reproductive Rights (20 February 2014) 'Center for Reproductive Rights' (*Center for Reproductive Rights*). https://www.reproductiverights.org/

Committee on Human Gene Editing: Scientific, Medical, and Ethical Considerations and others (2017) Human genome editing: science, ethics, and governance. National Academies Press. https://www.nap.edu/catalog/24623

Committee on Science, Technology, and Law, Policy and Global Affairs and National Academies of Sciences, Engineering, and Medicine (2016) In: Olson S (ed) International summit on human gene editing: a global discussion. National Academies Press. http://www.nap.edu/catalog/21913

Cyranoski D, Reardon S (2015) Chinese scientists genetically modify human embryos. Nature News. http://www.nature.com/news/chinese-scientists-genetically-modify-human-embryos-1.17378

Department of Health (England) and others, UK Policy Framework for Health and Social Care Research

Dickenson D (2007) Property in the body: feminist perspectives. Cambridge University Press

Diuana V et al (2016) Women's reproductive rights in the penitentiary system: tensions and challenges in the transformation of reality. Ciência & Saúde Coletiva 21:2041

Donnelly J (1984) Cultural relativism and universal human rights. Hum Rights Q 6:400

Drahos P (ed) (2017) Regulatory theory foundations and applications. ANU Press

Falck EJ (2014) Technology and the memetic self. In: Global issues and ethical considerations in human enhancement technologies. IGI Global

Freeman RE (1984) Strategic management: a stakeholder approach. Pitman

George TE, Yoon AH (2014) The gavel gap- who sits in judgment on state courts. American Constitution Society for Law and Policy. http://gavelgap.org/pdf/gavel-gap-report.pdf

Glendinning C (1991) Notes toward a Neo-Luddite Manifesto

Global Fund for Women, 'Global Fund for Women' (*Global Fund for Women*). https://www.globalfundforwomen.org/

Greenfield R (1 August 2012) Technology doesn't ruin our lives, we do. The Atlantic. https://www.theatlantic.com/technology/archive/2012/08/technology-doesnt-ruin-our-lives-we-do/325180/

Haraway DJ (2000) A cyborg manifesto: science, technology, and socialist-feminism in the late twentieth century. In: Badmington N (ed) Posthumanism. Macmillan Education UK

Harris J (2015) Why human gene editing must not be stopped|Science|The Guardian. https://www.theguardian.com/science/2015/dec/02/why-human-gene-editing-must-not-be-stopped

Hennette-Vauchez S (2015) More women - but which women? The rule and the politics of gender balance at the European Court of Human Rights. Eur J Int Law 26:195

History, Unabomber: Ted Kaczynski: facts and summary (HISTORY.com). http://www.history.com/topics/unabomber-ted-kaczynski

Inhorn M (2009) Right to assisted reproductive technology: overcoming infertility in low-resource countries. Int J Gynecol Obstet 106:172

Jasanoff S (2016) The ethics of invention: technology and the human future. WW Norton & Company

Kaczynski T (1991) Industrial society and its future

Krattiger A (September 2013) Promoting access to medical innovation. World Intellectual Property Organization (WIPO)|Magazine. http://www.wipo.int/wipo_magazine/en/2013/05/article_0002.html

Laurie G, Harmon SHE, Arzuaga F (2012) Foresighting futures: law, new technologies, and the challenges of regulating for uncertainty. Law Innov Technol 4:1

Mahalatchimy A et al (2012) The legal landscape for advanced therapies: material and institutional implementation of European Union rules in France and the United Kingdom. J Law Soc 39:131

Nuffield Council on Bioethics (2016) Genome editing: an ethical review. Nuffield Council on Bioethics

Nuffield Council on Bioethics (2018) Genome editing and human reproduction: social and ethical issues. Nuffield Council on Bioethics

Regulation (EU) 2016/679 of the European Parliament and Of the Council 2016

Roth R (2017) She doesn't deserve to be treated like this: prisons as sites of reproductive injustice. In: Ross LJ et al (eds) Radical reproductive justice: foundations, theory, practice, critique. The Feminist Press

RT Question More (2017) Pentagon revealed as top funder of controversial gene editing tech — RT US News. https://www.rt.com/usa/412019-pentagon-darpa-gene-drive/

Sándor J (2015) The ethical and legal analysis of embryo preimplantation testing policies in Europe. In: Sills ES (ed) Screening the single euploid embryo. Springer International Publishing

Schmeer K (1999) Stakeholder analysis guidelines. Policy toolkit for strengthening health sector reform 1

Sen A (1997) Human rights and Asian values. New Republic:33

Sunstein C (1993) After the rights revolution, reconceiving the regulatory state. Harvard University Press

Supreme Court of the United States, Current Members. https://www.supremecourt.gov/about/biographies.aspx

The Eleanor Roosevelt Papers Project, The Universal Declaration of Human Rights. https://www2.gwu.edu/~erpapers/humanrights/lessonplans/

United Nations Population Fund (2014) Reproductive Rights Are Human Rights. https://www.unfpa.org/publications/reproductive-rights-are-human-rights

World Health Organization, WHO|Maternal and Reproductive Health (*WHO*). http://www.who.int/gho/maternal_health/en/

World Health Organization, GHO|Global Strategy for Women's, Children's and Adolescents' Health (2016–2030) (*WHO*). http://apps.who.int/gho/data/node.gswcah

World Health Organization, WHO|World Health Organization. http://gamapserver.who.int/gho/interactive_charts/mdg5_mm/atlas.html

World Trade Organization, World Health Organization and World Intellectual Property Organization (2013) Promoting access to medical technologies and innovation: intersections between public health, intellectual property and trade. WTO

References

A, B and C v Ireland (Application No 25575/05) [2010] HUDOC (Grand Chamber, European Court of Human Rights)

Abdulaziz, Cabales and Balkandali v The United Kingdom (1985) European Court of Human Rights Application no. 9214/80; 9473/81; 9474/81, HUDOC

Addis A (2015) Human dignity in comparative constitutional context: in search of an overlapping consensus. J Int Comp Law 2:1

Adolphs R (2013) The biology of fear. Curr Biol 23:R79

African Charter on Human and Peoples Rights (21 October 1986). http://www.humanrights.se/wp-content/uploads/2012/01/African-Charter-on-Human-and-Peoples-Rights.pdf

Agamben G (2005) State of exception (Kevin Attell tr). The University of Chicago Press

Agar N (1998) Liberal eugenics. Public Aff Q 12:137

Agar N (2006) The debate over liberal eugenics. Hast Cent Rep 36:4

Age Discrimination Act 1992

Agence France-Presse (20 February 2018) Malaysian Artist Jailed for Clown Face Caricature of PM Najib Razak That Went Viral. South China Morning Post. https://www.scmp.com/news/asia/southeast-asia/article/2133990/malaysian-artist-jailed-clown-face-caricature-pm-najib

Agence France-Presse, Bangkok (9 June 2017) Man Jailed for 35 Years in Thailand for Insulting Monarchy on Facebook. The Guardian. http://www.theguardian.com/world/2017/jun/09/man-jailed-for-35-years-in-thailand-for-insulting-monarchy-on-facebook

Ahmad N, Lilienthal G, Hussain M (2016) Law of assisted reproductive surrogacy in Malaysia: a critical overview. Commonw Law Bull 42:355

Airedale NHS Trust v Bland [1993] AC 789

Alexy R (2010) A theory of constitutional rights. Oxford University Press, Oxford

American Congress of Obstetricians and Gynecologists, Committee on Ethics (2007) ACOG Committee Opinion No. 360: sex selection. Obstet Gynecol 109:475–478

American Constitution Society for Law and Policy, 'The Gavel Gap' (*The Gavel Gap*). http://gavelgap.org

American Society for Reproductive Medicine (2015) Assisted reproductive technology- a guide for patients. http://www.fertilityanswers.com/wp-content/uploads/2016/04/assisted-reproductive-technologies-booklet.pdf

American Society for Reproductive Medicine, Ethics Committee (2013) Use of preimplantation genetic diagnosis for serious adult onset conditions: a committee opinion. Fertil Steril 100:54

American Society for Reproductive Medicine, Ethics Committee (2015) Use of reproductive technology for sex selection for nonmedical reasons. Fertil Steril 103:1418

Andorno R (2002) Biomedicine and international human rights law: in search of a global consensus. Bull World Health Org 5

Andorno R (2009) Human dignity and human rights as a common ground for a global bioethics. J Med Philos 34:223

Andorno R (2013a) Principles of international biomedical law. In: Principles of international biolaw: seeking common ground at the intersection of bioethics and human rights. Editions Bruylant, Bruxelles

Andorno R (2013b) The dual role of human dignity in bioethics. Med Health Care Philos 16:967

Andrews LB, Elster N (2000) Regulating reproductive technologies. J Legal Med 21:35

Annas G, Andrews L, Isasi R (2002) Protecting the endangered human: toward an international treaty prohibiting cloning and inheritable alterations. Am J Law Med 28:151–178

Aroney N (2009) The constitution of a federal commonwealth: the making and meaning of the Australian Constitution. Cambridge University Press, Cambridge

AsiaNews.it (21 September 2005) Malaysia Malay converts to Christianity. http://www.asianews.it/news-en/Malay-converts-to-Christianity-cannot-renounce-Islam-4157.html

Assembly UG (1979) Convention on the elimination of all forms of discrimination against Women. 20 Retrieved April 2006

Assisted Reproductive Technology Act [NSW] 2007 (No 69) 43

Assisted Reproductive Technology Amendment Bill [NSW] 2016 17

Assisted Reproductive Treatment Act [SA] 1988 14

Assisted Reproductive Treatment Act [VIC] 2008 (No 76)

Assisted Reproductive Treatment Regulations [SA] 2010 4

Association CG Press (24 October 2017) Northern Ireland abortion law "Inhuman and Degrading", Supreme Court Told, The Irish News. http://www.irishnews.com/news/northernireland-news/2017/10/24/news/northern-ireland-abortion-law-inhuman-and-degrading-supreme-court-told-1170249/

Association of Southeast Asian Nations, ASEAN Human Rights Declaration and the Phnom Penh Statement On The Adoption of the ASEAN Human Rights Declaration (AHRD). http://www.asean.org/storage/images/ASEAN_RTK_2014/6_AHRD_Booklet.pdf

Australia Act 1986 (No. 142, 1985). https://www.legislation.gov.au/Details/C2004A03181/Html/Text

Australia and Australian Government Solicitor (2016) *Australia's constitution: with overview and notes by the Australian Government Solicitor*

Australian Associated Press (20 February 2015a) Thailand bans commercial surrogacy | World News, The Guardian. https://www.theguardian.com/world/2015/feb/20/thailand-bans-commercial-surrogacy

Australian Associated Press (20 February 2015b) Thailand bans commercial surrogacy, The Guardian. https://www.theguardian.com/world/2015/feb/20/thailand-bans-commercial-surrogacy

Australian Broadcasting Corporation v Lenah Game Meats Pty Ltd [2001] HCA 63

Australian Human Rights Commission 'Right to Life'. https://www.humanrights.gov.au/right-life

Australian Human Rights Commission Act 1986

Australian Trade and Investment Commission, Biotechnology A Powerhouse for Science and Innovation. https://www.austrade.gov.au/International/Buy/Australian-industry-capabilities/Biotechnology

Aveyard H (2002) Implied consent prior to nursing care procedures. J Adv Nurs 39:201

Ayres I, Braithwaite J (1992) Responsive regulation: transcending the deregulation debate. Oxford University Press, Oxford

Badham V (2017) We can beat the robots - with Democracy|Opinion|The Guardian. https://www.theguardian.com/commentisfree/2017/nov/03/we-can-beat-the-robots-with-democracy

Bahadur G (2001) The Human Rights Act (1998) and its impact on reproductive issues. Hum Reprod 16:785

Baharuddin SAB (2008) 7 Competing domains of control: Islam and Human Rights in Malaysia. Islam and human rights in practice: perspectives across the Ummah 8:108

Baldwin R, Black J (2008) Really responsive regulation. Modern Law Rev 71:59. https://doi.org/10.1111/j.1468-2230.2008.00681.x

Baldwin R, Cave M (1999) Understanding regulation: theory, strategy, and practice. Oxford University Press, Oxford

Baltimore D et al (2015) A prudent path forward for genomic engineering and germline gene modification. Science 348:36

Banyard K (2010) The equality illusion: the truth about men and women today. Faber and Faber, London, pp 181–182

Barak A (1996) Constitutional human rights and private law. Rev Const Stud III:218

Barak A (2012) Proportionality: constitutional rights and their limitations. Cambridge University Press, Cambridge

Barak A (2015) Human dignity: the constitutional value and the constitutional right. Cambridge University Press, Cambridge

Bari AA, Shuaib FS (2009) Constitution of Malaysia: text and commentary. Pearson Prentice Hall, Upper Saddle River

Barlow K (6 July 2017) Federal government accused of undermining reproductive rights. Huffington Post. http://www.huffingtonpost.com.au/2017/07/06/federal-government-accused-of-undermining-reproductive-rights_a_23018464/

Barnes M, Wallace N (2017) Laws and ethics affecting clinical trials in Africa. J Health Care Law Policy 19:25

Barnett RE (1986) Four senses of the public law-private law distinction. Harv J Law Public Policy 9:267

Barnhart MG (1997) Ideas of nature in an Asian context. Philos East West 47:417

Barr MD, Lee Kuan Yew: race, culture and genes. 18

Bartlett J (4 March 2018) Will 2018 be the year of the Neo-Luddite? The Guardian. http://www.theguardian.com/technology/2018/mar/04/will-2018-be-the-year-of-the-neo-luddite

Basas CG (2014) What's bad about wellness? What the disability rights perspective offers about the limitations of wellness. J Health Polit Policy Law 39:1035

Baumann F, Humanism and Transhumanism. 17

Bawden T (27 July 2017) Scientists call for new rules on GM designer babies. iNews. https://inews.co.uk/news/health/new-rules-will-be-needed-to-exploit-designer-baby-breakthrough-in-britain. Accessed 27 Jan 2018

Bayefsky M (2015) The regulatory gap for preimplantation genetic diagnosis. Hastings Center Rep 45:7

Bayefsky M (2016) Comparative preimplantation genetic diagnosis policy in Europe and the USA and its implications for reproductive tourism. Reprod Biomed Soc Online 3:41

Bayefsky M, Jennings B (2015) Regulating preimplantation genetic diagnosis in the United States: the limits of unlimited selection. Springer, Berlin

BBC News (11 September 2002) IVF Wrangle Cases Go to Court. http://news.bbc.co.uk/2/hi/health/2249442.stm

BBC News (24 June 2011) Profile: Thaksin Shinawatra. http://www.bbc.com/news/world-asia-pacific-13891650

BBC News (14 November 2012) Abortion "Would Have Saved Wife". http://www.bbc.com/news/uk-northern-ireland-20321741

BBC News (21 July 2017) First woman appointed as UK's top judge. https://www.bbc.com/news/uk-40679293

BBC News (6 September 2018a) Historic India Ruling Legalises Gay Sex. https://www.bbc.com/news/world-asia-india-45429664

BBC News (27 November 2018b) Terminally ill Noel Conway loses Supreme Court Appeal. https://www.bbc.com/news/uk-england-shropshire-46359845

BBC News (21 January 2019) China Turns on "Gene Editing" Scientist. https://www.bbc.com/news/world-asia-46943593

Beaney WM (1966) The right to privacy and American Law. Law Contemp Probl 31:253

Beauchamp T, Childress J (2011) Principles of biomedical ethics. Oxford University Press, Oxford

Bell AV (2016) The margins of medicalization: diversity and context through the case of infertility. Soc Sci Med 156:39

Bello W (21 July 1998) "Asian Values" Democracy. Focus on the Global South. https://focusweb. org/node/365

Belluck P (4 August 2017) Gene editing for "designer babies"? Highly unlikely, scientists say. The New York Times. https://www.nytimes.com/2017/08/04/science/gene-editing-embryos-designer-babies.html

Bennett B (12 July 2018) How Brett Kavanaugh Could Change the Supreme Court—and America. Time. http://time.com/5336621/brett-kavanaugh-supreme-court/

Bentham J, Burns JH, Hart HLA (1988) A fragment on government. Cambridge University Press, Cambridge

Beyleveld D, Brownsword R (2007) Consent in the law. Hart, Oxford

Biegel S (2001) Beyond our control?: confronting the limits of our legal system in the age of cyberspace. MIT Press

Bimber B (1990) Karl Marx and the three faces of technological determinism. Soc Stud Sci 20:333

Binsaleh S, Lo KC (2007) Varicocelectomy: microsurgical inguinal varicocelectomy is the treatment of choice. Can Urol Assoc J 1:277

Black J (1998) Regulation as facilitation: negotiating the genetic revolution. Mod Law Rev 61:621

Black J (2005) What is regulatory innovation? In: Black J, Lodge M, Thatcher M (eds) Regulatory innovation. Edward Elgar

Boer D, Boehnke K (2015) What are values? Where do they come from? A developmental perspective. In: Brosch T et al (eds) Handbook of value: perspectives from economics, neuroscience, philosophy, psychology and sociology. Oxford University Press, Oxford

Bogdanor V (2009) The new British Constitution. Hart, Oxford

Bognar G (2016) Is disability mere difference? J Med Ethics 42:46

Bolam v Friern Hospital Management Committee [1957] 1 WLR 582

Book of Genesis of the Bible, Chapter 18 & 19, Sodom & Gomorrah

Botkin J (1998) Ethical issues and practical problems in preimplantation genetic diagnosis. J Law Med Ethics 26:17

Bowcott O (7 June 2018) Northern Ireland Abortion Law Clashes with Human Rights, Judges Say. The Guardian. https://www.theguardian.com/world/2018/jun/07/supreme-court-dismisses-bid-to-overturn-northern-ireland-abortion-laws

Bowers v Hardwick [1986] Supreme Court 478 U.S. 186

Boyle RJ, Savulescu J (2001) Ethics of using preimplantation genetic diagnosis to select a stem cell donor for an existing person. Br Med J 323:1240

Braithwaite J (2002) Restorative justice and responsive regulation. Oxford University Press, Oxford

Braithwaite J (2017a) Types of responsiveness. In: Drahos P (ed) Regulatory theory: foundations and applications. Australian National University Press, Canberra

Braithwaite V (2017b) Closing the gap between regulation and the community. In: Drahos P (ed) Regulatory theory: foundations and applications. Australian National University Press, Canberra

Braithwaite J, Drahos P (2000) Global business regulation. Cambridge University Press, Cambridge

Breavington v Godleman (1988) 169 CLR 41

Brezina PR, Zhao Y (2012) The ethical, legal, and social issues impacted by modern assisted reproductive technologies. Obstet Gynecol Int 2012:1–7. https://www.hindawi.com/journals/ogi/2012/686253/

Brezina PR, Ke RW, Kutteh WH (2013) Preimplantation genetic screening: a practical guide. Clin Med Insights Reprod Health 7:37

Briggs H (24 January 2018) First monkey clones created in the lab. BBC News. http://www.bbc.com/news/health-42809445

Brock DW (2009) Is selection of children wrong? In: Human enhancement. Oxford University Press, Oxford

Brookings (30 November 2001) Event summary: a debate on the ethics of genetic engineering. Brookings. https://www.brookings.edu/opinions/event-summary-a-debate-on-the-ethics-of-genetic-engineering/

Brosch T et al (2015) Handbook of value: perspectives from economics, neuroscience, philosophy, psychology and sociology. Oxford University Press, Oxford

Brownsword R (2008) Rights, regulation, and the technological revolution. Oxford University Press, Oxford

Brownsword R (2011) Why I Wrote ... rights, regulation, and the technological revolution. Clin Ethics 6:207

Brownsword R, Goodwin M (2012) Law and the technologies of the twenty-first century. Cambridge University Press, Cambridge

Brownsword R, Yeung K (2008) Regulating technologies: legal futures, regulatory frames and technological fixes. Hart

Buchanan A et al (2001) From chance to choice: genetics and justice. Cambridge University Press, Cambridge

Buchitchon S (2016) The Protection of Children Born from Assisted Reproductive Technology Act 2015: scientific advances, ethics and concerns over the use of human embryo. Adv Sci Lett 22:1610

Buck v. Bell, 274 U.S. 200 (1927) (*Justia Law*). https://supreme.justia.com/cases/federal/us/274/200/case.html

Bull and another v Hall and another (2013) UKSC 73

Bulman M (24 October 2017) Supreme Court to Scrutinise Northern Ireland's "degrading and Humiliating" Abortion Laws. The Independent. http://www.independent.co.uk/news/uk/home-news/northern-ireland-abortion-law-uk-supreme-court-scrutiny-women-terminations-degrading-humiliating-a8016921.html

Burns JH (2005) Happiness and utility: Jeremy Bentham's equation. Utilitas 17:46

Burrus T (23 June 2011) One generation of Oliver Wendell Holmes, Jr. Is Enough. Cato Institute. https://www.cato.org/blog/one-generation-oliver-wendell-holmes-jr-enough

Busby M (2 July 2018) Marriage of 11-Year-Old Girl to a 41-Year-Old Man Provokes Backlash in Malaysia. The Independent. https://www.independent.co.uk/news/world/asia/child-marriage-malaysia-11-woman-man-41-wedding-underage-unicef-provokes-backlash-a8426311.html

BVerfGE 75 [1987] Verfassungsgericht 369 1 BvR 313/85

Byrne L (3 January 2019) Concerns as Irish Abortion Services Start. BBC News. https://www.bbc.com/news/world-europe-46737593

C-528/16 Fédération Nature & Progrès v Premier Ministre, Ministre de l'agriculture, de l'agroalimentaire et de la forêt [2018] Court of Justice of the European Union C-528/16, Curia

Caamano JM (2016) International, commercial, gestational surrogacy through the eyes of children born to surrogates in Thailand: a cry for legal attention. Boston Univ Law Rev 96:37

Callahan D (1986) How technology is reframing the abortion debate. Hastings Center Rep 16:33

Callahan S (2009) The ethical challenges of the new reproductive technologies. In: Morrison EE (ed) Health care ethics: critical issues for the 21st century, 2nd edn. Jones and Bartlett Publishers, Burlington

Campbell v MGN Ltd. [2004] UKHL 22

Canal G (7 March 2017) Everything You Need to Know About the Anti-Vaxxer Movement. Global Citizen. https://www.globalcitizen.org/en/content/everything-you-need-to-know-about-the-anti-vaxxer/

Cannon L (1991) President Reagan: the role of a lifetime. Public Affairs

Carleton N (2016) Fear of the unknown: one fear to rule them all? J Anxiety Disord 41:5

Carroll D (2011) Genome engineering with Zinc-Finger nucleases. Genetics 188:773

Case of A, B and C v Ireland [2010] Grand Chamber 25579/05

Center for Ethics and Law in Biomedicine (19 January 2016) Anniversary bioethics debate on gene editing. https://celab.ceu.edu/events/2016-01-19/anniversary-bioethics-debate-gene-editing

Center for Reproductive Rights (20 February 2014a) Malaysia. Center for Reproductive Rights. https://www.reproductiverights.org/our-regions/asia/malaysia

Center for Reproductive Rights (20 February 2014b) 'Center for Reproductive Rights' (*Center for Reproductive Rights*). https://www.reproductiverights.org/

Center for Reproductive Rights (Asia) (20 February 2014) Malaysia. https://www.reproductiverights.org/our-regions/asia/malaysia

Centre for Health, Law and Emerging Technologies (HeLEX) — Nuffield Department of Population Health. https://www.ndph.ox.ac.uk/research/centre-for-health-law-and-emerging-technologies-helex

Chaffin BC, Gosnell H, Cosens BA (2014) A decade of adaptive governance scholarship: synthesis and future directions. Ecol Soc 19:56

Chambers P, Waitoolkiat N (2016) The resilience of Monarchised Military in Thailand. J Contemp Asia 46:425

Chan CK (1985) Eugenics on the rise: a report from Singapore. Int J Health Serv 15:707

Chatham House (2015) The elusive consensus in international affairs. Chatham House. https://www.chathamhouse.org//node/17691

Chatterjee A (2004) Cosmetic neurology the controversy over enhancing movement, mentation, and mood. Neurology 63:968

Chen QB (1994) Chinese Constitutional Law. BLJ 26:77

Chen Z-H et al (2017) Targeting genomic rearrangements in tumor cells through Cas9-mediated insertion of a suicide gene. Nature Biotechnol 35:543

Chesterton GK (2000) Eugenics and other evils: an argument against the scientifically organized state. Inkling Books

Chial H (2008) Rare genetic disorders: learning about genetic disease through gene mapping, SNPs, and microarray data. Nat Educ 1:192

Christiansen K (14 November 2017) Genome editing: are we opening a back door to eugenics? Science Nordic. http://sciencenordic.com/genome-editing-are-we-opening-back-door-eugenics

Christine Goodwin v the United Kingdom (Application No 28957/95) [2002] HUDOC (Grand Chamber, European Court of Human Rights)

Chua A (2011) Battle hymn of the tiger mother. Penguin Group, London

Clark AI (1905) Studies in Australian constitutional law. CF Maxwell, Melbourne

Clarke AC (1985) Profiles of the future, 1st edn. Warner Books, New York

Cochrane K (7 February 2014) The truth about the tiger mother's family. The Guardian. http://www.theguardian.com/lifeandstyle/2014/feb/07/truth-about-tiger-mothers-family-amy-chua

Coco v AN Clark (Engineers) Ltd. [1969] RPC 41

Cody A, Nawaz M (9 November 2017) UN Slams Australia's Human Rights Record. The Conversation. http://theconversation.com/un-slams-australias-human-rights-record-87169

Coggon J, Miola J (2011) Autonomy, liberty, and medical decision-making. Camb Law J 70:523

Cohen IG (2014) What (if anything) is wrong with human enhancement? What (if anything) is right with it? Tulsa Law Rev 49:645

Cohen A (2016) Imbeciles, The Supreme Court, American Eugenics and the Sterilization of Carrie Buck. Penguin Press, London

Coke E (1628) The first part of the institutes of the Lawes of England. Or, a commentarie upon Littleton, not the name of a lawyer onely, but of the law it selfe. Societe of Stationers, London

Coke E (1642) The second part of the institutes of the lawes of England. Containing the exposition of many ancient, and other statutes; Whereof you may see the particulars in a table following. Miles Fletcher and Robert Young for Ephraim Dawson

Collingridge D (1982) The social control of technology. St Martin's Press, New York

Committee on Human Gene Editing: Scientific, Medical, and Ethical Considerations and others (2017) Human genome editing: science, ethics, and governance. National Academies Press. https://www.nap.edu/catalog/24623

Committee on Science, Technology, and Law, Policy and Global Affairs and National Academies of Sciences, Engineering, and Medicine (2016) In: Olson S (ed) International summit on human gene editing: a global discussion. National Academies Press, Washington, D.C. http://www.nap.edu/catalog/21913

Commonwealth of Australia Constitution Act (The Constitution). https://www.legislation.gov.au/Details/C2013Q00005/Html/Text

Commune de Morsang-sur-Orge v Societe Fun Production et MWackenheim [1995] Conseil d'Etat 136727, Cons Etat

Conseil de l'Europe (1953) Convention for the protection of human rights and fundamental freedoms

Conseil de l'Europe (1997) Convention for the protection of human rights and dignity of the human being with regard to the application of biology and medicine: convention on human rights and biomedicine. Editions du Conseil de l'Europe. http://193.205.211.30/lawtech/images/lawtech/law/convenzioneoviedo.pdf

Constitution of the Kingdom of Thailand (2007). https://www.samuiforsale.com/law-texts/thai-law-translations-constitution-of-thailand.html

Cook A (17 July 2002) Our post-human future: consequences of the biotechnology revolution by Francis Fukuyama. PopMatters. https://www.popmatters.com/our-posthuman-future-2496243108.html

Cook M (19 August 2017) China rushes into embryo selection. BioEdge. https://www.bioedge.org/bioethics/china-rushes-into-embryo-selection/12399

Cook R, Dickens BM (2003) Human rights dynamics of abortion law reform. Hum Rights Q 25(1):2–3

Cordelia T (2004) Pre-implantation testing and the protection of the savior sibling. Deakin Law Rev 5:121

Cosens BA et al (2017) The role of law in adaptive governance. Ecol Soc 22:30

Costa J-P (2013) Human Digity in the jurisprudence of the European Court of Human Rights. In: McCrudden C (ed) Understanding human dignity. The British Academy

Crimes Amendment (Zoe's Law) Bill 2017. https://www.parliament.nsw.gov.au/bills/Pages/bill-details.aspx?pk=2936

Crockin SL (2005) Reproduction, genetics and the law. Reprod BioMed Online 10:692

Cunningham M (16 February 2017) Well Done! Australia Adds More Funds for Sexual and Reproductive Health Program. Global Citizen. https://www.globalcitizen.org/en/content/australian-government-announces-further-funding-fo/

Cwik B (2017) Designing ethical trials of germline gene editing. New Engl J Med 377:1911

Cyranoski D, Reardon S (2015) Chinese scientists genetically modify human embryos. Nature News. http://www.nature.com/news/chinese-scientists-genetically-modify-human-embryos-1.17378

Dafoe A (2015) On technological determinism: a typology, scope conditions, and a mechanism. Sci Technol Human Values 40:1047

Dahlan R, Faudzi FS (9 March 2016) The Syariah Court: Its Position Under The Malaysian Legal System. Mondaq. http://www.mondaq.com/x/472794/trials+appeals+compensation/The+Syariah+Court+Its+Position+Under+The+Malaysian+Legal+System

Daniels CR et al (2016) Informed or misinformed consent? Abortion policy in the United States. J Health Politics Policy Law 41:181

Dasgupta S (15 February 2018) A New "ASEAN Way": finding a regional solution for human rights violations in Rakhine State. The Kenan Institute for Ethics at Duke University. https://kenan.ethics.duke.edu/a-new-asean-way-finding-a-regional-solution-for-human-rights-violations-in-rakhine-state/

Davidson D (27 February 2018) Muslims Guard Bishop Heckled by Zealous Youths at Kuching Court Complex. The Malaysian Insight. https://www.themalaysianinsight.com/s/40086/

Dawkins R (2016) The selfish gene, 4th edn. Oxford University Press, Oxford

de Anarclens P (2007) The United Nations as a social and economic regulator. In: de Sanarclens P, Kazancigli A (eds) Regulating globalization, critical approaches to global governance. United Nations University, Tokyo, pp 8–35

Dean R (2007) Erosion of access to abortion in the United States: lessons for Australia. Deakin Law Rev 12:123

Dearden L (18 January 2018) Noel Conway: terminally ill man wins right to challenge court ruling preventing "dignified Death". The Independent. http://www.independent.co.uk/news/uk/home-news/noel-conway-latest-terminally-ill-right-to-die-court-challenge-ruling-motor-neurone-suicide-a8165606.html

Deeney MS (2013) Bioethical considerations of preimplantation genetic diagnosis for sex selection. Wash Univ Jurisprud Rev 5:333

Deonandan R (2015) Recent trends in reproductive tourism and international surrogacy: ethical considerations and challenges for policy. Risk Manag Healthcare Policy 8:111

Department of Health (England) and others, UK Policy Framework for Health and Social Care Research

Department of Health, UK (9 November 2007) Human Fertilisation and Embryology Act 1990 - an Illustrative Text. http://webarchive.nationalarchives.gov.uk/content/20130107105354/http://www.dh.gov.uk/en/Publicationsandstatistics/Publications/PublicationsLegislation/DH_080205

Department of Health UK, Health and Safety Executive, 'Biological Agents: Managing the Risks in Laboratories and Healthcare Premises' 82

Department of Jurisconsult R and LD, 'Bioethics and the Case-Law of the Court' (Council of Europe / European Court of Human Rights 2016) Research Report

Department of Social Services, Australian Government, 'Values and Law'. https://www.dss.gov.au/our-responsibilities/settlement-and-multicultural-affairs/programs-policy/taking-the-initiative/resources/values-and-law

Deppe C (2010) Tao Te Ching: a window to the Tao through the words of Lao Tzu. Fertile Valley Publishing. https://terebess.hu/english/tao/Deppe.pdf

Dhammanada KS (2002) What Buddhists believe, 4th edn. Buddhist Missionary Society Malaysia

Dickenson D (2007) Property in the body: feminist perspectives. Cambridge University Press, Cambridge

Dickson v United Kingdom, Application No. 44362/04

Dietz T, Ostrom E, Stern PC (2003) The struggle to govern the commons. Science 302:1907

Disability Discrimination Act 1992

Diuana V et al (2016) Women's reproductive rights in the penitentiary system: tensions and challenges in the transformation of reality. Ciência & Saúde Coletiva 21:2041

Dodson M (1993) Social justice for indigenous peoples, 3rd edn. Aboriginal Research Institute Publications, Underdale

Doherty P, Sutton A (1997) Man-made man: ethical and legal issues in Genetics. Four Courts Press

Donnelly J (1984) Cultural relativism and universal human rights. Hum Rights Q 6:400

Douglas v Hello! Ltd [2005] EWCA Civ 595

Draft Constitution of Thailand 2016 (Unofficial English Translation). http://constitutionnet.org/sites/default/files/thailand-draft-constitution_englishtranslation_june_2016.pdf

Drahos P (ed) (2017) Regulatory theory foundations and applications. Australian National University Press, Canberra

Dudgeon MR, Inhorn MC (2004) Men's influences on women's reproductive health: medical anthropological perspectives. Soc Sci Med 59:1379

Dyson F (1997) Can science be ethical? N Y Rev Books 44:46

Elbarbary RA et al (2017) Tudor-SN–mediated endonucleolytic decay of human cell MicroRNAs promotes G_1/S phase transition. Science 356:859

Encyclopedia of Bioethics, vol 2, 3rd edn (Thomson Gale 2004)

Engeli I, Rothmayr CA (2016) When doctors shape policy: the impact of self-regulation on governing human biotechnology: when doctors shape policy. Regul Gov 10:248

Equality Act 2010 (merging the Equal Pay Act 1970, the Sex Discrimination Act 1975, the Race Relations Act 1976, the Disability Discrimination Act 1995, the Employment Equality (Religion or Belief) Regulations 2003, the Employment Equality (Sexual Orientation) Regulations 2003, the Employment Equality (Age) Regulations 2006, the Equality Act 2006 Part 2, and the Equality Act (Sexual Orientation) Regulations 2007)

Erdman JN (2015) The politics of global abortion rights. Brown J World Aff 39:22

Ettorre E (2000) Reproductive genetics, gender and the body: "Please Doctor, May I Have a Normal Baby?". Sociology 34:403

Etuk SJ (2009) Reproductive health: global infertility trend. Niger J Physiol Sci 24:85

European Commission JRC Science for Policy (2018) JRCF7- Knowledge Health and Consumer Safety, Overview of EU National Legislation on Genomics. European Commission. EUR29404EN http://publications.jrc.ec.europa.eu/repository/bitstream/JRC113479/policy_ report_-_review_of_eu_national_legislation_on_genomics_-_with_identifiers.pdf

European Medicines Agency (2018) Report of the EMA Expert Meeting on Genome Editing Technologies Used in Medicinal Product Development. European Medicines Agency. EMA/47066/2018

Evans v United Kingdom, Application No. 6339/05

Everton K (2014) Walking the edge with controversial use of Pre-Implantation Genetic Diagnosis (PGD): opinions and attitudes of genetic counsellors. University of South Carolina Scholar Commons

Exploring Constitutional Conflicts 'Levels of Scrutiny Under the Equal Protection Clause'. http://law2.umkc.edu/faculty/projects/ftrials/conlaw/epcscrutiny.htm

Falck EJ (2014) Technology and the memetic self. In: Global issues and ethical considerations in human enhancement technologies. IGI Global

Farokhmanesh M (12 December 2016) How a Trump Administration Threatens Women's Health. The Verge. https://www.theverge.com/2016/12/12/13904032/ trump-womens-reproductive-health-affordable-care-planned-parenthood

Farrell P (20 January 2015) Baby Gammy, Born into Thai Surrogacy Scandal, Granted Australian Citizenship | Australia News, The Guardian. https://www.theguardian.com/australia-news/2015/ jan/20/baby-gammy-born-into-thai-surrogacy-scandal-granted-australian-citizenship

Federal Constitution of Malaysia (Federal Constitution) 1957

Federal Constitution of Malaysia 1957

Feikert C (30 April 2012) Sex selection & abortion: Australia. Library of Congress. https://www. loc.gov/law/help/sex-selection/australia.php

Finnis J (2015) Grounding human rights in natural law. Am J Jurisprud 60:199

Fischer-Lescano A, Teubner G (2004) Regime collisions: the vain search for legal unity in the fragmentation of global law. Mich J Int Law 25:99–1045

Flanigan J (2013) Adderall for all: a defense of pediatric neuroenhancement. HEC Forum 25:325

Foley KE (23 January 2018) Chinese scientists already used Crispr gene editing on 86 human patients. Quartz. https://qz.com/1185488/ chinese-scientists-used-crispr-gene-editing-on-86-human-patients/

Foster C (2011) Human dignity in bioethics and law. Hart, Oxford

Foucault M (1963) Naissance de La Clinique Une Archéologie Du Regard Médical. Presses Universitaires de France

Foucault M (1976) The history of sexuality volume I: an introduction. Pantheon Books

Foucault M (1977) Discipline and punish: the birth of the prison. Vintage Books, Random House

Foucault M, Gordon C (1980) Power/knowledge: selected interviews and other writings, 1972–1977, 1st American edn. Pantheon Books

Fox D (2010) Retracing liberalism and remaking nature: designer children, research embryos, and featherless chickens. Bioethics 24:170

FPA, The Family Planning Association, A Sexual Health Charity, Abortion Law, England, Scotland and Wales. https://www.fpa.org.uk/sexual-and-reproductive-rights/abortion-rights/ abortion-law

Francioni F (2007a) Biotechnologies and international human rights. Hart Publishing, Oxford

Francioni F (2007b) Biotechnologies and international human rights. Bloomsbury Publishing

Franklin DL (29 June 2015) How the 1942 case of a one-footed chicken thief laid the foundation for marriage equality. Slate Magazine. http://www.slate.com/articles/news_and_politics/jurisprudence/2015/06/gay_marriage_supreme_court_ruling_how_skinner_v_oklahoma_laid_the_foundation.html

Fredman S et al (2015) The potential challenges to equality law in the UK. Oxford Human Rights Hub. https://ohrh.law.ox.ac.uk/wordpress/wp-content/uploads/2015/08/here.pdf

Freedland J (2 July 2015) 1984 by George Orwell, Book of a Lifetime: An Absorbing, Deeply Affecting Political Thriller. The Independent. http://www.independent.co.uk/arts-entertainment/books/reviews/1984-by-george-orwell-book-of-a-lifetime-an-absorbing-deeply-affecting-political-thriller-10360789.html

Freeman RE (1984) Strategic management: a stakeholder approach. Pitman

Fukuyama F (2003) Our posthuman future: consequences of the biotechnology revolution. Farrar, Straus and Giroux

Fuller LL (1969) The morality of law. Yale University Press

Gabbatis J (14 February 2018) Dolly the sheep: 15 years after her death, cloning still has the power to shock. The Independent. https://www.independent.co.uk/news/science/dolly-the-sheep-cloning-15-years-death-future-humans-monkeys-what-next-a8208896.html

Galton D (2002) Eugenics. The future of human life in the 21st century. Abacus

Ganesan JS (7 October 2016) A short history of the word "Kiasu". Esquire Singapore. https://www.esq.sg/lifestyle/culture/news/A-Short-History-Of-Kiasu

Gardbaum S (2008) Human rights as international constitutional rights. Eur J Int Law 19:749

Gauri V, Gloppen S (2012) Human rights-based approaches to development: concepts, evidence, and policy. Polity 44:485

Gavaghan C (2007) Defending the genetic supermarket: the law and ethics of selecting the next generation. Routledge-Cavendish, Abingdon

Geller T (2016) In privacy law, it's the U.S. vs. the world. Commun ACM 59:21

Gender Recognition Act 2004

GenomeWeb (4 December 2018) WHO to set up panel to develop gene-editing standards. GenomeWeb. https://www.genomeweb.com/scan/who-set-panel-develop-gene-editing-standards?utm_source=addthis_shares&fbclid=IwAR320wblX33e1znwYOf8bkOLtvjc9mOrGSaImu8ARQ22ST6k3U2xn-2injo#.XGcDwcnSMNY.facebook

George TE, Yoon AH (2014) The gavel gap- who sits in judgment on state courts. American Constitution Society for Law and Policy. http://gavelgap.org/pdf/gavel-gap-report.pdf

Ghaidan v Godin-Mendoza [2004] UKHL 30

Gibbs WW (2014) Biomarkers and aging: the clock-watcher. Nature 508:168

Gillon R (1985) Autonomy and the principle of respect for autonomy. Br Med J (Clinical research ed.) 290:1806

Gleicher N, Orvieto R (2017) Is the hypothesis of preimplantation genetic screening (PGS) still supportable? A review. J Ovarian Res 10:21

Gleicher N, Kushnir VA, Barad DH (2014) Preimplantation Genetic Screening (PGS) still in search of a clinical application: a systematic review. Reprod Biol Endocrinol 12:22

Gleicher N et al (2016) Accuracy of Preimplantation Genetic Screening (PGS) is compromised by degree of mosaicism of human embryos. Reprod Biol Endocrinol 14:54

Glendinning C (1991) Notes toward a Neo-Luddite Manifesto

Glenn LM (2003) Crossing species boundaries. Am J Bioethics 3:27–28

Global Fund for Women, 'Global Fund for Women' (*Global Fund for Women*). https://www.globalfundforwomen.org/

Gomez J (3 July 2017) One Year Later, Duterte Remains a Human Rights Nightmare. The Diplomat. https://thediplomat.com/2017/07/one-year-later-duterte-remains-a-human-rights-nightmare/

Goold I et al (2014) The human body as property? Possession, control and commodification. J Med Ethics 40:1

Goslinga-Roy GM (2000) Body boundaries, fiction of the female self: an ethnographic perspective on power, feminism, and the reproductive technologies. Fem Stud 26:113

Gotsis T, Ismay L. Abortion Law: A National Perspective, Briefing Paper No. 2/2017. NSW Parliamentary Research Service. https://www.parliament.nsw.gov.au/researchpapers/Documents/Abortion%20Law.pdf

Gozzetti A, Le Beau MM (2000) Fluorescence in situ hybridization: uses and limitations. Semin Hematol 37:320

Graham M et al (2016) Women's reproductive choices in Australia: mapping federal and state/territory policy instruments governing choice. Gend Issues 33:335

Grant v HM Land Registry and Another [2011] EWCA Civ 769

Grant v the United Kingdom (32570/03)

Greenfield R (1 August 2012) Technology doesn't ruin our lives, we do. The Atlantic. https://www.theatlantic.com/technology/archive/2012/08/technology-doesnt-ruin-our-lives-we-do/325180/

Gregg AR (2016) Noninvasive prenatal screening for fetal aneuploidy, 2016 update: a position statement of the American College of Medical Genetics and Genomics. Genet Med 18:10

Gribble J, Bremner J (26 November 2012) The challenge of attaining the demographic dividend. Population Reference Bureau (PRB). https://www.prb.org/demographic-dividend/

Griswold v. Connecticut, 381 U.S. 479 (1965) (*Justia Law*). https://supreme.justia.com/cases/federal/us/381/479/case.html

Groll D, Lott M (2015) Is there a role for "human nature" in debates about human enhancement? Philosophy 90:623

Gross M (2003) Dawn of the saviour sibling. Curr Biol 13:R541

Grugel J, Piper N (2009) Do rights promote development? Global Soc Policy Interdisciplinary J Public Policy Soc Dev 9:79

Gunningham N, Sinclair D (2017) Smart regulation. In: Drahos P (ed) Regulatory theory: foundations and applications. Australian National University Press

Gunningham N, Grabosky P, Sinclair D (1998) Smart regulation: designing environmental policy. Clarendon Press

Guttinger S (2017) Trust in science: CRISPR–Cas9 and the Ban on human Germline editing. Sci Eng Ethics

Gyngell C, Douglas T (2015) Stocking the genetic supermarket: reproductive genetic technologies and collective action problems: stocking the genetic supermarket. Bioethics 29:241

Habermas J (2003) The future of human nature. Polity Press

Hallich O (2017) Sperm donation and the right to privacy. New Bioethics 23:107

Handyside A (2010) Let parents decide. Nature 464:978

Harari YN (2015) Sapiens: a brief history of humankind, 1st edn. Harper

Haraway DJ (2000) A Cyborg Manifesto: science, technology, and socialist-feminism in the late twentieth century. In: Badmington N (ed) Posthumanism. Macmillan Education, London

Harding A, Leyland P (2011) The constitutional system of Thailand: a contextual analysis. Hart Publishing, Oxford

Harlow C (1980) "Public" and "Private" law: definition without distinction. Mod Law Rev 43:241

Harmon SHE (2016) Modernizing biomedical regulation: foresight and values in the promotion of responsible research and innovation. J Law Biosci 3:680

Harris J (2010) Enhancing evolution: the ethical case for making better people. Princeton University Press

Harris J (2015) Why human gene editing must not be stopped|Science|The Guardian. https://www.theguardian.com/science/2015/dec/02/why-human-gene-editing-must-not-be-stopped

Harryono M et al (2006) Thailand medical tourism cluster. Harvard Business School Microeconomics of Competitiveness

Harvard TH (2016) Chan School of Public Health and STAT, 'The Public and Genetic Editing, Testing and Therapy'. Harvard TH Chan School of Public Health. https://cdn1.sph.harvard.edu/wp-content/uploads/sites/94/2016/01/STAT-Harvard-Poll-Jan-2016-Genetic-Technology.pdf

Hassan S, Lopez C (2005) Human rights in Malaysia: globalization, national governance and local responses. In: Wah FLK, Ojendal J (eds) Southeast Asian responses to globalization: restructuring governance and deepening democracy. Institute of Southeast Asian Studies, Singapore

Hawking SW (2005) The theory of everything: the origin and fate of the universe. Phoenix Books

Hediger R (2016) Becoming with animals: sympoiesis and the ecology of meaning in London and Hemingway. Stud Am Nat 11:5

Heinemann T, Honnefelder L (2003) Principles of ethical decision making regarding embryonic stem cell research in Germany. Bioethics 16:530

Hennette-Vauchez S (2015) More women - but which women? The rule and the politics of gender balance at the European Court of Human Rights. Eur J Int Law 26:195

Hewison K (2014) Considerations on inequality and politics in Thailand. Democratization 21:846

Heydarian RJ (11 May 2018) A peaceful revolution in Malaysia. Al-Jazeera. https://www.aljazeera.com/indepth/opinion/peaceful-revolution-malaysia-180511140532987.html

Heyman SJ (1991) First duty of government: protection, liberty and the fourteenth amendment. Duke Law J 41:507

Hickling RH (1972) The legal system of Thailand. Hong Kong Law J 2:8

Hill G (2005) Resolving a true conflict between state laws: a minimalist approach. Melb Univ Law Rev 29:39

History, Unabomber: Ted Kaczynski: facts and summary (HISTORY.com). http://www.history.com/topics/unabomber-ted-kaczynski

History, 'The 1960s - Facts & Summary'. (*HISTORY.com*) http://www.history.com/topics/1960s

Hive Legal, Australia's Life Science Sector – Snapshot, Trends and Our Work. http://hivelegal.com.au/australias-life-science-sector-snapshot-trends-and-our-work/

Hobbes T. Leviathan or the Matter, Forme, & Power of a Common-Wealth Ecclesiasticall and Civill. Andrew Crooke at the Green Dragon in St Paul's Church-yard 1651. https://socialsciences.mcmaster.ca/econ/ugcm/3ll3/hobbes/Leviathan.pdf

Holmes O (29 October 2016) Thailand's Crackdown on "Insults" to the Monarchy Spreads Abroad. The Guardian. http://www.theguardian.com/world/2016/oct/29/thailand-bhumibol-monarchy-insults-law

Hong M et al (2017) Correction of a pathogenic gene mutation in human embryos. Nature 548:413

House of Representatives, The Parliament of the Commonwealth of Australia, 'Australian Bill of Rights Bill 2017'. https://www.legislation.gov.au/Details/C2017B00161/Html/Text

Hovhannisyan A (2018) Ōta Tenrei's defense of birth control, eugenics and euthanasia. Contemp Jpn 30:28

Hsu P, Lander E, Zhang F (2014) Development and applications of CRISPR-Cas9 for genome engineering. Cell 157:1262

Hudson KL (2006) Preimplantation genetic diagnosis: public policy and public attitudes. Fertil Steril 85:1638

Hughes JJ (2010) Humans should be free of all biological limitations including sex. Am J Bioethics 10:15

Human Fertilisation and Embryology (Mitochondrial Donation) Regulations 2015

Human Fertilisation and Embryology Act 1990 (Chapter 37) 48

Human Fertilisation and Embryology Act 2008 (Chapter 22) 120

Human Fertilisation and Embryology Authority S and ID, 'The HFE Act (and Other Legislation) - HFEA' http://hfeaarchive.uksouth.cloudapp.azure.com/www.hfea.gov.uk/134.html. Accessed 15 Jan 2018

Human Fertilization and Embryology Authority, Code of Practice. https://ifqlive.blob.core.windows.net/umbraco-website/2062/2017-10-02-code-of-practice-8th-edition-full-version-11th-revision-final-clean.pdf

Human Reproductive Technology Act [WA] 1991

Human Rights Commission of Malaysia (Amendment) Act 2009 (Act A1353)

Human Rights Commission of Malaysia Act 1999 (Act A597)

Human Rights Watch, World Report 2016: 'Thailand'. https://www.hrw.org/world-report/2016/country-chapters/thailand

Human Rights Watch, World Report 2018: Rights Trends in Australia. https://www.hrw.org/world-report/2018/country-chapters/australia

Human Rights Watch, World Report 2018: Rights Trends in Malaysia. https://www.hrw.org/world-report/2018/country-chapters/malaysia

Human Rights Watch, World Report 2018: Rights Trends in Thailand. https://www.hrw.org/world-report/2018/country-chapters/thailand

Human Tissues Act 1974 (Act 130) 1974 (Act 130) 9

Humanists UK (3 February 2013) Abortion and Sexual and Reproductive Rights. https://humanism.org.uk/campaigns/public-ethical-issues/sexual-and-reproductive-rights/

Humanists UK (9 October 2017) Humanists UK to Intervene in Northern Ireland Abortion Supreme Court Case. https://humanism.org.uk/2017/10/09/supreme-court-grants-humanists-uk-permission-to-intervene-in-northern-ireland-abortion-case/

Hume D (1896) A treatise of human nature. Oxford University Press, Oxford

Hunt L (15 November 2018) Thailand's politics heat up ahead of elusive election. The Diplomat. https://thediplomat.com/2018/11/thailands-politics-heat-up-ahead-of-elusive-election/

ICPD, 'Policy Recommendations for the ICPD Beyond 2014: Sexual and Reproductive Health & Rights for All' http://icpdtaskforce.org/resources/policy-recommendations-for-the-ICPD-beyond-2014.pdf

Illustrative Text: Human Embryology and Fertilisation Authority Act 1990. As Amended: An Illustrative Text

IMDb, *Science and the Swastika*. http://www.imdb.com/title/tt0808104/

Inglis-Arkell E (28 April 2013) Technology isn't magic: why Clarke's third law always bugged me. io9. http://io9.gizmodo.com/technology-isnt-magic-why-clarkes-third-law-always-bug-479194151

Ingram C (12 March 2003) State issues apology for policy of sterilization. Los Angeles Times. http://articles.latimes.com/2003/mar/12/local/me-sterile12

Inhorn MC (2003) Global infertility and the globalization of new reproductive technologies: illustrations from Egypt. Soc Sci Med 56:1837

Inhorn MC (2007) Masculinity, reproduction, and male infertility surgery in the Middle East. J Middle East Women's Stud 3:1

Inhorn M (2007a) Reproductive disruptions and assisted reproductive technologies in the Muslim world. In: Reproductive disruptions: gender, technology and biopolitics in the new millennium

Inhorn M (2007b) Reproductive disruptions: gender, technology and biopolitics in the new millennium, vol 11. Berghan Books

Inhorn MC (2009) Right to assisted reproductive technology: overcoming infertility in low-resource countries. Int J Gynecol Obstet 106:172

Inhorn MC (2013) Why Me? Male infertility and responsibility in the Middle East. Men Masculinities 16:49

Inhorn MC, Birenbaum-Carmeli D (2008) Assisted reproductive technologies and culture change. Ann Rev Anthropol 3:177

Inhorn MC, Patrizio P (2015) Infertility around the globe: new thinking on gender, reproductive technologies and global movements in the 21st century. Hum Reprod Update 21:411

International Women's Development Agency (IWDA) (26 April 2015) Reproductive Rights, Abortion & Zoe's Law: Why Freedom of Choice Is Still Feminism's Biggest Fight. https://iwda.org.au/reproductive-rights-abortion-zoes-law-why-freedom-of-choice-is-still-feminisms-biggest-fight/

Jabour B (10 November 2015) UN examines Australia's forced sterilisation of women with disabilities. The Guardian. http://www.theguardian.com/australia-news/2015/nov/10/un-examines-australias-forced-sterilisation-of-women-with-disabilities

Jacobs B (2 August 2017) Donald Trump proposes law to cut immigration numbers by half in 10 years. The Guardian. http://www.theguardian.com/us-news/2017/aug/02/trump-immigration-law-reduction-10-years

Jaede M (2017) The concept of the common good. Working Paper, University of Edinburgh, p 18

Jalloh v Germany (Application 54810/00) [2006] HUDOC (Grand Chamber, European Court of Human Rights)

Jasanoff S (2016) The ethics of invention: technology and the human future. WW Norton & Company

Jefferson T (2002) The Declaration of Independence. Scholastic Inc

Jensen EM (2012) The individual mandate, taxation, and the constitution. J Tax Invest 30:31

John Pfeiffer Pty Ltd v Rogerson (2000) 203 CLR 503

Jomo KS (2004) The new economic policy and interethnic relations in Malaysia. UNRISD, Geneva

Juslaws & Consult, 'Unofficial Translation- Act Providing for the Protection for Children Born Through Assisted Reproductive Technologies'

Kaczynski T (1991) Industrial society and its future

Kaelber L, Eugenics: compulsory sterilization in 50 American states. https://www.uvm.edu/~lkaelber/eugenics/

Kamel RMA (2013) Assisted reproductive technology after the birth of Louise Brown. Gynecol Obstet 3:156. https://www.omicsonline.org/assisted-reproductive-technology-after-the-birth-of-louise-brown-2161-0932.1000156.php?aid=16043

Kango-Singh M (2010) In: Speicher M, Antonarakis SE, Motulsky AG (eds) Vogel and Motulsky's human genetics-problems and approaches. BioMed Central. https://humgenomics.biomedcentral.com/articles/10.1186/1479-7364-5-1-73

Kant I (2003) The critique of pure reason. JMD Meiklejohn tr, The Project Gutenberg

Karovska-Andonovska B (2014) Right to privacy in the constitutions of the European Countries and the US constitution. Vizione 22:127

Kass L (1997) The wisdom of repugnance. New Republic:17

Kater MH (1987) The burden of the past: problems of a modern historiography of physicians and medicine in Nazi Germany. Ger Stud Rev 10:31

Kaye J et al (2015) Dynamic consent: a patient interface for twenty-first century research networks. Eur J Hum Genet 23:141

Kelly J (8 April 2016) Why are Northern Ireland's abortion laws different? BBC News. http://www.bbc.com/news/magazine-35980195

Kendler K, Greenspan R (2006) The nature of genetic influences on behaviour: lessons from simpler organisms. Am J Psychiatry 163:1683–1694

Kerr K (2014) Queensland abortion laws: criminalising one in three women. QUT Law Rev 14:24

Keszthelyi C (2 May 2017) Government's "Stop Brussels" Campaign Revs Up. Budapest Bus J. http://bbj.hu/politics/governments-stop-brussels-campaign-revs-up_132259

Kevles DJ (1999) Eugenics and human rights. Br Med J 319:435

Kirk J (2003) Conflicts and choice of law within the Australian Constitutional Context. Federal Law Rev 31:247

Kleinlein T (2012) Constitutionalization of international law. Das Max-Planck-Institut für ausländisches öffentliches Recht und Völkerrecht 231:703

Knowles LP, Kaebnick GE (eds) (2007) Reprogenetics: law, policy and ethical issues. John Hopkins University Press, Baltimore

Koskenniemi M, Leino P (2002) Fragmentation of international law? Postmodern anxieties. Leiden J Int Law 15(3):553–579

Krajewska A (2012) International biomedical law in search for its normative status. Revista De Derecho Y Genoma Humano = Law and the Human Genome Review 36:121

Krattiger A (September 2013) Promoting access to medical innovation. World Intellectual Property Organization (WIPO)|Magazine. http://www.wipo.int/wipo_magazine/en/2013/05/article_0002.html

Kurzweil R (2014) The singularity is near. In: Sandler RL (ed) Ethics and emerging technologies. Palgrave Macmillan

Lander ES et al (2019) Adopt a moratorium on heritable genome editing. Nature 567:165

Lando H (2017) Alf Ross and the functional analysis of law. SSRN Electronical J 17

Lau PL (2018) The Genius & The Imbecile: disentangling the "Legal" framework of autonomy in modern liberal eugenics, from non-therapeutic gene enhancement use in gene editing technologies. In: Current debates in international relations and law, vol 4. IJOPEC, London

Laurie G, Harmon SHE, Arzuaga F (2012) Foresighting futures: law, new technologies, and the challenges of regulating for uncertainty. Law Innov Technol 4:1

Lawrence v Texas [2003] Supreme Court 539 U.S. 558

Lee HP (2004) Competing conceptions of rule of law in Malaysia. In: Asian discourses of rule of law: theories and implementation of rule of law in twelve Asian countries, France and the U.S. Routledge Taylor & Francis Group, London

Lee MYK (2008) Universal human dignity: some reflections in the Asian context. Asian J Comp Law 3:1

Lee HP (2010) A Federal Human Rights Act and the reshaping of Australian Constitutional Law. UNSW Law J 33(1):88

Lee MYH (31 March 2016a) Donald Trump's Claim He Evolved into "pro-Life" Views, like Ronald Reagan. The Washington Post. https://www.washingtonpost.com/news/fact-checker/wp/2016/03/31/donald-trumps-claim-he-evolved-into-pro-life-views-like-ronald-reagan/?noredirect=on&utm_term=.a98e5a4e62f8

Lee H-A (2016b) Affirmative action regime formation in Malaysia and South Africa. J Asian Afr Stud 51:511

Leem SY (2017) Gangnam-style plastic surgery: the science of westernized beauty in South Korea. Med Anthropol 36:657

Lembaga Tatatertib Perkhidmatan Awam, Hospital Besar Pulau Pinang v Utra Badi A/L K Perumal (2000) Malayan Law J 3:281 (Kuala Lumpur Court of Appeal)

Lessig L (1997) The constitution of code: limitations on choice-based critiques of cyberspace regulation. Commlaw Conspectus: J Commun Law Technol Policy 5:181

Lessig L (1999) Code and other laws of cyberspace. Basic Books

Lessig L (2001) The future of ideas: the fate of the commons in a connected world, 1st edn. Random House

Lessig L (2006) Code: Version 2.0, 2nd edn. Basic Books

Level Crossings, Law Commission Consultation Paper No. 194, Scottish Law Commission and Law Commission Discussion Paper No. 143, 'Regulatory Theory' (2010). https://www.scot-lawcom.gov.uk/files/5312/8024/5698/regulatory_theory.pdf

Levett C (13 January 2007) Toothless Tiger ASEAN hopes to replace polite silence with a roar. Sydney Morning Herald. https://www.smh.com.au/news/world/toothless-tiger-asean-hopes-to-replace-polite-silence-with-a-roar/2007/01/12/1168105177847.html

Lewis R (2017) Human genetics: concepts and applications, 12th edn. McGraw-Hill Publishing Company

Lewis v HSBC Bank plc UKEAT/0364/06/RN and UKEAT/0412/06/RN

Lie W (2004) Equality and non-discrimination under international human rights law. Research Notes, Norwegian Centre for Human Rights. University of Oslo, Oslo

Liebert W, Schmidt JC (2010) Collingridge's Dilemma and technoscience: an attempt to provide a clarification from the perspective of the philosophy of science. Poiesis Praxis 7:55

Lincoln R (16 May 2018) CRISPR pioneer Jennifer Doudna explains gene-editing technology in Prather lectures. Harvard Gazette. https://news.harvard.edu/gazette/story/2018/05/crispr-pioneer-jennifer-doudna-explains-gene-editing-technology-in-prather-lectures/

Lipohar v The Queen (1999) 200 CLR 485

Liu CK (2007) "Saviour Siblings"? The distinction between PGD with HLA tissue typing and pre-implantation HLA tissue typing: winner of the Max Charlesworth Prize Essay 2006. J Bioeth Inq 4:65

Ly KD, Agarwal A, Nagy ZP (2011) Preimplantation genetic screening: does it help or hinder IVF treatment and what is the role of the embryo? J Assist Reprod Genet 28:833

Ma Y, Zhao Y, Liao M (2015) The values demonstrated in the constitution of the People's Republic of China. In: Ladikas M et al (eds) Science and technology governance and ethics. Springer

Macklin R (2003) Dignity is a useless concept. BMJ 327:1419

Mahalatchimy A et al (2012) The legal landscape for advanced therapies: material and institutional implementation of European Union Rules in France and the United Kingdom. J Law Soc 39:131

Mahlmann M (2012) Human dignity and autonomy in modern constitutional orders. In: Rosenfeld M, Sajo A (eds) The Oxford handbook of comparative constitutional law. Oxford University Press, Oxford

Maier KE (1989) Pregnant women: fetal containers or people with rights? Affilia 4:8

Malaysian Medical Council (14 November 2006) Guideline of the Malaysian Medical Council MMC Guideline 003/2006 assisted reproduction

Malm S (19 February 2018) Malaysian Rapper Probed over Lunar New Year Dog Video. Mail Online. http://www.dailymail.co.uk/news/article-5407651/Malaysian-rapper-probed-Lunar-New-Year-dog-video.html

Mandel GN (2009) Regulating emerging technologies. Law Innov Technol 1:75

Marshall J, Transport of biological materials. University of Edinburgh, p 51

Matsubara Y (1998) The enactment of Japan's sterilization laws in the 1940s: a prelude to postwar eugenic policy. Historia Scientiarum 8:187

Maturo A (2012) Medicalization: current concept and future directions in a bionic society. Mens Sana Monogr 10:122

McCrudden C (2008) Human dignity and judicial interpretation of human rights. Eur J Int Law 19:655

McDonald H (23 October 2017) Supreme Court to hear challenge to Northern Ireland Abortion Law. The Guardian, http://www.theguardian.com/world/2017/oct/23/supreme-court-to-hear-challenge-to-northern-ireland-abortion-law

McDonald H, Graham-Harrison E, Baker S (26 May 2018) Ireland votes by landslide to legalise abortion. The Guardian. https://www.theguardian.com/world/2018/may/26/ireland-votes-by-landslide-to-legalise-abortion

Mehlman MJ (1999) How will we regulate genetic enhancement. Wake Forest Law Rev 34:671

Meidinger E (1987) Regulatory culture: a theoretical outline. Law Policy 9(4):355–386

Mendel as the Father of Genetics:: DNA from the Beginning. http://www.dnaftb.org/1/bio.html

Metzl JF (10 October 2014) The genetics epidemic. Foreign Affairs. https://www.foreignaffairs.com/articles/united-states/2014-10-10/genetics-epidemic. Accessed 14 Jan 2017

Michaels R (2006) The functional method of comparative law. In: Reimann M, Zimmermann R (eds) The Oxford handbook of comparative law. Oxford University Press, Oxford

Miko I (2008) Gregor Mendel and the principles of inheritance. Nat Educ:134

Miller J (2014) The influence of human rights and basic rights in private law in the United States. Am J Comp Law 62:133

Minari J et al (2014) The emerging need for family-centric initiatives for obtaining consent in personal genome research. Genome Med 6:118

Ministry of Foreign Affairs of Denmark, Denmark for the UN Human Rights Council. http://um.dk/en/foreign-policy/denmark-for-the-un-human-rights-council/

Ministry of Health Malaysia, Medical Development Division (October 2012) Standards for assisted reproductive technology facility - embryology laboratory and operation theatre

Mohamad Bin Senik v Public Prosecutor (2005) Malay Law J 4:164 (Kuala Lumpur Court of Appeal)

Mohd Mutalip SS (2012) Promoting Malaysia through "fertility" tourism. J Tourism Hosp Culinary Arts 4:1

Montgomery v Lanarkshire Health Board [2015] UKSC 11

Morningstar News (23 February 2018) Highest Court in Malaysia to hear appeal of Christian Converts from Islam. https://morningstarnews.org/2018/02/highest-court-malaysia-hear-appeal-christian-converts-islam/

Morrison M (2016) Overdiagnosis, medicalisation and social justice: commentary on Carter et al (2016) 'A definition and ethical evaluation of overdiagnosis. J Med Ethics 42:720

Morrison M (2017) "A good collaboration is based on unique contributions from each side": assessing the dynamics of collaboration in stem cell science. Life Sci Soc Policy 13

Morrison M, Dickenson D, Lee SS-J (2016) Introduction to the article collection "Translation in healthcare: ethical, legal, and social implications". BMC Med Ethics 17

Müller J-W (11 February 2016) The problem with Poland. The New York Review of Books. http://www.nybooks.com/daily/2016/02/11/kaczynski-eu-problem-with-poland/

Mulvenna B (2004) Pre-implantation genetic diagnosis, tissue typing and beyond: the legal implications of the Hashmi Case. Med Law Int 6:163

Munné S et al (1994) Chromosome Mosaicism in human embryos. Biol Reprod 51:373

Munro K (10 March 2017) Fred Nile gives renewed push to Zoe's law to criminalise harm to a fetus. The Sydney Morning Herald. http://www.smh.com.au/nsw/fred-nile-gives-renewed-push-to-zoes-law-to-criminalise-harm-to-a-fetus-20170309-guup40.html

Muntarbhorn V (2005) Rule of law and aspects of human rights in Thailand- from conceptualization to implementation? In: Peerenboom R (ed) Asian discourses of rule of law. Routledge, Abingdon

Murphy T (2013) Health and human rights. Hart Publishing, Oxford

Mutua M (2001) Savages, victims, and saviors: the metaphor of human rights. Harv Int Law J 42:201

Nalbandian E (2011) Sociological jurisprudence: Roscoe Pound's discussion on legal interests and jural postulates. Mizan Law Rev 5:9

NaRanong A, NaRanong V (2011) The effects of medical tourism: Thailand's experience. Bull World Health Org 89:336

National Health and Medical Research Council (2017) Ethical guidelines on the use of assisted reproductive technology in clinical practice and research. Australian Government NHMRC, Canberra

National Human Genome Research Institute. (National Human Genome Research Institute (NHGRI)) https://www.genome.gov/

National Human Genome Research Institute (NHGRI) 'An Overview of the Human Genome Project'. https://www.genome.gov/12011238/an-overview-of-the-human-genome-project/

Natipodhi P (2014) Practice of sex selection in Asian Region. Working Paper Series Asian Law Institute

Nature (1998) China's "eugenics" law still disturbing despite relabelling. Nature 394:707

Nayal MB (2013) Preimplantation genetic diagnosis. http://emedicine.medscape.com/article/273415-overview

Nazlina M (14 January 2016) Prosecution wraps up case in Azmi Sharom Sedition Trial. The Star. https://www.thestar.com.my/news/nation/2016/01/14/azmi-sharom-trial-prosecution/

Nelson TE, Oxley ZM (1999) Issue framing effects on belief importance and opinion. J Polit 61:1040

Nemudryi AA et al (2014) TALEN and CRISPR/Cas genome editing systems: tools of discovery. Acta Naturae 6:22

Newland K (July 2015, Issue No. 13) Irregular Maritime migration in the Bay of Bengal: the challenges of protection, management and cooperation. International Organization for Migration, Migration Policy Institute. http://publications.iom.int/system/files/pdf/mpi-iom_brief_no_13.pdf

Nianias H (2 November 2015) It's 2015 and British women's right to safe abortions is still uncertain. The Debrief. https://thedebrief.co.uk/news/real-life/2015-british-womens-right-safe-abortions-still-uncertain/

Nielsen VL, Parker C (2009) Testing responsive regulation in regulatory enforcement. Regul Gov 3:376

Nolan MA (2012) The constitutional system of Thailand: a contextual analysis. Aust J Asian Law 13:1

Noonan JT (1977) Abortion in the American context. Human Life Rev 3:29

Norberg J (20 August 2016) 'Why can't we see that we're living in a golden age? The Spectator. https://www.spectator.co.uk/2016/08/why-cant-we-see-that-were-living-in-a-golden-age/

North Carolina Administration, NC DOA: Welcome to the Office of Justice for Sterilization Victims. https://ncadmin.nc.gov/about-doa/special-programs/welcome-office-justice-sterilization-victims

Norwitz ER, Levy B (2013) Noninvasive prenatal testing: the future is now. Rev Obstet Gynecol 6:48

Noyes J (21 November 2013) 'On Zoe's law, and the accidental/on purpose erosion of your reproductive rights. Junkee. http://junkee.com/on-zoes-law-and-the-accidentalon-purpose-erosion-of-your-reproductive-rights/21659

Nozick R (1974) Anarchy, State, and Utopia. Basic Books

Nuffield Council on Bioethics (2016) Genome editing: an ethical review. Nuffield Council on Bioethics

Nuffield Council on Bioethics (2017) Non-invasive prenatal testing: ethical issues. Nuffield Council on Bioethics

Nuffield Council on Bioethics (2018) Genome editing and human reproduction: social and ethical Issues. Nuffield Council on Bioethics

Nunan D, Di Domenico M (2017) Big data: a normal accident waiting to happen? J Bus Ethics 145:481

Nyamu-Musembi C, Cornwell A (2004) What is the 'right-based approach' all about?: Perspectives from international development agencies. Institute of Development Studies, Brighton

O'Cinneide C (14 September 2011) Equality: a constitutional principle? (UK Constitutional Law Association) https://ukconstitutionallaw.org/2011/09/14/colm-ocinneide-equality-a-constitutional-principle/

O'Cinneide C (2012) The human rights act and the slow transformation of the UK's political constitution. Annales U. Sci. Budapestinensis Rolando Eotvos Nominatae 53:239

O'Connor A (28 October 2017) How the death of Savita Halappanavar changed the abortion debate. The Irish Examiner. http://www.irishexaminer.com/analysis/how-the-death-of-savita-halappanavar-changed-the-abortion-debate-461787.html

O'Mahony C (2012) There is no such thing as a right to dignity. Int J Const Law 10:551

Obergefell v Hodges [2015] Supreme Court 14–556

OECD (2010) Biomedicine and Health Innovation: Synthesis Report. http://www.oecd.org/health/biotech/46925602.pdf

Osborn F (1937) Development of a eugenic philosophy. Am Sociol Rev 2:389

Owigar JWB (2002) Ethics and living values in constitution. Const Kenya Rev Comm, 4. http://www.commonlii.org/ke/other/KECKRC/2002/4.html

Panaspornprasit C (2017) Thailand: the historical and indefinite transitions. Southeast Asian Aff:351

Pardes A (14 January 2016) How commercial surrogacy became a massive international business. Vice. https://www.vice.com/en_us/article/how-commercial-surrogacy-became-a-massive-international-business

Parens E (2013) On good and bad forms of medicalization. Bioethics 27:28

Parker C (2013) Twenty years of responsive regulation: an appreciation and appraisal: twenty years of responsive regulation. Regul Gov 7:2

Parliament of Malaysia, 'Official Portal of The Parliament of Malaysia' http://www.parlimen.gov.my/index.php?lang=en

Parry v the United Kingdom (Application No. 42971/05)

Parsa-Parsi RW (2017) The revised declaration of Geneva: a modern-day physician's pledge. J Am Med Assoc 318:1971

Patterson D et al (2015) The dark future of constitutionalism: the cosmopolitan constitution. Const Commentary 30:667

Pellissier H (22 June 2015) Do you fear eugenics? China does not, and that's a problem - interview with Chad White. Institute for Emerging Technologies and Ethics. https://ieet.org/index.php/IEET2/more/pellissier20150622

Penal Code 1936 (Act 574) 314

Pennings G (2004) Saviour siblings: using preimplantation genetic diagnosis for tissue typing. Int Congr Ser 1266:311

Pennings AJ (1 July 2012) Arthur C. Clarke's three laws of innovation. Writings on Digital Strategies, ICT Economies, and Global Communications. http://apennings.com/political-economies-in-sf/arthur-c-clarkes-three-laws-of-innovation/

Peter F (2008) Pure epistemic proceduralism. Episteme: A J Soc Epistemol 5:33

Peters PG (2004) How safe is safe enough? Obligations to the children of reproductive technology. Oxford University Press, Oxford

Petersen K (2014) Decriminalizing abortion- the Australian experience. In: Rowlands S (ed) Abortion care. Cambridge University Press, Cambridge

Phillips DC (1995) The good, the bad, and the ugly: the many faces of constructivism. Edu Res 24:5

Pitkin H (1981) Justice: on relating public and private. Polit Theory 9:327–352

Planned Parenthood of Southeastern Pa. v. Casey, 505 U.S. 833 (1992) (*Justia Law*) https://supreme.justia.com/cases/federal/us/505/833/

Polya R (2008) Chronology of Genetic Engineering Regulation in Australia: 1953–2008. https://www.aph.gov.au/About_Parliament/Parliamentary_Departments/Parliamentary_Library/pubs/BN/0809/ChronGeneticEngineeringA

Poort L, van Beers B, van Klink B (2016) Introduction: symbolic dimensions of biolaw. In: Symbolic legislation theory and developments in biolaw. Springer, Basel

Portnoy J (27 February 2015) Va. General Assembly agrees to compensate eugenics victims. The Washington Post. https://www.washingtonpost.com/local/virginia-politics/va-general-assembly-agrees-to-compensate-eugenics-victims/2015/02/27/b2b7b0ec-be9e-11e4-bdfa-b8e8f594e6ee_story.html?noredirect=on&utm_term=.129bf4f66cb1

Posner RA (2006) The role of the judge in the twenty-first century. BUL Rev 86:1049

Powell T (20 December 2017) May Faces Brexit Grilling by Powerful Committee of MPs. Evening Standard. https://www.standard.co.uk/news/politics/brexit-latest-theresa-may-to-be-grilled-on-eu-withdrawal-negotiations-by-committee-of-powerful-mps-a3723596.html

Powell CMH. Being human: how should we define life and personhood? Enrichment Journal. http://enrichmentjournal.ag.org/201002/201002_134_define_person.cfm

Pretty v the United Kingdom (Application No 2346/02) [2002] HUDOC (Fourth Section, European Court of Human Rights)

Prohibition of Human Cloning for Reproduction Act 2002 2016 (No 144, 2002)

Pronto AN (2008) Some thoughts on the making of international law. Eur J Int Law 19:601

Protection for Children Born Through Assisted Reproductive Technologies Act 2015 (167/2553)

Public Prosecutor v Azmi Bin Sharom (Criminal Ref. No. 06-5-12/2014(W)) (6 October 2015). http://www.kehakiman.gov.my/judgment/file/PRESS_SUMMARY_PP_V_AZMI_SHAROM.pdf

R (on the application of Quintavalle) v Human Fertilisation and Embryology Authority [2005] UKHL 28

R and F v the United Kingdom (Application No. 35748/05)

R v Brennan and Leach [2010] QDC 329

R v Sood [2006] NSWSC 1141

Racial Discrimination Act 1975

Rantsev v Cyprus & Russia (Application No 25965/04) HUDOC (First Section, European Court of Human Rights)

Rath J (2018) Safety and security risks of CRISPR/Cas9. In: Schroeder D et al (eds) Ethics dumping. Springer International Publishing, New York

Rauch BJ et al (2017) Inhibition of CRISPR-Cas9 with bacteriophage proteins. Cell 168:150

Rawlinson MC, Donchin A (2005) The quest for universality: reflections on the universal draft declaration on bioethics and human rights. Dev World Bioeth 5:258

Reaney P (25 September 2004) In search of baby perfect. Center for Genetics and Society. http://www.geneticsandsociety.org/article.php?id=1469

Reardon S (2015) Global summit reveals divergent views on human gene editing. Nature 528:173

Regulation (EU) 2016/679 of the European Parliament and Of the Council 2016

Reiss MJ, Straughan R (1996) Improving nature? The science and ethics of genetic engineering. Cambridge University Press, Cambridge

Research and Library Division Department of Jurisconsult (2016) Bioethics and the Case-Law of the Court. Council of Europe / European Court of Human Rights, Research Report

Research Involving Human Embryos Act 2002 2016 (No 145, 2002)

Resnik DB, Vorhaus DB (2006) Genetic modification and genetic determinism. Philos Ethics Humanit Med 1

Rhode H (15 June 2012) Can Muslims reopen the gates of Ijtihad? Gatestone Institute. http://www.gatestoneinstitute.org/3114/muslims-ijtihad

Ribeiro GL (2001) Cosmopolitanism. Int Encycl Soc Behav Sci 4:2842

Richardson M (2002) Whither breach of confidence: a right of privacy for Australia. Melb Univ Law Rev 26:381

Riggan K (31 December 2009) G12 country regulations of assisted reproductive technologies. The Centre for Bioethics and Human Dignity. https://cbhd.org/content/g12-country-regulations-assisted-reproductive-technologies

Robertson JA (2003a) Extending preimplantation genetic diagnosis: medical and non-medical uses. J Med Ethics 29:213

Robertson JA (2003b) Extending preimplantation genetic diagnosis: the ethical debate: ethical issues in new uses of preimplantation genetic diagnosis. Hum Reprod 18:465

Robertson J (2010) Eugenics in Japan: Sanguinous repair. In: Bashford A, Levine P (eds) The Oxford handbook of the history of eugenics

Robinson M (2007) The value of a human rights perspective in health and foreign policy. Bull World Health Org 85:241

Roe v. Wade, 410 U.S. 113 (1973) (Justia Law) https://supreme.justia.com/cases/federal/us/410/113/

Romanow L (2012) The women of Thailand. Glob Majority E-J 3:44

Rorty MV (2003) The future of human nature. Notre Dame Philos Rev

Rosen M (2012) Dignity: its history and meaning. Harvard University Press, Cambridge

Roth R (2017) She doesn't deserve to be treated like this: prisons as sites of reproductive injustice. In: Ross LJ et al (eds) Radical reproductive justice: foundations, theory, practice, critique. The Feminist Press, New York

Rothstein A (Winter 2015, Number 44) Vaccines and their critics, then and now. The New Atlantis. https://www.thenewatlantis.com/publications/vaccines-and-their-critics-then-and-now

Royal College of Nursing of the United Kingdom v Department of Health and Social Security [1981] AC 800

RT Question More (2017) Pentagon revealed as top funder of controversial gene editing tech — RT US News. https://www.rt.com/usa/412019-pentagon-darpa-gene-drive/

Ruiz-Chiriboga OR (2013) The American Convention and the protocol of San Salvador: two intertwined treaties-non-enforceability of economic, social and cultural rights in the Inter-American System. Neth Quart Human Rights 31:159

Saetz SB, Court MV, Henshaw, MW (1985) Eugenics and the third Reich. Eugen Bull

Sajó A, Ryan C (2016) Judicial reasoning and new technologies: framing, newness, fundamental rights and the internet. In: Pollicino O, Romeo G (eds) The internet and constitutional law. The protection of fundamental rights and constitutional adjudication. Routledge, Abingdon

Saliba I (June 2011) What Is Sharia Law? I Law Library of Congress. https://www.loc.gov/law/help/sharia-law.php

Saltman Engineering Co. Ltd. V Campbell Engineering Co. [1984] 65 RPC 203

Sandberg R (2016) The failure of legal pluralism. Eccles Law J 18:137

Sandel M (2004) The case against perfection. Atl Mon 293:51

Sanders L (2017) 40 more genes linked to intelligence. Sci News 191:14

Sándor J (2012) Bioethics and basic rights: persons, humans and boundaries of life. In: Rosenfeld M, Sajos A (eds) The Oxford handbook of comparative constitutional law. Oxford University Press, Oxford, pp 1142–1161

Sándor J (2015) The ethical and legal analysis of embryo preimplantation testing policies in Europe. In: Sills ES (ed) Screening the single euploid embryo. Springer International Publishing, New York

Sands MT (2013) Saviour siblings: a relational approach to the welfare of the child in selective reproduction. Routledge, Abingdon

Savulescu J (2001) Procreative beneficence: why we should select the best children. Bioethics 15:413

Savulescu J (2007a) Genetic interventions and the ethics of enhancement of human beings. In: The Oxford handbook of bioethics. Oxford University Press, Oxford

Savulescu J (2007b) In defence of procreative beneficence. J Med Ethics 33:284

Savulescu J (2009) Genetic interventions and the ethics of enhancement of human beings. Read Philos Technol:417

Savulescu J, Kahane G (2011) Disability: a welfarist approach. Clin Ethics 6:45

Schaefer GO (2 August 2016) The future of genetic enhancement is not in the west. The Conversation. http://theconversation.com/the-future-of-genetic-enhancement-is-not-in-the-west-63246

Schloendorff v The Society of the New York Hospital (1914) 211 NY 125 (The Court of Appeals of New York)

Schmeer K (1999) Stakeholder analysis guidelines. Policy toolkit for strengthening health sector reform 1

Schwartz CM et al (2016) Synthetic RNA polymerase III Promoters facilitate high-efficiency CRISPR–Cas9-mediated genome editing in *Yarrowia Lipolytica*. ACS Synthetic Biol 5:356

Scott R (2006) Choosing between possible lives: legal and ethical issues in preimplantation genetic diagnosis. Oxf J Leg Stud 26:153

Selgelid MJ (2014) Modern eugenics and human enhancement. Med Healthcare Philos 17:3

Sen A (1997) Human rights and Asian values. New Republic:33

Serour GI (2005) Religious perspectives of ethical issues in ART: Islamic perspectives of ethical issues in ART. Middle East Fertil Soc J 10:185–190

Sex Discrimination Act 1984

Shanks P (2005) Human genetic engineering: a guide for activists, skeptics, and the very perplexed. Nation Books

Sharp T (12 June 2013) Right to privacy: constitutional rights & privacy laws. Live Science. https://www.livescience.com/37398-right-to-privacy.html

Shear MD (28 June 2018) Supreme Court Justice Anthony Kennedy Will Retire. The New York Times. https://www.nytimes.com/2018/06/27/us/politics/anthony-kennedy-retire-supreme-court.html

Sheehan M (2009) Making sense of the immorality of Unnaturalness. Camb Q Healthc Ethics 18:177

Sheffield and Horsham v the United Kingdom (Application No: 31-32/1997/815-816/1018-1019) and

Sheldon S (2004) Hashmi and Whitaker: an unjustifiable and misguided distinction? Med Law Rev 12:137

Sholley JB (1951) Constitution of the United States of America, *Cases on Constitutional Law*. Bobbs-Merrill

Shuaib FS (2012) The Islamic legal system in Malaysia. Pacific Rim Law Policy J 21:85

Shuster E (1997) Fifty years later: the significance of the Nuremberg code. N Engl J Med 337:1436

Sifris R, Belton S (2 June 2017) Australia: abortion and human rights. Health and Human Rights Journal. https://www.hhrjournal.org/2017/06/australia-abortion-and-human-rights/

Siliadin v France (Application No 73316/01) [2005] HUDOC (Second Section, European Court of Human Rights)

Silver L (1997) Remaking Eden: how genetic engineering and cloning will transform the American family. Avon Books Inc

Silver LM (2000) Reprogenetics: third millennium speculation: the consequences for humanity when reproductive biology and genetics are combined. EMBO Rep 1:375

Simson RS (2014) What does the right to life really entail-a framework for depolarizing the abortion debate. Connecticut Public Int Law J 14:107

Singapore Democratic Party, Eugenics in Singapore. http://yoursdp.org//news/eugenics_in_singapore/2008-11-09-558

Singer P (2009) Parental choice and human improvement. In: Human enhancement. Oxford University Press, Oxford

Singer P (2016) The human genome and the genetic supermarket. In: Ethics in the real world: 82 brief essays on things that matter. Princeton University Press, Princeton

Skinner v. Oklahoma Ex Rel. Williamson, 316 U.S. 535 (1942) (Justia Law). https://supreme.justia.com/cases/federal/us/316/535/case.html

Smiley S (2005) Genetic modification: study guide (exploring the issues). Independence Educational Publishers

Smith MK (2013) The Human Fertilisation and Embryology Act 2008: restrictions on the creation of "Saviour Siblings" and the relevance of the harm principle. New Genet Soc 32:154

Smith J (3 August 2014) Mother Country: the harrowing truth behind Thai "fertility Tourism". The Independent. http://www.independent.co.uk/voices/comment/mother-country-the-harrowing-truth-behind-thai-fertility-tourism-9644517.html

Solomon JM (2010) New governance, preemptive self-regulation and the blurring of boundaries in regulatory theory and practice. Wis Law Rev:591

Somek A (2014) The cosmopolitan constitution. Oxford University Press, Oxford

Spriggs M, Savulescu J (2002) Saviour siblings. J Med Ethics 28:289

Stankovic B (2005) "It's a Designer Baby!": Opinions on regulation of preimplantation genetic diagnosis. UCLA J Law Technol:3

Stanton-Ife J (2000) Should equality be a constitutional principle? King's Law J 11:133

Stasi A (2015) Maternal surrogacy and reproductive tourism in Thailand: a call for legal enforcement. Ubon Ratchathani Law J 8:17–36

Stock G (2003) Redesigning humans: choosing our genes, changing our future. Mariner Books

Stock G (2005) Germinal choice technology and the human future. Reprod BioMed Online 10:27

Stockholm Resilience Centre 'Adaptive Governance' (6 December 2010). http://www.stockholm-resilience.org/research/research-streames/stewardship/adaptive-governance-.html

Stone G et al (2005) Constitutional law, 5th edn. Aspen Publishers

Stone GR et al (2013) (Chapter 7) Constitutional law, 7th edn. Wolters & Kluwer, Alphen aan den Rijn

Strategy and Information Directorate Human Fertilisation and Embryology Authority, 'The HFE Act (and Other Legislation) - HFEA' http://hfeaarchive.uksouth.cloudapp.azure.com/www.hfea.gov.uk/134.html

Strawson G (11 September 2014) Sapiens: a brief history of humankind by Yuval Noah Harari – review. The Guardian. https://www.theguardian.com/books/2014/sep/11/sapiens-brief-history-humankind-yuval-noah-harari-review

Sunstein C (1993) After the rights revolution, reconceiving the regulatory state. Harvard University Press, Cambridge

Supiot A (2007) Homo Juridicus: on the anthropological function of the law. Verso, Brooklyn

Supreme Court of the United States, Current Members. https://www.supremecourt.gov/about/biographies.aspx

Suter SM (2002) The routinization of prenatal testing. Am J Law Med 28:233

Sweet AS, Mathews J (2008) Proportionality balancing and global constitutionalism. Columbia J Transntl Law 72:47

Szoke H, Neame L, Johnson L (2006) Old technologies and new challenges: assisted reproduction and its regulation. In: Freckelton I, Petersen K (eds) Disputes & dilemmas in health law. Federation Press

Tan Tek Seng v Suruhanjaya Perkhidmatan Pendidikan & Anor (1996) Malay Law J 1:261 (Kuala Lumpur Court of Appeal)

Taranto S (22 January 2018) How abortion became the single most important litmus test in American Politics. The Washington Post. https://www.washingtonpost.com/news/made-by-history/wp/2018/01/22/how-abortion-became-the-single-most-important-litmus-test-in-american-politics/?noredirect=on&utm_term=.b4015648272b

Tarrant-Cornish T (26 December 2017) Richest country in the World: China to overtake the US as most powerful economy. Express. https://www.express.co.uk/news/world/896869/China-economy-US-richest-country-world-Donald-Trump-trade-GDP-research

Taylor AL (2004) Governing the globalization of public health. J Law Med Ethics 32:500

Taywaditep KJ, Coleman E, Dumronggittigule P. The International Encyclopedia of Sexuality: Thailand. http://www.sexarchive.info/IES/thailand.html

Teare HJA et al (2017) The RUDY study: using digital technologies to enable a research partnership. Eur J Hum Genet 25:816. http://www.nature.com/doifinder/10.1038/ejhg.2017.57

ten Have HAMJ, Jean MS (2009) The UNESCO universal declaration on bioethics and human rights: background, principles and application. UNESCO Publishing

Thai Law Forum, 'Thailand Medical Council Regulations on Surrogacy and IVF | Thailand Law Forum' http://www.thailawforum.com/medical-surrogacy-regulations/

Thailand Penal Code, Thailand Criminal Law (Unofficial English Translation) https://www.samui-forsale.com/law-texts/thailand-penal-code.html

Thananithichot S, Satidporn W (2016) Political dynasties in Thailand: the recent picture after the 2011 general election. Asian Stud Rev 40:340

The Economist (22 September 2016) Taking the rap - religious freedom in Malaysia. https://www.economist.com/news/asia/21707565-malaysias-culture-tolerance-under-threat-taking-rap

The Eleanor Roosevelt Papers Project, The Universal Declaration of Human Rights. https://www2.gwu.edu/~erpapers/humanrights/lessonplans/

The Guardian (4 March 2011) Profiles of the future by Arthur C Clarke – review. The Guardian. https://www.theguardian.com/science/2011/mar/04/profiles-future-arthur-clarke-review

The Nation (28 November 2014) Thailand Portal 'Commercial Surrogacy Bill Passes First Reading with 177 to 2 Votes' http://www.nationmultimedia.com/national/Commercial-surrogacy-bill-passes-first-reading-wit-30248734.html

The Star, 'Pakatan Harapan 100 Days' https://www.thestar.com.my/ph100/

The Straits Times (18 May 2018) "Cash Is King": The Fall of Malaysia's First Couple. https://www.straitstimes.com/asia/se-asia/cash-is-king-the-fall-of-malaysias-first-couple

Thepgumpanat P, Tanakasempipat P (21 May 2017) Three years after Coup, Junta is deeply embedded in Thai Life. Reuters. https://www.reuters.com/article/us-thailand-military/three-years-after-coup-junta-is-deeply-embedded-in-thai-life-idUSKCN18G0ZJ

Thomas A (31 July 2017) Super-intelligence and eternal life: transhumanism's faithful follow it blindly into a future for the elite. The Conversation. http://theconversation.com/super-intelligence-and-eternal-life-transhumanisms-faithful-follow-it-blindly-into-a-future-for-the-elite-78538

Thompson C (2016) IVF global histories, USA: between rock and a marketplace. Reprod Biomed Soc Online 2:128

Thomson M (2003) Confidence, privacy and damages: Hello! To Clarity. New Law Journal

Tien L (2005) Architectural regulation and the evolution of social norms. Yale J Law Technol 7:23

Tisdall S (23 July 2017) Europe seeks a long-term answer to a refugee crisis that needs a solution now. The Observer. http://www.theguardian.com/world/2017/jul/22/divided-europe-refugee-crisis-italy-serbia-greece

TMC Fertility Centre (21 February 2013) 'Home' (TMC Fertility Centre) http://www.tmcfertility.com/

TMC Fertility Centre (23 March 2015) 'Pre-Implantation Genetic Diagnosis (PGD)' (TMC Fertility Centre) http://www.tmcfertility.com/treatment-options/pre-implantation-genetic-diagnosis-pgd/

Torii T (1997) The new economic policy and the United Malays National Organization- with special reference to the restructuring of Malaysian Society. Dev Econ XXXV(3):209–239

Travis J (4 December 2015) Inside the Summit on human gene editing: a reporter's notebook. Science. http://www.sciencemag.org/news/2015/12/inside-summit-human-gene-editing-reporter-s-notebook

Trebilcock MJ, Iacobucci EM (2009) Designing competition law institutions: values, structure, and mandate. Loyola Univ Chicago Law J 41:455

Trstenjak V, Weingerl P (eds) (2016) The influence of human rights and basic rights in private law, vol 15. Springer

Tsaknis L (1993) The jurisdictional basis, elements and remedies in the action for breach of confidence- uncertainty abounds. Bond Law Rev 5:18

Turriziani J (2014) Designer babies: the need for regulation on the quest for perfection. http://scholarship.shu.edu/cgi/viewcontent.cgi?article=1595&context=student_scholarship. Law School Student Scholarship 595

Tysiac v Poland (Application No 5410/03) [2007] HUDOC (Fourth Section, European Court of Human Rights)

Uchiyama M, Nagai A, Muto K (2018) Survey on the perception of germline genome editing among the general public in Japan. J Human Genet 63:745

Uitz R (2015) Can you tell when an illiberal democracy is in the making? An appeal to comparative constitutional scholarship from hungary. Int J Const Law 13:279

Umeda S (6 April 2015) Thailand: new surrogacy law | global legal monitor. Library of Congress. http://www.loc.gov/law/foreign-news/article/thailand-new-surrogacy-law/

UNESCO (11 November 1997) Universal Declaration on the Human Genome and Human Rights. http://portal.unesco.org/en/ev.php-URL_ID=13177&URL_DO=DO_TOPIC&URL_SECTION=201.html

UNESCO (16 October 2003) International Declaration on Human Genetic Data. http://unesdoc.unesco.org/images/0013/001361/136112e.pdf

UNESCO (2006) The Universal Declaration on Bioethics and Human Rights. http://unesdoc.unesco.org/images/0014/001461/146180E.pdf

UNESCO (5 October 2015) UNESCO Panel of Experts Calls for Ban on "Editing" of Human DNA to Avoid Unethical Tampering with Hereditary Traits. https://en.unesco.org/news/unesco-panel-experts-calls-ban-editing-human-dna-avoid-unethical-tampering-hereditary-traits

United Nations (1979) Convention on the elimination of all forms of discrimination against women. http://www.un.org/womenwatch/daw/cedaw/cedaw.htm

United Nations, In opening debate on human cloning ban, some speakers urge outright prohibition, others favour partial ban to allow for medical advances | Meetings Coverage and Press Releases. https://www.un.org/press/en/2002/l2995.doc.htm

United Nations, Office of the High Commissioner, Human Rights, Sexual and Reproductive Health and Rights. http://www.ohchr.org/EN/Issues/Women/WRGS/Pages/HealthRights.aspx

United Nations, Development Policy and Analysis Division, Department of Economic and Social Affairs, 'Country Classification- Data Sources, Country Classification and Aggregation Methodology' (United Nations Secretariat) http://www.un.org/en/development/desa/policy/wesp/wesp_current/2014wesp_country_classification.pdf

United Nations General Assembly, Report of the Sixth Committee, 'International Convention against the Reproductive Cloning of Human Beings (A/59/516/Add.1). https://digitallibrary.un.org/record/542699/files/A_59_516_Add.1-EN.pdf

United Nations Population Fund (2014) Reproductive Rights Are Human Rights. https://www.unfpa.org/publications/reproductive-rights-are-human-rights

United States v. Carolene Products Co. 304 U.S. 144 (1938) (*Justia Law*) https://supreme.justia.com/cases/federal/us/304/144/case.html

US National Library of Medicine, Genetics Home Reference, 'What Are the Different Ways in Which a Genetic Condition Can Be Inherited?' (Genetics Home Reference) https://ghr.nlm.nih.gov/primer/inheritance/inheritancepatterns

van Gossum O, Arts B, Verheyen K (2010) From "smart regulation" to "regulatory arrangements". Policy Sci 43:245–261. https://doi.org/10.1007/s11077-101-9108-0

van Klink B (2016) Symbolic legislation: an essentially political concept. In: van Klink B, van Beers B, Poort L (eds) Symbolic legislation theory and developments in biolaw, vol 4. Springer, Berlin

VAWnet, National Resource Centre on Domestic Violence, Reproductive justice, reproductive health and reproductive rights: a framework. https://vawnet.org/sc/reproductive-justice-frameworks

Vetlesen AJ (2005) The future of human nature. Scand J Disabil Res 7:232

Vo v France (Application No. 53924/00)

Wade N (21 December 2017) Scientists seek ban on method of editing the human genome. The New York Times. https://www.nytimes.com/2015/03/20/science/biologists-call-for-halt-to-gene-editing-technique-in-humans.html

Wade N (19 January 2018) Scientists seek moratorium on edits to human genome that could be inherited. The New York Times. https://www.nytimes.com/2015/12/04/science/crispr-cas9-human-genome-editing-moratorium.html

Warbrick C (1989) Federal aspects of the European Convention on Human Rights. Mich J Int Law 10:698

Watson M (29 December 2013) Battle of life and death never ends for anti-abortion campaigners. The Age. http://www.theage.com.au/victoria/battle-of-life-and-death-never-ends-for-antiabortion-campaigners-20131228-300tv.html

Watts J (15 October 2017) UK Government Watchdog Pushes for New British "right to Equality" to Stop Brexit Leading to More Discrimination. The Independent. http://www.independent.co.uk/news/uk/politics/brexit-discrimination-laws-right-to-equality-uk-equalities-watchdog-eu-a7999461.html

Watts G (31 January 2018) "Eugenics" case highlights dark chapter in Japanese history. Asia Times. https://www.asiatimes.com/2018/01/article/eugenics-case-highlights-dark-chapter-japanese-history/

Weihua L, Xinwu Z (2000) Harvard Girl Liu Yiting: a character training record. Writers Publishing House

Weisberg SM, Badgio D, Chatterjee A (2017) A CRISPR new world: attitudes in the public toward innovations in human genetic modification. Front Public Health 5

Weltgesundheitsorganisation (ed) (2005) Sexually transmitted and other reproductive tract infections: a guide to essential practice

West R (1988) Jurisprudence and gender. Univ Chic Law Rev 55:1

Whitman JQ (2003) The two western cultures of privacy: dignity versus liberty. Yale Law J 113:1151. 73

Whittaker A (2002a) The struggle for abortion law reform in Thailand. Reprod Health Matters 10:45

Whittaker A (2002b) "The Truth of Our Day by Day Lives": abortion decision making in rural Thailand. Cult Health Sex 4:1

Whittaker A, Speier A (2010) "Cycling Overseas": care, commodification, and stratification in cross-border reproductive travel. Med Anthropol 29:363

Wiesenthal DL, Wiener NI (1999) Ethical questions in the age of the new eugenics. Sci Eng Ethics 5:383

Wilkie M. 'Australian Bill of Rights Bill 2017 No., 2017' 38

Winslow ER, Kodner IJ (2004) Ethics and genetic testing. Semin Colon Rectal Surg 15:186

Women's Agenda (16 August 2017) Reproductive freedom is unfinished business in Australia: Tanya Plibersek. https://womensagenda.com.au/politics/reproductive-freedom-unfinished-business-australia-tanya-plibersek/

Women's Electoral Lobby, Reproductive Rights (*Women's Electoral Lobby*) http://www.wel.org.au/reproductive_rights

Workplace Gender Equality Act 2012

World Bank (2018) Global economy to edge up to 3.1 percent in 2018 but future potential growth a concern. World Bank. http://www.worldbank.org/en/news/press-release/2018/01/09/global-economy-to-edge-up-to-3-1-percent-in-2018-but-future-potential-growth-a-concern

World Bank Global Knowledge and Research Hub (December 2017) Malaysia economic monitor: turmoil to transformation 20 years after the Asian Financial Crisis. https://openknowledge.worldbank.org/bitstream/handle/10986/28990/MalaysiaEconomicMonitor2017.pdf?sequence=6&isAllowed=y

World Health Organization, 'WHO | Infertility Is a Global Public Health Issue' (WHO) http://www.who.int/reproductivehealth/topics/infertility/perspective/en/

World Health Organization, GHO|Global Strategy for Women's, Children's and Adolescents' Health (2016–2030) (*WHO*). http://apps.who.int/gho/data/node.gswcah

World Health Organization, WHO Expert Advisory Committee on Developing Global Standards for Governance and Oversight of Human Genome Editing. https://www.who.int/ethics/topics/human-genome-editing/committee-members/en/

World Health Organization, WHO|Maternal and Reproductive Health (*WHO*). http://www.who.int/gho/maternal_health/en/

World Health Organization, WHO|World Health Organization. http://gamapserver.who.int/gho/interactive_charts/mdg5_mm/atlas.html

World Medical Association (19 October 2013) WMA Declaration of Helsinki - Ethical Principles for Medical Research Involving Human Subjects. https://www.wma.net/policies-post/wma-declaration-of-helsinki-ethical-principles-for-medical-research-involving-human-subjects/

World Medical Association (2018) WMA declaration of Helsinki - ethical principles for medical research involving human subjects. https://www.wma.net/policies-post/wma-declaration-of-helsinki-ethical-principles-for-medical-research-involving-human-subjects/

World Trade Organization, World Health Organization and World Intellectual Property Organization (2013) Promoting access to medical technologies and innovation: intersections between public health, intellectual property and trade. WTO

Wright T (12 September 2016) 1MDB Scandal Around Malaysian Prime Minister Najib Puts Spotlight on Wife. Wall Street Journal. http://www.wsj.com/articles/1mdb-scandal-around-malaysian-prime-minister-najib-puts-spotlight-on-wife-1473606895

X, Y and Z v The United Kingdom (1997) Grand Chamber, European Court of Human Rights Application No. 21830/93, HUDOC

Yap MT (2003) Fertility and population policy: the Singapore experience. J Popul Soc Secur (Popul) 1(Suppl):643

Yatim R (1995) Freedom under executive power in Malaysia: a study of executive supremacy. Endowment Publications, Washington, D.C.

Yong E (2 August 2017) The designer baby era is not upon us. The Atlantic. https://www.theatlantic.com/science/archive/2017/08/us-scientists-edit-human-embryos-with-crispr-and-thats-okay/535668/

Yuehtsen JC (2010) Eugenics in China and Hong Kong: nationalism and colonialism, 1890s–1940s. In: Bashford A, Levine P (eds) The Oxford handbook of the history of eugenics

Yuen M (29 November 2015) Act to ensure country has regulations on artificial reproduction – Nation. The Star. https://www.thestar.com.my/news/nation/2015/11/29/birth-of-a-new-law-soon-act-to-ensure-country-has-regulations-on-artificial-reproduction/

Zakhiri M et al (2016) Legal position of Fatwa: observations from selected jurisdictions. Seminar on Law and Society (SOLAS 2016), School of Law, Universiti Utara Malaysia

Zeidman LA (2011) Neuroscience in Nazi Europe Part I: eugenics, human experimentation, and mass murder. Can J Neurol Sci/Journal Canadien des Sciences Neurologiques 38:696

Zubedy A (21 June 2012) NEP: the good and the bad. Malaysia Today. https://www.malaysia-today.net/2012/06/21/the-nep-the-good-and-the-bad/

Zurairi AR (4 July 2018) Kelantan Deputy MB: child marriage "Common", LGBT a Bigger Issue. Malay Mail. https://www.malaymail.com/s/1648815/kelantan-deputy-mb-child-marriage-common-lgbt-a-bigger-issue

Printed by Printforce, the Netherlands